Metals and Micro-organisms

by

Martin N. Hughes
Reader in Chemistry

and

Robert K. Poole
Professor of Microbiology

King's College London
(University of London)

LONDON NEW YORK

Chapman and Hall

First published in 1989 by Chapman and Hall Ltd
11 New Fetter Lane, London EC4P 4EE

Published in the USA by Chapman and Hall
29 West 35th Street, New York NY 10001

Typeset in 10/12pt Palatino
by Best-Set Typesetter Ltd, Hong Kong

Printed in Great Britain by St. Edmundsbury Press,
Bury St. Edmunds, Suffolk.

ISBN 0 412 244004

British Library Cataloguing in Publication Data

Hughes, Martin N.
Metals and Micro-organisms.
1. Organisms. Metals
I. Title II. Poole, Robert K.
574.19'214

ISBN 0-412-24400-4

Library of Congress Cataloging in Publication Data

Hughes. Martin N. (Martin Neville), 1938-
Metals and Micro-organisms/Martin N. Hughes and
Robert K. Poole.
p. cm.
Bibliography: p.
Includes index.
ISBN 0-412-24400-4
1. Metals—Physiological effect. 2. Micro-organisms—Effect of metals on. 3. Metals—Metabolism. I. Poole, Robert K. II. Title.
QR92.M45H84 1989
576'.119214—dc20 89-32096
CIP

Contents

Acknowledgements

We are indebted to a number of colleagues, collaborators and friends and hope that their involvement or the publication of this book will not jeopardize these relationships. We have been helped, inspired and encouraged. We acknowledge the efforts of Professor S. John Pirt and Dr Harry Rosenberg in reading and improving Chapters 2 and 3 respectively, but absolve them of all blame for errors and omissions. Our sincere gratitude goes to Jenny and Joy for their tolerance and assistance with some of the less interesting aspects of authorship.

Preface

This book was written in the belief that one of the interfaces between microbiology and chemistry had become too important, and perhaps too diffuse, to be consigned to texts in either subject area alone, and that an integrated view was timely. Our title, aside from any alliterative appeal, perfectly summarizes this area.

The book is about the effects of metals on micro-organisms and, in turn, the effects of micro-organisms on metals and their compounds. Some of these phenomena, such as the necessity of metal ions in cell nutrition, the antimicrobial activities of metal compounds, or the distinctive biological roles of metals such as molybdenum, are well known and have become part of the fabric of undergraduate texts in our respective disciplines. Other interactions, such as the biosorption of metal ions to cell surfaces, metal transport systems, and the confirmation that metals such as vanadium and nickel have essential biological functions, are only just beginning to be studied, understood or discovered. Pointers to future developments exist. Thus, the contributions of molecular biology will be increasingly felt, particularly in the elucidation of the structure of proteins that bind, transport or have their function modulated by metals. Interest in biotechnological aspects is gaining in momentum particularly in the 'new field' of biohydrometallurgy. Advances in instrumental techniques, particularly in magnetic resonance methodology, can be expected to provide new structural and functional information on roles for metals and facilitate measurements of intact cells.

Our ordering of material in the book is perhaps formal. We begin with coverage of those aspects of coordination chemistry central to an understanding of biological function and discuss the difficulties of analysing metals in biological systems. An account of the nutritional requirements of micro-organisms for growth also highlights experimental difficulties, in this case the provision of adequately chelated metals, and speciation and metal limitation. We continue with detailed accounts of how these essential elements are taken up by cells and the functions of metals, main group and transition, in a wide range of cellular activities. Later chapters deal with the toxic effects of metals and metal compounds, particularly organometallics, on micro-organisms, the mechanisms by which these effects are sometimes resisted, and the considerable biotechnological significance of metal–microbe interactions.

These topics have been considered separately in several specialist reviews and multi-author treatises, but not, to our knowledge, in a single volume that integrates chemical and biological viewpoints. Appropriately, the authors' backgrounds are, respectively, in inorganic chemistry and microbiology, and we have collaborated for many years. Despite this, our individual approaches may have resulted in some seams, which we trust will not be too evident or distracting. We are conscious that the area of geomicrobiology may not have received here the attention it deserves but this exciting area has been quite recently surveyed by W.E. Krumbein (1983) in *Microbial Geochemistry* (Blackwell). With this exception, we have attempted to provide a comprehensive coverage of topics but, necessarily, citations to the original research literature are illustrative rather than encyclopaedic. References to material appearing in 1988 are for the most part restricted to reviews and books.

This book is intended to be of value to specialists in apparently disparate aspects of microbiology and of inorganic chemistry, unified by an interest in metal–microbe interactions. We hope that this contribution to the literature will be a timely statement of what has been achieved and will provide a stimulus to this rapidly developing aspect of 'inorganic microbiology'.

Glossary of terms

ADP	adenosine 5′-diphosphate
δALA	δ amino laevulinic acid
AMP	adenosine 5′-monophosphate
ATP	adenosine 5′-triphosphate
ATPase	adenosine triphosphatase
CCCP	carbonylcyanide m-chlorophenylhydrazone
CD	circular dichroism
DCCD	NN′-dicyclocarbodi-imide
DNA	deoxyribonucleic acid
E_m	mid-point potential (v)
E^o	standard electrode potential (v)
E_o	standard electrode potential at a given pH (v)
EDTA	ethylenediaminetetraacetic acid
EGTA	ethyleneglycolbis(aminoethylether)tetraacetic acid
ESR	electron spin resonance
EXAFS	extended X-ray absorption fine structure
F	Faraday's constant = 96487 JV^{-1}
FAD	flavin adenine dinucleotide
FMN	flavin mononucleotide
FTIR	Fourier Transform Infra-red
g	g = hν/βH, where β = the Bohr magneton, H = magnetic field, h = Planck's constant, ν = frequency
HEPES	*N*-2-hydroxyethylpiperazine-*N*′-2-2-ethanesulphonic acid
K	equilibrium constant or formation constant
kb	kilobase
K_m	Michaelis constant
MCD	magnetic circular dichroism
M_r	relative molecular mass
MQ	menaquinone
NBD	7-chloro-4-nitrobenzo-2-oxa-1,3-diazole
NEM	N-ethylmaleimide
NMR	nuclear magnetic resonance

NTA	nitrilotriacetic acid
Q	ubiquinone
RNA	ribonucleic acid
SOD	superoxide dismutase
SDS	sodium dodecyl sulphate
XPS	X-ray photoelectron spectroscopy
$Y_{sucrose}$ Y_o	molar growth yield, with substrate indicated by subscript (g mol^{-1})
Δp	proton motive force or pmf (v), $= \Delta\,\tilde{\mu}_{H^+}/F$
ΔpH	pH difference between two sides of a membrane
$\Delta\psi$	membrane potential (v); the electrical potential difference between two bulk phases separated by a membrane
$\Delta\tilde{\mu}H^+$	the electrochemical potential difference for protons between two bulk phases separated by a membrane (kJ mol^{-1})

1

The functions of metals in micro-organisms

Why should we fear to be crushed by savage elements, we who are made up of the same elements?

Emerson
The Conduct of Life (1860)

1.1 INTRODUCTION

Metal ions are involved in all aspects of microbial life. The cations K^+ and Mg^{2+} are bulk intracellular species, while Na^+, Ca^{2+}, a number of transition elements and Zn^{2+} are all essential metals, although they are present only at trace or ultratrace levels. These cations have a variety of functions, although in many cases their roles are not yet fully understood. However, in general terms, they are known to be involved in the stabilization of a range of biological structures, from cell walls to protein conformations. They are often highly effective catalysts of a range of diverse biochemical processes and in some cases are able to trigger, moderate or inhibit reactions. The subject of metals and micro-organisms is, then, one of considerable significance, and many fundamental questions have to be addressed. Why is a particular metal needed for a certain function? How is the metal taken up into the cell and transported to a specific site? In what way are other metals excluded from the site? What is the detailed role of the metal in a particular biochemical process?

In addition to these fundamental issues, there is currently a great deal of interest in other aspects of metal–microbe interaction. A number of metals, including some essential metals at high concentration, are toxic to micro-organisms. Much research is concerned with the mechanisms underlying these toxic effects and with the way in which some organisms develop resistance to toxic metals. Such toxicity has, of course, been exploited in the production of antimicrobial metal compounds. There is, too, increasing awareness of the potentially important role of micro-organisms in solving major industrial and environmental problems associated with metals. The exploitation of chemolithotrophic organisms has allowed the development of pollution-free methods for the extraction of metals from low-grade or refractory ores. These methods may well offer the best solution to the problem of maintaining adequate supplies of certain key metals in the future. Micro-organisms and microbiological material have also been used in the recovery or removal of valuable or toxic metals from industrial process streams or effluents, and appear to have advantages over conventional methods, including ion exchange resins.

The important, but often unappreciated, role of micro-organisms in geological processes is manifested in the laying down of mineral sulphides and the precipitation of calcium carbonate deposits through the action of sulphate-reducing bacteria and certain marine algae respectively. It also appears that some rare minerals have crystallized through initial precipitation on a cell surface matrix.

Inorganic chemists have been increasingly involved in the last two decades in the study of the roles of metals in biological systems. They have moved from the investigation of model systems involving coordination compounds to studies on the structure, electronic properties and reactivity of isolated metalloenzymes, metal-containing cofactors and biologically-relevant metal cluster compounds. Future developments are likely to centre on the behaviour of metal ions in whole cells under physiologically relevant conditions. Micro-organisms offer the opportunity to study the function of metals in intact biological systems, which have well-studied nutritional requirements and growth characteristics, which are amenable to genetic manipulation, and which respond to external pressures.

Micro-organisms also offer a challenge to inorganic chemists to prepare model systems that duplicate the reactivity and selectivity displayed in microbial coordination chemistry. Inorganic chemists sometimes need to synthesize simple ligands that bind metal ions selectively and strongly, but the best examples of such ligands are often those synthesized by micro-organisms, such as the ligands secreted under certain conditions to solubilize and transport iron(III). Similarly the nitrogen-fixing organisms were a great stimulus to inorganic chemists in their successful attempts to prepare metal dinitrogen complexes that allow the reduction of dinitrogen under mild conditions.

1.2 THE ESSENTIAL METALS

A metal may be demonstrated to be essential by setting up conditions where the growth of the organism is limited by the supply of the metal. Such conditions are achieved by reducing the concentration of the metal in the medium to very low levels and showing that the amount of biomass obtained under these conditions is dependent on the amount of metal present. These matters are discussed fully in Chapter 2.

Table 1.1 gives estimates for the amount of cation required to give 100 g dry biomass when that cation is growth-limiting (Pirt, 1975). These data, taken for a range of organisms, demonstrate the high requirements for K^+ and Mg^{2+} and the low requirements for the transition elements. It should be stressed that growth yields for trace elements vary considerably with growth conditions, and that it is often difficult to show a requirement for a trace metal such as sodium because it is present in sufficient amounts as a contaminant in the components of the medium and cannot be easily removed. The essentiality of several other metals, such as nickel and vanadium, has only recently been acknowledged. Nickel was well known to be a component of the enzyme urease, but was not recognized as an essential growth requirement for several micro-organisms. In one case this was due to the adventitious supply of nickel from stainless steel needles used in syringes. The role of vanadium in the nitrogen-fixing organism *Azotobacter chroococcum* was only established in 1986 (Robson *et al.*), despite a report in 1936 that vanadium could substitute for molybdenum as a trace element essential for nitrogen fixation in *Azotobacter vinelandii* (Bortels, 1936). However, the demonstration that a trace metal enhances growth of an organism cannot always be unambiguously interpreted as evidence for essentiality, as the

Table 1.1 Growth yields* under metal-limited conditions

Approximate amount of cation (g) to give 100 g dry biomass			
K^+	1.7	Co^{2+}	0.001
Mg^{2+}	0.1–0.4	Cu^{n+}	0.001
Ca^{2+}	0.1	Zn^{2+}	0.005
Mn^{2+}	0.005	Mo^{n+}	0.001
Fe^{n+}	0.015		

* Growth yields vary considerably with growth conditions.

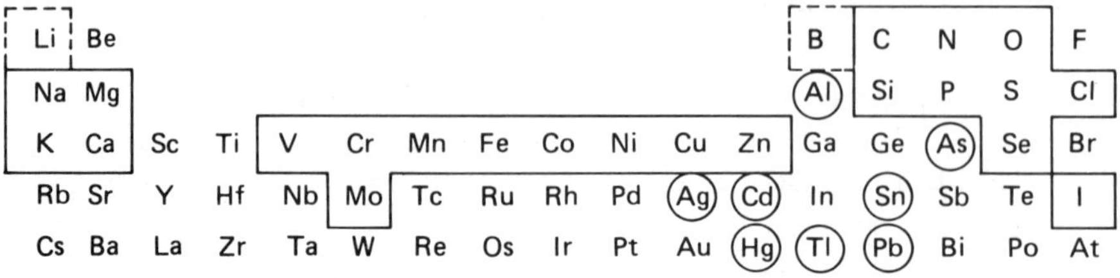

Figure 1.1 The Periodic Table: hydrogen, the noble gases, the lanthanides and the actinides are not shown. □ elements known to be essential; ⬚ elements that may be essential; ○ elements that are toxic.

element may function as a growth stimulant by substituting (if inefficiently) for another metal of similar chemistry. The methanogen *Methanobacterium wolfei* has a growth requirement for tungsten (Winter *et al.*, 1984), but the tungsten may be substituting for molybdenum or possibly nickel. A further complication centres on the speciation of the metal in the growth medium. It may be present in several forms, depending upon the components and pH of the medium, and not all of these may be taken up by the organism.

Figure 1.1, the Periodic Table of the elements, shows the elements known to be essential in biology, those which may be essential, and some well-known toxic elements. It is important to note that essential elements may become toxic at high concentration, and that elements notorious for their toxicity may exert beneficial effects at very low concentration levels.

1.2.1 Classes of metals showing biological activity

The biologically important metals shown in Figure 1.1 fall into well-defined sections of the Periodic Table, each of which will be discussed.

(a) The s-block elements

The essential *s*-block elements are the monovalent cations of sodium and potassium and the divalent cations of magnesium and calcium, from Groups I and II respectively. These cations have noble gas configurations and their chemistry is essentially ionic. They interact weakly with ligands, particularly in the cases of Na^+ and K^+, due to their poor polarizing power (Section 1.3.3). All four cations prefer to bind to oxygen donor ligands. Many calcium salts are insoluble, and the ready precipitation of calcium as the carbonate, phosphate or oxalate should be noted in particular.

The biological roles of these cations are difficult to explore, partly because they interact weakly with ligands and partly because their properties do not allow the exploitation of instrumental techniques other than nuclear magnetic resonance (NMR). It should also be noted that while calcium and sodium have radioactive isotopes of convenient half-life, the isotope ^{42}K has a rather short half-life of 12.4 hours. An attempt has been made to overcome these problems by the use of probe metal cations to replace the native cation, so allowing the characterization of the coordination environments of the metal-binding site through the exploitation of the electronic and nuclear properties of the probe (Hughes, 1981). Table 1.2 lists data for the four cations and for some probes. Ionic radii, preferred coordination number and geometry, and flexibility in the coordination environment are important factors in considering binding properties of these cations. There is an interesting contrast between Mg^{2+} and Ca^{2+}: magnesium prefers to form six-coordinate compounds of regular octahedral geometry, but Ca^{2+}, a larger cation, is able to accept higher coordination numbers and irregular geometries. The greater flexibility of Ca^{2+} for binding sites allows it to function more effectively than Mg^{2+} in cross-linking negatively charged ligand groups in cell walls, as these groups provide an irregular site. This explains why calcium ions rather than magnesium ions have the role of stabilizing the structure of cell walls.

A notable feature of these four cations is their selective distribution in biological systems. In general, K^+ and Mg^{2+} are intracellular cations

Table 1.2 Some properties of the *s*-block cations and probe replacements

Native cations			Probes		
	Ionic* radius ($\times 10^{-10}$m)	Common coordination numbers		Ionic* radius ($\times 10^{-10}$m)	Common coordination numbers
Na^+	0.95	6	Li^+	0.60	4
K^+	1.33	6, 8	Rb^+	1.48	8
Mg^{2+}	0.65	6	Tl^+	1.40	6
Ca^{2+}	0.99	6, 8, higher	Mn^{2+}	0.80	6
			Ni^{2+}	0.69	6
			La^{3+}	1.15–0.93	

* Pauling ionic radii, taken from Cotton and Wilkinson (1980). These values relate to six-coordination; ionic radii for other coordination numbers will differ slightly (Sinha, 1976).

while Na^+ and Ca^{2+} are rejected by the cell. It is not surprising, therefore, that K^+ and Na^+ often have different biological functions from each other and may be mutually inhibitory. The halobacteria do tolerate higher internal concentrations of NaCl, but this is necessary to maintain osmotic balance with the saline medium in which they grow. In this case certain enzymes are activated by both sodium and potassium cations. The selective distribution of magnesium and calcium cations is exploited in various trigger and control processes. The exclusion of calcium from the cell cytoplasm means that the controlled appearance of calcium ions in the cell (either by translocation from outside the cell or by release from internal stores) can trigger a range of cellular processes via binding of Ca^{2+} to receptor proteins, which are not activated by magnesium ions.

The ability of cells to distinguish between Na^+ and K^+, between Mg^{2+} and Ca^{2+}, and between Na^+ and Ca^{2+} (which have similar ionic radii) illustrates well the selectivity incorporated into biological binding sites for metals. Such selectivity has been duplicated in simple model ligands such as the 'crown' cyclic ethers, the 'cryptand' bicyclic ethers and the ionophoric antibiotics (Hughes, 1981, 1987; Sigel, 1985).

(b) The transition metals and zinc

The transition elements have partly filled *d* orbitals, and with zinc, bind ligands much more strongly than do the *s*-block elements. Zn(II) and Cu(I) have d^{10} electronic configurations, and so their complexes are diamagnetic and, having no $d \rightarrow d$ absorption spectra, usually colourless. The 3*d* metals from vanadium to zinc are all essential elements, while molybdenum appears to be the only essential second row transition element, a point of great interest in itself. A considerable advantage arises in the study of biological molecules containing transition metals as their electronic properties can be exploited to help characterize their binding sites.

Zinc is not redox-active and so will have different functions in biology from those associated with the transition elements. Manganese is redox-active in certain enzymes, where it is present as Mn(III), but is usually found as Mn(II) in a range of enzymes where it has a non-redox role similar to that of Mg(II).

Transition metals, notably iron and copper, are important catalysts of biological redox processes, and may be involved in the catalysis of electron transfer and the oxidation-reduction of substrates. In the latter instance, the metal centre usually has an open site to allow coordination of substrate molecules. There are many reasons why these inorganic elements are such effective catalysts. Transition metals can adopt alternative coordination numbers and geometries, and so undergo changes in coordination number during the course of a redox process. Ligand substitutions at the transition metal centre are usually very fast, but can be controlled if necessary. Redox reactions can be effected by group transfer to or from the metal. Transition metals sometimes have a range of accessible oxidation states to allow the ready transfer of electrons, while the redox potential of transition metal couples can be varied considerably in a controlled manner by changing the coordination environment. Major variation in potential can result from the preferential stabilization of high or low oxidation state of the metal by the presence of a particular type of ligand. Fine tuning of the potential may be effected by the exploitation of more subtle interactions, such as a change in the spin state of the metal or hydrogen bonding interactions. This means that the potential of a cytochrome can be determined precisely by the choice of axial ligand and the overall environment provided by the protein.

Table 1.3 gives some typical coordination numbers and geometries for these metals. Simple coordination complexes often involve regular geometries, with four or six-coordination, but the much more complicated

Table 1.3 Stereochemistries, coordination numbers and oxidation states of some 3*d* elements (important geometries are indicated by*)

Metal	Ionic[+] radius ($\times 10^{-10}$m)	Coordination numbers and geometries
Mn(II) (d^5)	0.80	6 (octahedral*)
Fe(III)(d^5)	0.64	4 (tetrahedral*), 6 (octahedral*)
Fe(II)(d^6)	0.76	4 (tetrahedral*), 6 (octahedral*)
Co(III)(d^6)	0.63	5 (square pyramidal), 6 (octahedral*)
Co(II)(d^7)	0.74	4 (tetrahedral*), 5 (trigonal bipyramidal, square pyramidal), 6 (octahedral*)
Ni(II)(d^8)	0.69	4 (square planar*, tetrahedral*), 5 (trigonal bipyramidal), 6 (octahedral*)
Cu(II)(d^9)	0.69	4 (square planar*, tetrahedral), 5 (square pyramidal, trigonal bipyramidal), 6 (distorted octahedral*)
Cu(I)(d^{10})	0.96	2 (linear), 3 (planar), 4 (tetrahedral*)
Zn(II)(d^{10})	0.74	4 (tetrahedral*), 5 (trigonal bipyramidal), 6 (octahedral)

[+] High-spin configurations.
See footnote to Table 1.2.

ligands found in biology, such as proteins, may, as a result of their own internal structural constraints, impose distorted geometries upon metals. The willingness of certain metals, notably Zn(II), to accommodate distorted sites is a central aspect of their efficiency as catalysts. Zinc enzymes sometimes catalyse reactions involving a succession of intermediates: some flexibility around the zinc is useful in accommodating the slightly different requirements of these intermediates.

Structural studies have frequently shown the presence of distorted geometries around the metal in a metalloenzyme. It has been suggested (Vallee and Williams, 1968) that this may be related to the catalytic efficiency of the enzyme, in that the protein forces the metal to adopt a geometry corresponding to the transition state of the reaction being catalysed. Many illustrations of this 'entatic state' hypothesis are found in transition metal biochemistry. Azurin is a blue copper protein, isolated from *Pseudomonas aeruginosa*, whose function is the catalysis of electron transfer. The copper centre cycles between Cu(I) and Cu(II), and the preferred geometries for these oxidation states are tetrahedral and square planar respectively. Nuclear movement is slow compared to the rate of electron transfer, so electron transfer in copper compounds might be expected to generate Cu(I) in a square planar geometry or Cu(II) in a tetrahedral one, which is energetically unfavoured as vibrationally excited copper species are formed. This problem is overcome in azurin, and in other blue electron-transfer copper proteins, as the protein enforces a distorted tetrahedral structure upon the copper centre, in which the bond angles differ by up to 50° from normal tetrahedral values. The structure is best described as being intermediate between tetrahedral and square planar geometries, and so is a compromise between the preferred geometries of Cu(I) and Cu(II). This facilitates electron transfer. The ligand groups also represent a compromise between the requirements of Cu(I) and Cu(II) (Adman *et al.*, 1978).

(c) Toxic elements

The most common examples of toxic elements are the $4d^{10}$ and $5d^{10}$ metals cadmium and mercury and the *p*-block elements thallium, tin, lead, and arsenic. Silver is less well known in this context but is probably the most toxic element to micro-organisms (Trevors, 1987a). These elements are all readily polarized, and may be classified as 'soft' in a chemical sense (Section 1.3.3(a)). The cations bind strongly to many ligands, particularly soft ligands such as thiols. They exert toxic effects in several ways, for example by displacing native metals from their normal binding sites or by binding to proteins and nucleic acids and altering their conformation. They also affect oxidative phosphorylation (Gibson et al., 1980) and membrane permeability. The chemical form of these elements must be considered in assessing their toxicity. In some cases, such as lead, tin or mercury, the toxicity of organoderivatives, in which the metal is bound to an organo group through a metal-carbon bond, is much greater than that of simple inorganic compounds. This results from the greater liposolubility of organoderivatives and the resulting enhancement of uptake into the cell. Certain metal cations undergo biomethylation in the environment (Thayer, 1984).

Concern is now being expressed over the toxicity of aluminium, and the increasing extent to which it is appearing in the environment through the solubilizing effect of acid raid on aluminosilicates in soil (Martin, 1986; Birchall and Chappell, 1988; MacDonald and Martin, 1988). Several reports in recent years have cited enzymes which require fluoride, citrate or phosphate as activators. In practice, these anions serve only to complex or precipitate aluminium, often present as a contaminant in commercial preparations of ATP, and so to remove an inhibitory factor. The toxicity of Al^{3+} usually results from inhibition of Mg^{2+}-dependent enzymes. Aluminium (Al^{3+}) can readily replace Mg^{2+} in biological systems because of its much higher affinity for ligands. For example, Al^{3+} binds to ATP^{4-} almost 10^7 times more strongly than does Mg^{2+}; thus, in enzymes which form ternary complexes with ATP and a metal cation, less than nanomolar amounts of free Al^{3+} can compete with millimolar levels of Mg^{2+} (Martin, 1986). However, Al^{3+} cannot maintain the activity of the enzyme because ligand exchange at Al^{3+} is about 10^5 times slower than ligand exchange at Mg^{2+}. Such toxicity resulting from lack of lability is an often-unappreciated complication. In addition to toxic effects arising from displacement of Mg^{2+}, aluminium may also compete with Fe^{3+} and Ca^{2+}.

1.3 METAL–BINDING SITES IN BIOLOGICAL MOLECULES

The biological function of metals is intimately linked to the nature of the ligands present in the coordination sphere. In some cases, the coordination positions around the metal may be fully filled by donor atoms from one multidentate molecule such as a protein. In others, one or more coordination positions may be available for the binding of a substrate molecule, either by expansion of the coordination number or, more usually, by having readily-displaceable water molecules coordinated to the metal centre.

This section will review the types of binding site available to metal ions in biological systems and discuss the ligand preferences of biologically important metals.

1.3.1 Possible metal-binding groups

The three major classes of biopolymer, proteins, nucleic acids and polysaccharides, all provide sites at which metal ions will bind. The ligand groups available include negatively charged groups such as carboxylate, thiolate or phosphate, and groups such as the amine function which coordinate to the metal centre through lone pairs of electrons. Metal ions are also bound by specialized macrocyclic ligands such as the porphyrins, while transition metals, notably iron, are often present in enzymes in dimers or clusters with bridging sulphide or oxide ligands.

(a) Ligand groups on peptides and proteins

Peptides and proteins offer a range of metal-binding groups, namely the amino and carboxylate termini, the deprotonated peptide link, and, especially, the side chain substituents (Figure 1.2). However, the picture is simplified because two amino acid residues are especially important; these are the histidine and cysteine residues which bind metals through the imidazole and thiol side chains respectively (Hughes, 1981). The carboxylate function is an excellent ligand for calcium, often acting as a bidentate group. The binding sites for calcium in a number of proteins have been determined by X–ray crystallography. Oxygen donor ligands only are involved, largely carboxylate groups of aspartic acid and glutamic acid residues, but also hydroxyl groups of serine and threonine, peptide carbonyl oxygen atoms or water. Coordination numbers are often higher than six. These structures are listed by Hughes (1987). Two special amino acid residues are of notable significance as ligands for calcium. Some calcium-binding proteins of mammalian origin contain γ-carboxyglutamate residues (the 'gla' proteins: Furie *et al.*, 1982) and occasionally β-hydroxyaspartate residues. Both provide sites of high affinity for calcium.

The amino acid residues acting as ligands may come from widely separated parts of the protein's primary structure which are brought together as a result of protein folding.

(b) Ligand groups on nucleotides and nucleic acids

The binding sites for metal ions available on nucleotides and nucleic acids are less diverse than in proteins. Metal ions bind readily to the

Residue	Substituent	Metals
Histidine	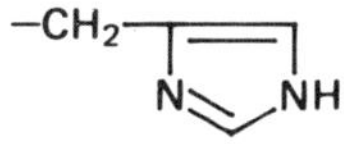	Zn, Fe (axial), Cu
Cysteine	$-CH_2SH$	Zn, Fe, Cu
Methionine	$-CH_2CH_2SCH_3$	Fe, Cu
Glutamic acid	$-CH_2CH_2COOH$	Ca, Zn, Fe
Aspartic acid	$-CH_2COOH$	Ca, Zn, Fe
γ-carboxyglutamic acid	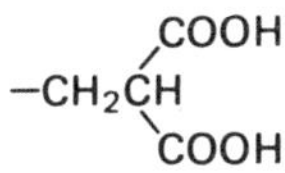	Ca
β-hydroxyaspartic acid	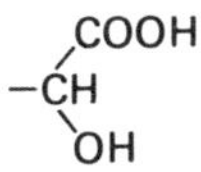	Ca
Tyrosine	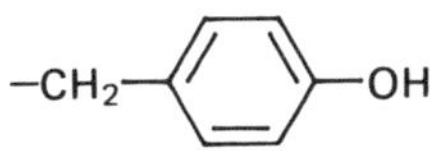	Zn, Fe, Mn, Ca
Serine	$-CH_2OH$	Ca
Asparagine	$-CH_2CONH_2$	Ca
Glutamine	$-CH_2CH_2CONH_2$	Fe
Threonine	$-CH(OH)CH_3$	Ca

Figure 1.2 Metal-binding groups in proteins.

phosphate and nucleotide base moieties and only weakly to the sugar groups. Binding of metal to the phosphate stabilizes the nucleic acid helical structure as negative charge is neutralized (with decreased repulsion) while binding of metals to the bases results in destabilization of the structure as hydrogen bonding is prevented (Eichhorn, 1981).

The five heterocyclic bases present in the nucleic acids are good ligands, and are shown in Figure 1.3 together with the important nucleotide adenosine triphosphate. Metal complexes of ATP often serve as substrates for enzymes (Cleland, 1982): thus, binding of Mg^{2+} to ATP

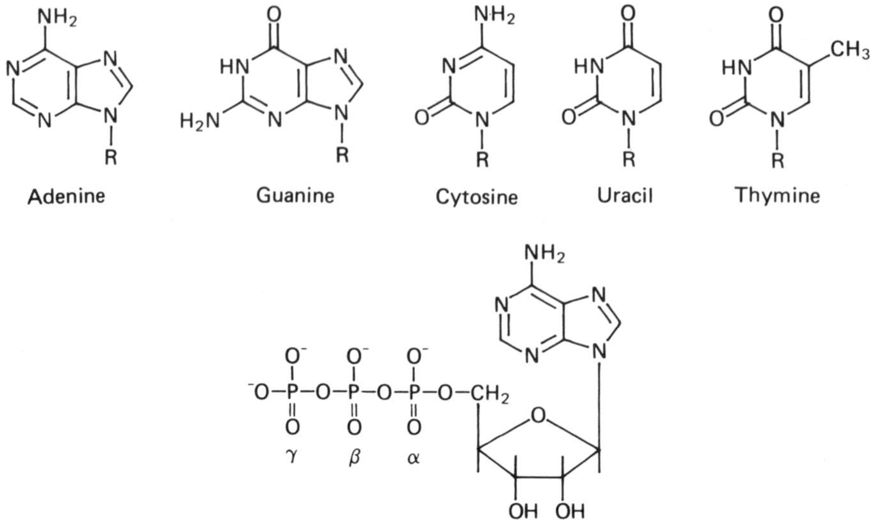

Figure 1.3 The nucleic acid bases and the structure of ATP.

results in the activation of the terminal phosphoryl group towards hydrolysis. Indeed both ATP and ADP exist in cells largely as the Mg(II) chelate complexes since intracellular [Mg^{2+}] is about 1 mmol dm^{-3}. It is clear that the $MgATP^{2-}$ complex can exist as a number of coordination isomers, for example the α, β and β, γ bidentate complexes (Figure 1.3). Furthermore, each of these bidentate complexes can exist in two conformations, as shown schematically in Figure 1.4 for the α, β complex. Nucleotide-utilizing enzymes are highly specific for one of these two forms, as demonstrated by the use of β,γ-bidentate complexes of Cr(III) or Co(III), which are inert to ligand substitution and so allow the stereospecificity of MgATP-utilizing enzymes to be probed (Cleland, 1982).

(c) Ligand groups on polysaccharides and analogues

The polysaccharides are an extremely diverse group of biopolymers. Many are simple, linear repeat polymers of a monosaccharide or a

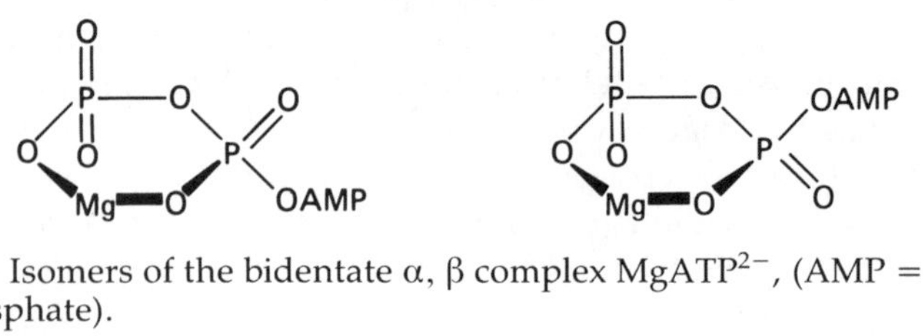

Figure 1.4 Isomers of the bidentate α, β complex $MgATP^{2-}$, (AMP = adenosine monophosphate).

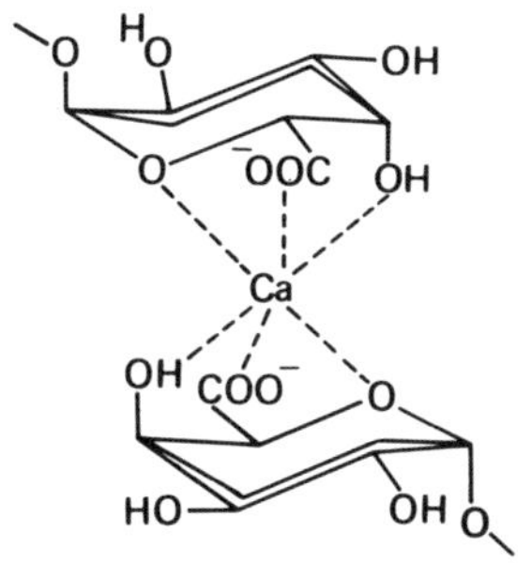

Figure 1.5 The D-galactopyranosiduronate–Ca^{2+} complex.

disaccharide, but more complicated structures may occur. A wide range of metal-binding compounds based upon polysaccharides or their analogues is found in micro-organisms (Rees *et al.*, 1982; Painter, 1983), often as important components of cell walls and membranes. Teichoic acid is a linear polymer of glycerol phosphate, where the phosphate group constitutes the metal-binding site, as it does in a range of lipopolysaccharides. Other typical compounds are peptidoglycan, chitosan and chitin. Peptidoglycan is found in bacterial cell walls and is made up of repeat dimers of *N*-acetylglucosamine and *N*-acetylmuramic acid cross-linked by a small peptide, which has a glutamic acid residue and may bind metals. Chitosan is an amino sugar, while in chitin this group is acetylated so accounting for the lower affinity of chitin for metals (Chapter 7). Polysaccharides thus offer a range of ligand groups in addition to the carbohydrate oxygen donors. Binding of metals is sometimes cooperative and there may be selectivity towards metals even though all the ligand groups involve oxygen coordination to the metal. The phosphate group in teichoic acid has a higher affinity for Mg^{2+} than Ca^{2+}, while in many cases the carbohydrate oxygen groups provide a metal-binding site which is specific for calcium. The latter behaviour is illustrated in Figure 1.5 which shows the binding of Ca^{2+} to the D-galactopyranuronic acid-containing glycans of Chrysophyte algae. This structure is based upon the use of lanthanide-shift NMR techniques, and does not distinguish between monodentate and bidentate carboxylate groups (Painter, 1983).

(d) Macrocyclic ligands

A range of tetrapyrrole ligands are available for binding metals, usually iron (Figure 1.6). These include porphyrins, hydroporphyrins (in which pyrrole double bonds are reduced), cobalt-binding corrins and nickel-binding corphins (Factor 430M). In each case, the macrocycle provides

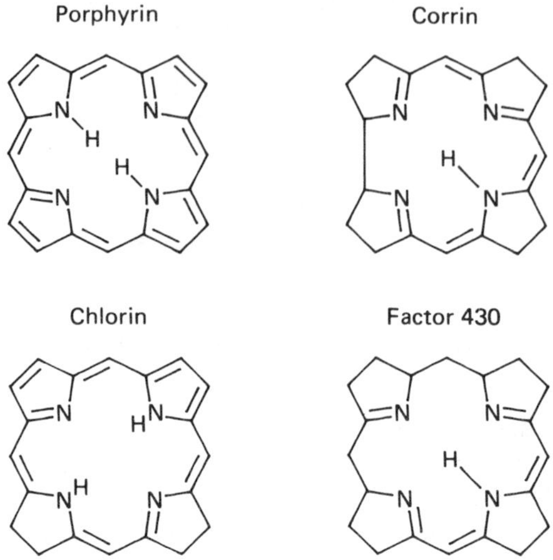

Figure 1.6 Tetrapyrrole macrocycles.

four in-plane nitrogen groups to bind the metal, while the two axial positions are usually filled with ligand groups provided by the associated protein. Iron, as both Fe(II) and Fe(III), is able to move in and out of the central cavity with respect to the plane of the porphyrin ring. This feature is an important aspect of the properties of iron porphyrins.

The functions of a diverse range of metalloenzymes containing porphyrin and other metal-substituted tetrapyrroles will be considered in detail in Chapter 5.

(e) Metal clusters with bridging sulphide or oxide

Iron-sulphur clusters are found in the iron-sulphur proteins discussed in Chapter 5. The clusters (Figure 1.7) are bound to the protein via cysteine thiolate groups, while the iron centres are bridged by sulphide which is released as H_2S on treatment with mineral acid. Clusters containing one, two, three and four iron atoms are known. Each iron atom is in an approximately tetrahedral environment of sulphur atoms. The one-iron protein rubredoxin, originally isolated from *Clostridium pasteurianum*, has no acid-labile, bridging sulphide, but is included in this discussion for convenience; in this case the iron is coordinated to four cysteine sulphur atoms. Aspects of the iron-sulphur proteins have been reviewed on many occasions (for example Lovenberg, 1973; Yoch and Carithers, 1979; Xavier *et al.*, 1980; Spiro, 1982). In a small number of examples, ligands other than sulphur donors appear to be included in

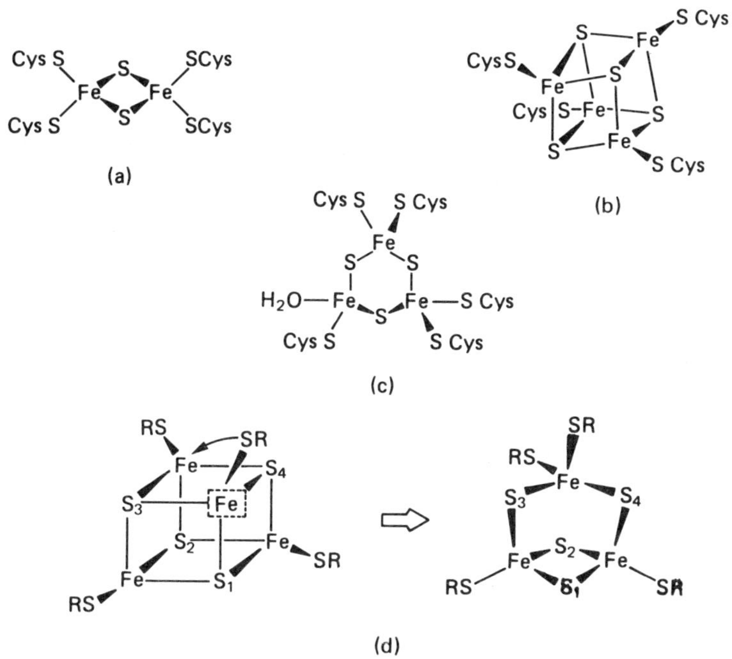

Figure 1.7 The structures of iron-sulphur clusters: (a) Two-iron centre; (b) Four-iron cluster; (c) The 3Fe.3S cluster in ferredoxin 1 (*Azotobacter vinelandii*). The aqua ligand has been queried (Malikayil *et al.*, 1985)*. (d) The conversion of a 4Fe.4S cluster into a 3Fe.4S cluster (R = Cys).

* Note added at proof: structure (c) is incorrect: *J. Am Chem. Soc.* (1988), **110**, 8721.

the coordination environment, as noted in Chapter 5.

The presence of iron clusters with the polarizable sulphide group as the bridging ligand has important implications for the redox and electronic properties of the electron-transfer protein. Each cluster transfers only one electron, despite the multi-metal character of the protein. The oxidized form of the 2Fe.2S protein putidaredoxin, isolated from *Pseudomonas putida*, is diamagnetic, due to antiferromagnetic coupling between two high-spin d^5 Fe(III) centres, while the reduced form of putidaredoxin is paramagnetic as one electron has been added to give (formally) Fe(II) and Fe(III). A 4Fe.4S cluster will have several overall oxidation state levels available to it, each level corresponding to a different number of Fe(II) and Fe(III) centres. This has been invoked to explain the great difference in redox potential between the 4Fe.4S ferredoxins (−420 to −240 mV) and the high potential iron proteins (HiPIP) found in some photosynthetic bacteria (around +350 mV). The structures of the clusters in *Chromatium* HiPIP and *Peptococcus aerogenes* ferredoxin are strikingly similar so the protein must exert some control over the redox potentials. The 4Fe.4S cluster appears to have three oxidation levels available, with HiPIP cycling between the upper two and ferredoxin (Fd) between the lower two: $HiPIP_{red}$ and Fd_{ox} share a common oxidation level, and are diamagnetic:

$$\begin{array}{ccccc} [Fe_4S_4(Cys)_4]^{3-} & \rightleftharpoons & [Fe_4S_4(Cys)_4]^{2-} & \rightleftharpoons & [Fe_4S_4(Cys)_4]^{-} \\ 3Fe(II)Fe(III) & & 2Fe(II)2Fe(III) & & Fe(II)3Fe(III) \\ Fd_{red} & \rightleftharpoons & Fd_{ox} & & \\ & & HiPIP_{red} & \rightleftharpoons & HiPIP_{ox} \end{array}$$

It is noteworthy that Fd_{ox} cannot normally be oxidized further and $HiPIP_{red}$ cannot be reduced further, despite being indistinguishable in structure. This must be due to some subtle influence of the protein (Carter, 1977), in accordance with the observation that $HiPIP_{red}$ can be reduced to the super-reduced form in the presence of 80% dimethylsulphoxide, which causes unfolding of the protein. Ferredoxin 1 from *Azotobacter vinelandii* has also been converted to a super-reduced state, which corresponds to all iron centres being reduced to Fe(II) (Stephens *et al.*, 1985).

The structures and function of the three iron clusters are not fully clarified (Beinert and Thompson, 1983). The cluster may exist either as 3Fe.3S or as 3Fe.4S, and in some cases the 3Fe cluster may be formed by damage to 4Fe.4S clusters during the isolation procedure (Beinert, Emptage and Dryer, 1983). Ferredoxin 1 (Fd1) from *Azotobacter vinelandii* contains a 4Fe.4S cluster and a 3Fe cluster. At 2.0×10^{-10}m resolution, the X-ray structure of the 3Fe centre is reported to be a nearly planar 3Fe.3S ring, with one iron atom bound by a water molecule or hydroxide group as well as a cysteinyl sulphur group, as shown in Figure 1.7 (Ghosh *et al.*, 1982). However, the use of electron spin echo techniques (Section 2.6.1 (c)) does not support the presence of a solvent-accessible hydroxide or water ligand (Malikayil *et al.*, 1985), so the structure of this 3Fe cluster does not yet appear to be conclusively resolved.

The interconversion of three-iron and four-iron clusters has been followed in ferredoxins from *Desulfovibrio gigas* and *Bacillus stearothermophilus* (Kent *et al.*, 1982; Bell *et al.*, 1982), and the 7Fe Fd1 has been converted to the 8Fe form (Stephens *et al.*, 1985).

The iron- and molybdenum-containing cofactor from nitrogenase is a cluster in which an iron atom has been replaced by molybdenum, although the exact structure is not known (Hughes, 1987).

Another type of iron binuclear site has recently been recognized (Theil, Eichhorn and Marzilli, 1983) in methane monooxygenase from

Methylococcus capsulatus and in one of the subunits of ribonucleotide reductase from *Escherichia coli* (Thelander and Reichard, 1979). The two iron centres are bridged by oxide to give a structure that is very similar to the binuclear iron cluster in the oxygen carrier haemerythrin. A non-haem iron protein, recently isolated from the periplasmic fraction of *Desulfovibrio vulgaris*, contains two rubredoxin-like FeS_4 centres and a haemerythrin-like binuclear iron cluster: this combination of prosthetic groups is reflected in the name of the protein – rubrerythrin (LeGall *et al.*, 1988). The physiological function of rubrerythrin is unknown at present.

(f) Specialized ligands

Ligands of microbiological origin that bind metals specifically include (a) the iron-carriers or siderophores (Chapter 3), which are usually phenolic or hydroxamic acid ligands, (b) the ionophoric antibiotics, which bind the IA and IIA cations (Sigel, 1985), and (c) the pterins which bind molybdenum by sulphur donor atoms in the carbon monoxide oxidase from *Pseudomonas carboxydovorans* (Johnson, 1980). Other specialized ligands will undoubtedly be discovered, probably from transport systems.

1.3.2 The strength of metal–ligand interaction: formation constants

The extent to which a metal will complex with a particular ligand is characterized by a formation constant. If a monodentate ligand is used, that is a ligand that binds to a metal by one donor atom only, then a series of stepwise equilibria exist for the formation of metal–ligand complexes of stoichiometry 1:1, 1:2, 1:3 up to the maximum coordination number of the metal atom (Equations 1.1–1.3, where charges and water molecules are omitted for simplicity and N is the maximum coordination number).

$$M + L \rightleftharpoons ML \qquad K_1 = [ML]/[M][L] \tag{1.1}$$

$$ML + L \rightleftharpoons ML_2 \qquad K_2 = [ML_2]/[ML][L] \tag{1.2}$$

$$ML_{(N-1)} + L \rightleftharpoons ML_N \qquad K_N = [ML_N]/[ML_{(N-1)}][L] \tag{1.3}$$

Alternatively, these reactions can be described in terms of overall formation constants (Equations 1.4–1.6). The two types of formation constant are related by $\beta_N = K_1 \times K_2 \times K_3 \times \ldots K_N$.

Table 1.4 Formation constants for ligands of biological importance at 25°C†

	log formation constant (K_1)									
Ligand	Mg^{2+}	Ca^{2+}	Mn^{2+}	Fe^{3+}	Fe^{2+}	Co^{2+}	Ni^{2+}	Cu^{2+}	Cu^{+}	Zn^{2+}
Glycine	3.44	1.38	3.44		4.3	5.23	5.77	8.62		5.52
Cysteine			4.1			9.3			19.2	9.86
Aspartic acid	2.43	1.60	3.74			5.94	7.12	8.57		5.84
Histidine			7.74*		9.3*	13.9*	15.9*	18.3*		12.9*
Glutamic acid	1.9	2.05	3.3		4.6	5.06	5.90	7.85		5.45
Citric acid	3.6	3.3	3.7	11.4	4.4	5.0		5.9		5.0
Polyphosphate	3.0	3.2	5.5*	6.5*	3.0	3.0		5.5*		6.0*
EDTA	9.12	11.0	14.0	25.1	13.9	16.2	18.6	18.8		16.3

* Values of β_2.
† See *Stability Constants*, Special Publication of the Chemical Society, No. 6 (1957) and Supplements.

$$M + L \rightleftharpoons ML \qquad \beta_1 = [ML]/[M][L] \tag{1.4}$$

$$M + 2L \rightleftharpoons ML_2 \qquad \beta_2 = [ML_2]/[M][L]^2 \tag{1.5}$$

$$M + ML \rightleftharpoons ML_N \qquad \beta_N = [ML_N]/[M][L]^N \tag{1.6}$$

Formation constants are usually measured at fixed temperatures and ionic strengths. Compilations of formation constants for a wide range of metals and ligands are available (Sillen and Martell, 1971): a selection of values is shown in Table 1.4. Such data can be used to check, for example, the extent of complexation of metal ions to components of growth media (Pirt, 1975) and to determine which metal would be most likely to bind at a particular ligand site.

Metal ions must compete with protons for ligands, and so metals bind less well at lower pH values as the potential donor group will be protonated to an increasing extent. Competition for a metal from two ligands could well be determined by the pK_a values of the protonated ligands. Formation constants show that ammonia and imidazole have similar affinity for Zn^{2+}, with log K_1 values of 2.37 and 2.58 respectively. But, at pH 7, less than 1% of the ammonia is present as NH_3 while over 50% of the imidazole is unprotonated so zinc will bind to imidazole. As the pH is increased above 7 so the extent of coordination of ammonia to zinc will increase.

(a) Conditional formation constants

Growth media or other environments may provide more than one ligand capable of binding metal ions. Competition between two metal ions for one particular ligand cannot then be assessed straightforwardly by the use of formation constant data, as one of them may be bound more firmly by other ligands. This type of situation requires the use of conditional formation constants, as developed originally by Schwarzenbach (1957) and subsequently by Ringbom (1963). The latter work describes methods for calculating the overall effect on metal binding of a range of competing equilibrium reactions, including protonation.

1.3.3 Factors that affect metal–ligand interaction

The complexing power of a metal ion depends upon its polarizing power, that is, on the charge/radius ratio of the ion. A cation of high polarizing power is 'seen' by the ligand as a centre of high density of positive charge with resulting strong interaction.

Ionic size decreases from left to right across the Periodic Table due to the effect of increased nuclear charge. This means that the alkali metal cations interact very weakly with ligands, but the alkaline earth cations Mg^{2+} and Ca^{2+} interact more strongly. In general, for small ligands, Mg^{2+} will interact more strongly than does Ca^{2+}, as shown by the data in Table 1.4. Other factors, including geometry, can disturb this sequence.

The divalent cations of the first transition series decrease in size down to Cu(II), as shown in Table 1.3. The result is an increase in polarizing power. Thus, formation constants for complexes of the 3*d* divalent cations with nitrogen-donor ligands increase in the series Mn(II) < Fe(II) < Co(II) < Ni(II) < Cu(II) > Zn(II). This is the Irving-Williams series. This series is not invariably followed. Change in spin-state of Fe(II) to a low-spin configuration results in a smaller radius and stronger interactions, while the inability of Cu(II) to form regular octahedral complexes may result in less strong interactions with rigid ligands.

(a) Hard and soft acids and bases

The relationship between the polarizing power of a cation and the extent of complex formation is not always straightforward. The cations K^+ and Tl^+ have similar radii and might, therefore, be expected to form complexes to a roughly equivalent extent. In practice, formation constants show that Tl^+ interacts about 10^3 times more strongly with ligands than does K^+. This situation arises because Tl^+ has more electrons than K^+ and is more easily polarized. In other words, Tl^+ is a 'soft' cation and is able to bind more strongly than K^+ through the existence of mutual polarization effects with ligands. Hard cations like K^+ are small, usually have a high charge and so are not easily polarized. Ligands with highly

Table 1.5 Hard and soft acids and bases*

Acids			Bases		
Hard	Intermediate	Soft	Hard	Intermediate	Soft
Na^+	Fe^{2+}	Cu^+	H_2O	Pyridine	CN^-
K^+	Co^{2+}	Pb^{2+}	ROH		RSH
Mn^{2+}	Ni^{2+}	Cd^{2+}	$-CO_2^-$		R_2S
Fe^{3+}	Cu^{2+}	Tl^+	NH_3		$S_2O_3^{2-}$
Co^{3+}	Zn^{2+}	Hg^{2+}	RNH_2		
			Cl^-		
			PO_4^{3-}		
			SO_4^{2-}		

* Pearson, 1965.
R = alkyl or aryl group.

electronegative donor atoms are hard bases, while ligands with polarizable donor atoms are soft. As a general rule, hard cations (or acids) form most stable complexes with hard ligands (or bases) while soft acids form most stable complexes with soft bases. Table 1.5 classifies biologically important metals and ligands as hard, soft and borderline acids and bases (Pearson, 1965), and provides a useful indication of likely combinations of metals and ligands.

Inspection of Table 1.5 shows that metal ions of microbiological significance are hard or borderline hard. Fundamental cellular constituents and potential binding groups are generally hard ligands, usually with N or O donor atoms. In contrast to the 'hardness' of living systems, the common toxic heavy metals associated with environmental pollution are soft, such as Co^{2+}, Hg^{2+}, Pb^{2+} and Tl^+. Hard metals cannot compete with these for binding sites (as noted above for K^+ and Tl^+) and so are displaced from their sites by soft cations (Ho, Ho and Hamilton, 1978). It must be emphasized that some biological ligands are soft, notably the sulphur-containing side chains of the cysteine and methionine residues. These groups are needed to moderate the overall hardness/softness of a metal binding site and so to control binding selectivity more effectively. They are also needed in the coordination environment of redox-active metals to stabilize the lower oxidation state.

(b) Chelation

Chelate ligands coordinate to a metal ion through two donor atoms to give a cyclic complex. Chelate complexes are more stable than analogous complexes with unidentate ligands of similar chemical type, as illustrated by the greater stability of $[Ni(en)_3]^{2+}$ (where en = $NH_2CH_2CH_2NH_2$) compared to $[Ni(NH_3)_6]^{2+}$. The magnitude of the chelate effect is often overemphasized as comparisons are made between formation constants of different units. Dimensionless formation constants may be calculated using mole-fractions to express concentrations, and these constants show the chelate effect to be much less marked.

(c) Back-bonding in transition metal complexes

Gaseous molecules such as O_2, N_2, CO, NO and N_2O are involved in microbial metabolism. Metal sites are involved in the binding and activation of these small molecules, a phenomenon that is well known in inorganic chemistry.

Both CO and N_2 readily form complexes with transition metals even though they have no basic character, as shown by their failure to undergo protonation. Complex formation arises from the synergic effect in which two bonding components reinforce each other strongly. These are a ligand-to-metal sigma (σ) bond and a 'back-bonding' metal-to-ligand pi (π) bond, represented in Figure 1.8. Sigma donation to the metal occurs to a substantial extent only because the electron density donated to the metal is transferred back to the antibonding π orbitals of the ligand through overlap with the *d* orbitals of the transition metal. The back-bonding only occurs because electron density is supplied to the metal from the ligand. The metal–ligand bond is thus created at the expense of the N–N or C–O bond, as bonding electron density is transferred via the metal into antibonding orbitals. This results in enhanced reactivity of the coordinated molecule.

The dioxygen molecule is activated by binding to Fe(II) or Cu(I). Electron transfer to dioxygen takes place via overlap of a metal 3*d* orbital

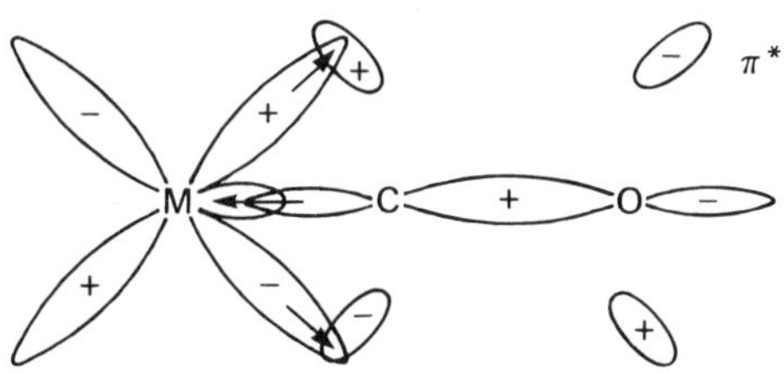

Figure 1.8 The synergic effect: bonding in a transition metal complex of carbon monoxide.

and the antibonding π molecular orbitals of dioxygen. The extent to which such electron transfer occurs is crucial in determining whether a haem protein acts as a carrier of O_2, an oxygenase or an oxidase. This is controlled by several factors, including the nature of the axial ligand supplied by the protein and the strength of its σ-donor and π-acceptor properties.

1.4 ELECTRONIC PROPERTIES OF TRANSITION METAL COMPLEXES

The reactivity of transition metals in biology is moderated by their ligands which control the geometry of the metal site and the electronic properties of the metal centre. In this section the significance of terms such as 'high-spin' and 'low-spin' will be explained.

1.4.1 Crystal field theory

Crystal field theory represents a simple approach to the bonding in transition metal compounds, and accounts semiquantitatively for many aspects of transition metal chemistry.

The orientation of the five 3*d* orbitals is shown in Figure 1.9. It may be seen that they fall into two groups. One group is made up of the $d_{x^2-y^2}$ and d_{z^2} orbitals whose electron density lies along the axes, while the

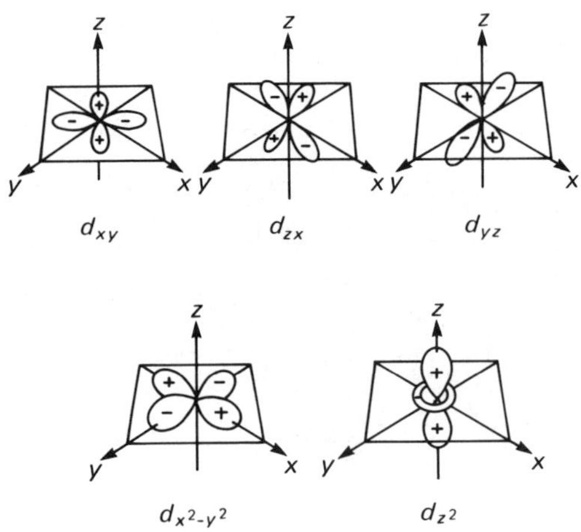

Figure 1.9 The *d* orbitals.

second group involves the other three orbitals which have their electron density lying between the axes.

Crystal field theory views the ligands as point charges and considers the effect of these charges on the energy of the five 3*d* orbitals. In the formation of an octahedral complex, six ligands will be brought up to the metal ion along the axes. The *d* orbitals will rise in energy as a result, but will not all be affected to the same extent. The orbitals lying along the axes will be most affected because electrons in these orbitals will experience the repulsive effects of the entering ligands most strongly. The orbitals lying between the axes will be less affected, as they do not point towards the incoming ligands.

The crystal field of the ligands thus causes the set of five degenerate *d* orbitals (that is, five *d* orbitals of equal energy) to split into two groups. One of them is a set of three degenerate orbitals (t_{2g}) and the other is a set of two degenerate orbitals (e_g). The energy difference between these two sets of orbitals is termed the crystal field splitting (Δ_0) and its magnitude is dependent upon the identity of the ligand. This is shown in Figure 1.10, where the energy given for the unsplit set of five *d* orbitals corresponds to the hypothetical situation in which the ion is surrounded by a spherically symmetrical crystal field without splitting

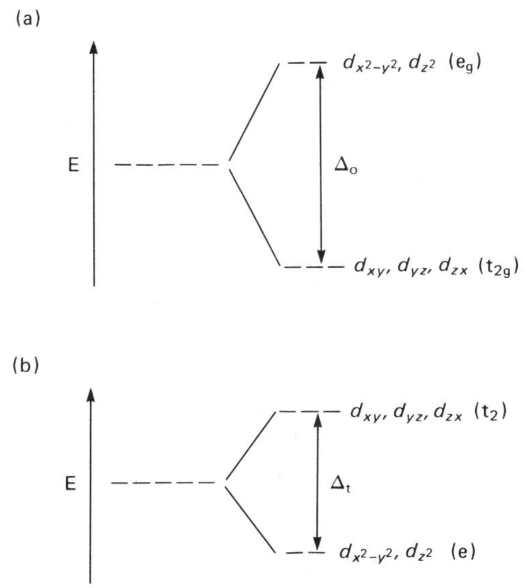

Figure 1.10 The splitting of the *d* orbitals by crystal fields. (a) Octahedral geometry. (b) Tetrahedral geometry.

of the *d* orbitals. Figure 1.10 also shows the splitting of the *d* orbitals brought about by the crystal field of four ligands arranged tetrahedrally about the metal ion. The splitting is reversed compared to the case for octahedral fields, as three of the four ligands are brought up between the axes and so the three orbitals pointing between the axes will be more destabilized than the orbitals pointing along the axes. The value of Δ_t is lower than that of Δ_0, with $\Delta_t = -4/9\Delta_0$.

Geometries around metal ions bound at biological sites are usually irregular. As a result of this, further splitting of the e_g and t_{2g} sets of *d* orbitals may take place. This presents a major complication in the detailed interpretation of the electronic properties of transition metal ions, and additional parameters will be needed to describe these effects fully, as described in standard texts in inorganic chemistry (Cotton and Wilkinson, 1980). Nevertheless, the simplified account presented here for regular geometries is sufficient to develop some basic concepts necessary for further chapters.

1.4.2 The number of unpaired electrons: high-spin and low-spin complexes

The way in which the *d* electrons of a transition metal are arranged in the two sets of *d* orbitals is highly significant. It should be noted that the number of *d* electrons available will depend upon the position of the metal in the transition metal series of the Periodic Table and the oxidation state of the metal. The transition metal iron has an electronic configuration of $3d^6 4s^2$, so iron(III) will have five *d* electrons and a d^5 configuration; Fe(II) is d^6. In general, the filling of orbitals with electrons is governed by Hund's rules, which state that the most stable arrangement is the one having the maximum number of unpaired electrons, arranged with parallel spins. The arrangement of electrons in the *d* orbitals is shown for the configurations d^1 to d^{10} in Figure 1.11 for octahedral complexes. For configurations d^1, d^2 and d^3 there is only one arrangement of electrons, so that these transition metal species have one, two and three unpaired electrons respectively.

For the d^4 configuration there are two possible ways of arranging the electrons. The fourth electron may be placed in the higher energy set of *d* orbitals, obeying Hund's rules, or it may be energetically more favourable to pair this electron than to place it in the higher energy *d* orbitals. In the former case there will be four unpaired electrons and in the latter case only two. The configuration actually found will be the more stable of these two arrangements, which will be decided by the

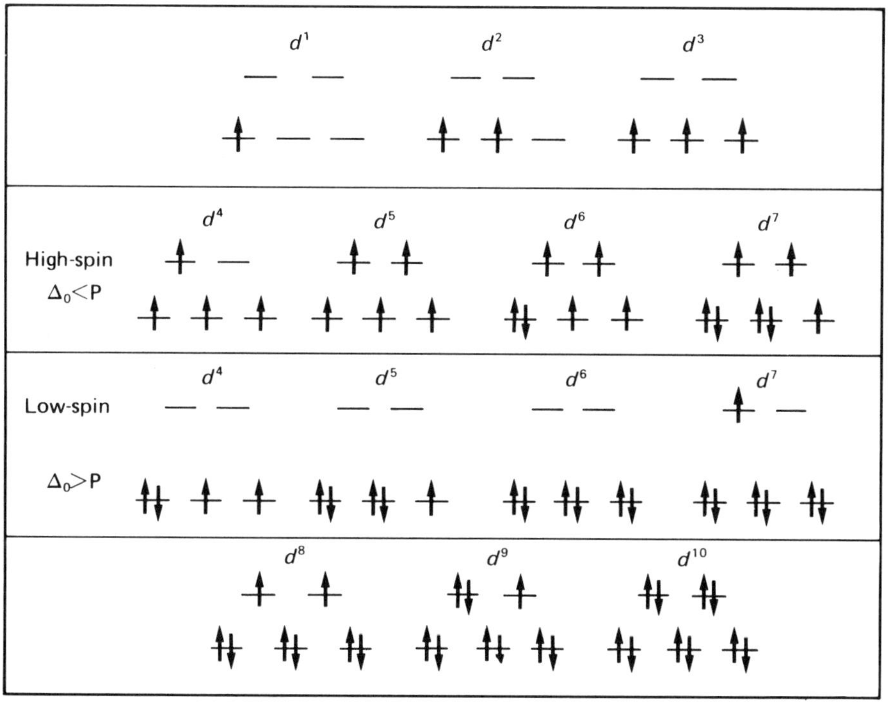

Figure 1.11 Electronic configurations d^1 to d^{10} for octahedral complexes showing high-spin and low-spin complexes.

relative magnitudes of Δ_0, the crystal field splitting and P, the pairing energy of the electrons. Two general cases hold:

1. The strong-field case where $\Delta_0 > P$, and so electrons are paired, giving **low-spin complexes,** and
2. The weak-field case where $\Delta_0 < P$, and it is more favourable to have unpaired electrons, giving **high-spin complexes.**

Figure 1.11 shows that high-spin and low-spin complexes are possible for d^4, d^5, d^6 and d^7 configurations in octahedral complexes. Only one arrangement is possible for d^8, d^9, d^{10} (and for d^1, d^2, d^3).

A similar argument may be applied to tetrahedral complexes, but as Δ_t values are low, the possibility of $\Delta_t > P$ is minimal and only high-spin complexes need be considered. Indeed, there are only two examples in inorganic chemistry of low-spin tetrahedral complexes. So, whenever Fe(II) and Fe(III) are found in iron-sulphur clusters, which offer tetrahedral environments to the metal, it is certain that they will be in the high-spin electronic configuration. In contrast, the iron centres in cytochromes or other haem molecules may be either high-spin or low-spin, depending on the nature of the axial ligands. Spin-state properties may be linked to the biological properties of the haem compound. Low-spin haems may be more effective in electron-transfer reactions, as the electron density is concentrated in orbitals that point towards the ligands.

A compound may be demonstrated to be low-spin or high-spin from its magnetic properties, which are related to the number of unpaired electrons present in the transition metal centre. Compounds with no unpaired electrons are diamagnetic. Compounds with unpaired electrons are paramagnetic, the value of the magnetic moment depending upon the number of unpaired electrons. As may be seen from Figure 1.11, low-spin Fe(II) (d^6) is diamagnetic and high-spin Fe(II) is paramagnetic, with four unpaired electrons. Low-spin Fe(III) and high-spin Fe(III) (d^5) are both paramagnetic, but have one and five unpaired electrons respectively. A fuller account of the magnetic properties of transition metal complexes may be found in the standard texts of inorganic chemistry (Cotton and Wilkinson, 1980).

1.4.3 Electronic spectra

Compounds containing transition metals are nearly always coloured. A number of metalloenzymes and metalloproteins also owe their colour to the presence of the transition metal, for example the blue copper electron transfer proteins such as azurin and plastocyanin (Hughes, 1987). Such colours arise from two sources, charge-transfer or *d–d* electronic transitions. Charge-transfer bands are of high intensity and usually involve the transfer of an electron between a metal orbital and a ligand orbital, which may occur in either direction. Charge-transfer bands are usually found in the u.v. region of the spectrum, but may tail off into the visible region. Occasionally, the charge-transfer band is of lower energy and is found in the visible spectrum. The energy at which charge-transfer bands occur will depend on the relative ease of oxidation or reduction of metal or ligand, and so, for a fixed metal ion, is diagnostic of the ligand, as illustrated in Section 2.6.1 (a).

The *d–d* bands are much weaker in intensity than charge-transfer bands. They arise from electronic transitions between the sets of *d* orbitals formed by the splitting of the degenerate *d* orbitals by the ligands. As a result, *d–d* spectra give a great deal of information about the geometry of the transition metal site, including indications of distortions from regular stereochemistry. Standard texts should be consulted for detailed discussion of these topics.

(a) Magnitude of crystal field splitting: the spectrochemical series

Values of the crystal field splitting parameter Δ may be obtained from *d–d* spectra, since these result from electronic transitions between the two sets of *d* orbitals. If all other factors are constant, the value of Δ will depend upon the ligand, that is upon the ability of the ligand to split the *d* orbitals. The Spectrochemical Series lists ligands in order of increasing values, for example, $I^- < Br^- < Cl^- < F^- < OH^-$ < oxalate $\leqslant H_2O < NH_3 \approx$ pyridine < ethylenediamine < bipyridyl $< NO_2^- < CN^- \approx CO$. Some ligands of biological interest lie in the sequence $–SH^- < –CO_2^-$ < amide < imidazole. The halides are at the weak end of the series because they are π-donors to the metal, while the ligands at the strong end are those with empty π orbitals which can accept electron density from the metal (as illustrated for carbon monoxide in Figure 1.8).

Values of Δ for a transition metal centre are an approximate average of the values for the individual ligands coordinated to the metal. It is possible, therefore, to draw limited conclusions about the types of ligand around a metal in biological sites from the value of Δ in association with the Spectrochemical Series.

1.5 ROLES FOR METALS

The roles of metal ions must ultimately reflect their fundamental properties, namely charge, size, electronic configuration (and hence redox properties), and rates of ligand exchange. Some points have been made earlier in this chapter. The IA cations do not bind ligands strongly, and so in solution behave as mobile hydrated cations, whose distribution is controlled by membranes. This gives rise to charge and concentration gradients across membranes which can be exploited in various ways. Other metal ions bind to ligands more tightly. This can be done selectively. A metal cation can complex with two ligands, and so cross-link groups, with stabilization of a structure or control of orientation of reactants. Coordination of the ligand to the metal may affect the charge distribution in the ligand and hence its reactivity. Finally, the nature of the ligand bound to a transition metal species will affect its redox potential. Thus, metal–ligand interaction is associated with selectivity, stabilization, cross-linking, orientation, reactivity and tuned redox potential (E^0) values – all terms that relate positively to important functions in biology. These roles will be discussed further in subsequent sections, although it must be emphasized that the relationship between the metal ion and its binding groups is a reciprocal one, in that these groups help control the behaviour of the metal ion by providing certain ligands in particular irregular geometries (Williams, 1985). It should also be noted that a metal ion may fulfil alternative roles under different sets of conditions and that, in general, metal ions serve as neutralizers of negative charge on a range of organic biopolymers and help to maintain osmotic balance inside the cell.

1.5.1 Trigger and control processes

The cations Na^+ and Ca^{2+} are usually associated with this type of process. The normal exclusion of Ca^{2+} from the cell cytoplasm means that the appearance of Ca^{2+} can trigger a range of biochemical and physiological processes. In general this does not involve direct activation of enzymes by binding of Ca^{2+}, as greater specificity can be achieved by the use of a calcium-binding protein such as calmodulin to mediate the trigger action of the cation. Activation of the enzyme takes place by binding the Ca^{2+}–calcium binding protein complex. Specific cases will be discussed in Chapter 4, but, for example, it may be noted that reversal of direction of movement of bacteria, as in chemotaxis, involves the influx of Ca^{2+}. This influx follows the concentration gradient passively from high to low concentrations of calcium, and is probably initiated by the opening of a gate that controls entry to the channel. The presence of Ca^{2+} in the cell gives rise to a reversal of the direction of movement, which will continue until the calcium level is reduced back to the original value by the operation of the efflux pump for calcium. Forward movement is then resumed.

The concentration gradients of alkali metal ions across membranes are exploited in several ways. The Na^+-gradient is involved in the transport of amino acids, which enter the cell as complexes with Na^+. Once inside the cell Na^+ dissociates from the complex and is pumped out. This and other examples are discussed in Section 4.2.1. Other metal ions known to function as catalysts in enzymatic processes may also serve as moderators of activity. Thus Zn^{2+} may have binding sites of high and low affinity on enzymes. Binding to high-affinity sites results in catalytic activity, but binding to low-affinity sites (at high concentrations of Zn^{2+}), gives rise to inhibition. Thus, high concentrations of Zn^{2+} cannot cause over-stimulation of enzyme activity.

1.5.2 Structural roles

Metal ions stabilize a variety of structures, usually by acting as a bridge between two groups. For example, thermolysin is a heat-stable, zinc-containing endopeptidase from *Bacillus thermoproteolyticus*. The heat stability is due directly to the presence of four calcium ions, which prevent the unfolding and denaturation of the protein. Matthews *et al.* (1974a,b) have established the structure of this enzyme. Zinc and potassium cations have similar structural roles in other enzymes. In some cases addition of the cation induces a conformational change in an enzyme and the generation of the active site configuration. Such behaviour has been established for potassium in yeast aldehyde dehydrogenase (Bostian *et al.*, 1982). It is possible that the importance of potassium ions in this type of process has not been fully appreciated, as enzymes may never have been studied under potassium-free conditions. In these examples, metal cations affect structures by cross-linking groups in different parts of the protein primary structure, in a manner analogous to the function of disulphide bridges. Metal cations can also serve to bridge an enzyme and its substrate, serving as a template to bring reacting groups into the correct relative orientation for reaction.

The interactions of the cations K^+ and Mg^{2+} with nucleic acids are of particular importance. Both cations play essential roles in the stabilization of nucleic acids and ribosomes, and in the activation of key enzymes in the biochemistry of the nucleic acids, as discussed in Chapter 4.

Divalent metal ions also help to stabilize cell walls by cross-linking carboxylate groups. This stabilization is reversed on removal of the cation, so allowing the permeability of the cell wall to be controlled. Reference was made in Section 1.2.1 (a) to the merits of Ca^{2+} over Mg^{2+} in carrying out this role in view of the greater affinity of calcium ions for binding sites of irregular geometry.

1.5.3 Metal ions as Lewis acids

Metal ions act as Lewis acids by accepting electron pairs from ligands. The strengths of Lewis acids increase with the polarizing power of the cation and, for the divalent transition metal ions, follow the Irving Williams series. If the group IIA cations are included, the series is $Ca^{2+} < Mg^{2+} < Mn^{2+} < Fe^{2+} < Co^{2+} < Ni^{2+} < Cu^{2+} > Zn^{2+}$.

The important Lewis acids in biology are Zn^{2+} and Mg^{2+}, although Ni(II) probably has this function in urease, and Fe(III) and Mn(III) may act as Lewis acids in the acid phosphatases. Zinc is a strong Lewis acid but can exercise this function in two distinct ways in the catalysis of hydrolysis reactions (Sigel, 1983). In one case, a substrate molecule such as a peptide is coordinated to the metal through the carbonyl group, resulting in polarization of the group and enhanced reactivity towards hydrolysis (Equation 1.6). In the second case, the zinc enhances the deprotonation of a Zn^{2+}-bound water molecule, generating a coordinated hydroxide group which is an effective nucleophile towards the substrate (Equation 1.7). In both equations the ligands to zinc are not shown. The reactivity of Zn^{2+} as a Lewis acid is much enhanced in enzymes, compared with simple coordination complexes, by the action of the protein in providing a site of appropriately distorted geometry and specific ligand groups.

$$\begin{matrix} X \\ Y \end{matrix}\!\!>\!C{=}O + Zn^{2+} \rightarrow \begin{matrix} X \\ Y \end{matrix}\; C{=}O{-}Zn^{2+} \xrightarrow{H-OH} XH + YCOOH + Zn^{2+} \qquad (1.6)$$

$$Zn{-}OH_2 \rightleftharpoons Zn{-}OH + H^+;\ Zn{-}OH + \text{substrate} \rightarrow \text{products} \qquad (1.7)$$

The fact that Zn^{2+} is a strong Lewis acid makes it an unsuitable catalyst for reactions involving weaker base centres. It is not unexpected, therefore, that a much weaker Lewis acid is also available in biology. This is Mg^{2+}, which is bound only weakly to enzymes and substrates, and which is associated particularly with the catalysis of reactions involving the hydrolysis and transfer of phosphates.

It is interesting to enquire why these two cations have been selected to function as Lewis acids, when, for example, Cu(II) is a stronger Lewis acid than Zn(II). It is noteworthy that magnesium and zinc are not transition elements and so do not undergo redox reactions. It is appropriate that they have been selected to catalyse reactions involving compounds such as proteins and nucleic acids because these compounds contain groupings which can be oxidized readily or damaged by free radicals. This in turn leads to the fundamental question of selectivity. How is it possible to distinguish between Cu(II) and Zn(II), and to place Zn(II) in the active sites of enzymes that catalyse hydrolysis reactions?

1.5.4 Transition metals as redox catalysts

Changes in the oxidation state of a transition metal in a redox enzyme may be linked either to the catalysis of the transfer of an electron or to the direct oxidation or reduction of a substrate (Section 1.2.1 (b)).

Proteins that catalyse electron transfer include cytochromes, iron-sulphur proteins and the blue copper electron transfer proteins. Both iron and copper are also involved in oxidases, oxygenases and hydrogenases. Molybdenum, vanadium, nickel and manganese (as Mn(III)) have a more limited role in biological redox processes. The function of all these transition metals is discussed in Chapter 5.

(a) Coupled redox centres

Iron and copper are each able to catalyse the transfer of one electron only. Despite this, they are often involved in redox reactions where the substrate undergoes two-, four- or six-electron redox reactions. One example is the reduction of the oxygen molecule to water, which is a four-electron process involving the intermediacy of reactive, toxic species such as superoxide and peroxide. The reaction must either involve firmly bound superoxo or peroxo intermediates or be carried out rapidly to minimize the opportunity for build up of these reactive species, particularly superoxide ion. This is accomplished through the participation of multi-metal redox centres or two covalently linked redox proteins, which can rapidly transfer two electrons to the substrate. Examples of this behaviour include (i) cytochrome oxidases of the aa_3 type, which have four metal centres made up of haems a and a_3 and two copper ions, in which case the coupled a_3-Cu site transfers two electrons to the bound dioxygen species; (ii) bacterial cytochrome c peroxidase, which has two haems coupled together to effect the two-electron reduction of hydrogen peroxide to water; and (iii) the NADPH-sulphite reductase of *Escherichia coli*, which also catalyses the six-electron reduction of nitrite to ammonia in two-electron steps through sirohaem covalently linked to a 4Fe.4S cluster (Christner *et al.*, 1983).

(b) Control of redox potential

One of the great advantages of a transition metal redox centre is that the value of the redox potential can be adjusted to specific values within a wide overall range. The factors that determine the magnitude of the reduction potential of a biological transition metal redox centre have been the subject of vigorous debate over many years, and a full analysis is beyond the scope of this chapter. Factors of major significance must include the chemical character of the ligands, which may stabilize either the oxidized or the reduced form of the metal, the electrostatic interactions between the redox centre and other polar groups, and the accessibility of the solvent to the metal site in the protein. Discussions are available of the various factors which influence the reduction potentials for cytochrome *c* (Moore and Williams, 1977), blue copper proteins (Roberts *et al.*, 1984), iron-sulphur proteins (Carter, 1977; Sheridan, Allen and Carter, 1981) and haem proteins (Stellwagen, 1978). A more recent general review is available (Moore, Pettigrew and Rogers, 1986), which assesses critically some of this earlier work.

(c) Mechanisms of redox reactions involving transition metal centres

This discussion is concerned with the oxidation or reduction of a substrate by transition metal complexes and, particularly, with electron–transfer reactions between transition metal centres. Redox reactions between two transition metal complexes are classed as **inner sphere** or **outer sphere**. Inner sphere reactions are those in which the oxidant and reductant are connected by a bridging ligand which is common to the coordination shell of both metal centres. Electron transfer occurs through the bridge, and is followed by bridge breaking, so that group transfer between the transition metal centres may occur, depending upon which metal-bridging ligand bond is broken. Outer sphere reactions involve electron transfer through the intact coordination shell of both metal ions, and require polarizable ligands for fast reaction. Redox reactions between transition metal centres and non-metallic compounds often involve coordination of the substrate, and possibly group transfer to or from the transition metal. Oxo-group transfer is probably an important feature of certain molybdoenzymes catalysing redox reactions of oxoanions.

Several factors contribute to the enhancement of the rate of electron transfer between metal centres in biology. The metal-binding sites provided by the protein ensure that the metal centres undergo minimal structural reorganization during the oxidation-reduction process, as noted in the discussion of the entatic state hypothesis for copper (Section 1.2.1 (b)). This makes them excellent electron-transfer catalysts, while the existence of polarizable and conjugated ligands ensures that electrons can move rapidly to and from the metal centre. In general, non-complementary redox reactions are slow, but this problem is overcome in biology by the use of multi-metal, coupled centres (Section 1.5.4 (a)). Electron transfer rates are also enhanced as the differences in redox potentials between transition metal centres are small.

Electron transfer between components of a membrane-bound respiratory chain is very fast, with an activation energy as low as 30 kJ mol^{-1}. This is due to the highly organized arrangement of electron carriers, the use of entatically-adjusted metal centres and highly conjugated ligand systems. The redox centres may be $15–25 \times 10^{-10}$m apart from each other. This distance is spanned partly by the use of porphyrins in cytochromes, but there will also be the need for entry channels through the protein for the electron to obtain access to the porphyrin system in an outer sphere type reaction. The role of protein residues in facilitating such electron movement is now thought to be of minor importance. In some cases, where metalloenzymes have a partially exposed metal centre, direct binding of a ligand may be important.

The ligand which receives an electron from the metal centre must have a low-energy, empty, acceptor orbital and a conjugated system to transfer the electron away before recombination with the metal can occur. If the ligands interact covalently with the metal, this will result in a decreased effective positive charge on the metal and the expansion of the *d*-electron density out from the nucleus. This will then favour electron transfer to the ligand, particularly if the metal is in a low-spin state, with the electron density accessible to the ligand.

1.6 SELECTIVE BINDING OF METALS

The question of selectivity is central to a full understanding of the roles of metal ions in biology. The correct functioning of an enzyme or some biological structure may depend upon the placing of a particular metal ion at a specific site, from which other metal ions must be excluded. Although it is not yet possible to give detailed assessments of how this can be accomplished for each metal ion, some generalizations seem possible at this stage. These generalizations are embodied in the well-established analytical chemistry of metals, where specific tests have been developed for the selective determination of one metal in the presence of others.

1.6.1 Selective binding of metals in analytical chemistry

Textbooks of analytical chemistry give a range of specific tests for metal ions which depend upon the formation of coloured coordination complexes with ligands that bind exclusively to a particular metal ion. These methods may then be used to analyse, specifically and quantitatively, for one metal in the presence of other competing metals. Several strategies have been developed to accomplish this.

One important method involves 'masking' other metal ions by adding a ligand which complexes them more strongly than does the ligand to be used in analysis. These metals cannot then participate in the analytical reaction. An example of this approach involves the use of

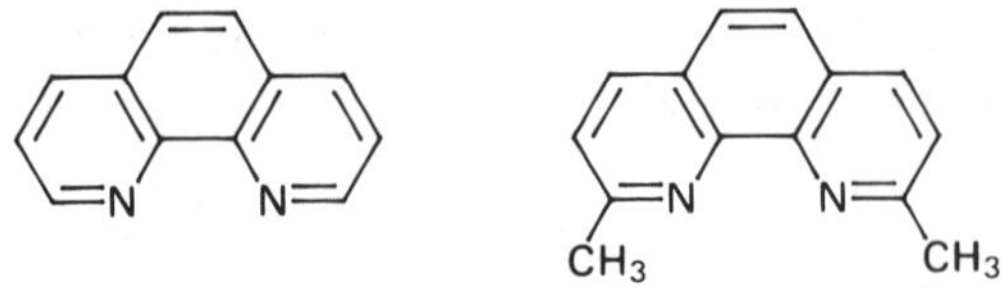

Figure 1.12 The structure of some 1, 10-phenanthrolines.

dimethylglyoxime to determine nickel, which is precipitated as an insoluble dimethylglyoximato complex in gravimetric analysis. Cobalt and iron interfere with this determination because they form slightly insoluble complexes with dimethylglyoxime which will be precipitated along with the nickel complex. The problem is overcome by adding *N*, *N*–bis (hydroxyethyl)glycine. Iron and cobalt form stable soluble complexes with this ligand and so are not available for reaction with dimethylglyoxime. In contrast, nickel forms a weaker complex with the substituted glycine than with dimethylglyoxime, and so the former ligand does not prevent the quantitative precipitation of nickel by dimethylglyoxime.

Selective analysis of metal ions may also exploit the preferences of the metals for certain geometries. This requires the use of bulky ligands which impose steric constraints for some geometries and not for others. The ligand 1,10-phenanthroline binds Fe(II) and Cu(I) well, as it is able to accommodate their preferences for octahedral and tetrahedral geometries respectively. The addition of methyl groups to positions adjacent to the nitrogen donor atoms in the 1,10-phenanthroline molecule (Figure 1.12) makes this molecule a selective ligand for Cu(I) in the presence of Fe(II). The presence of the methyl groups prevents the ligand from satisfying the stereochemical requirements of Fe(II) and complexation does not occur.

These examples show that a metal ion may be bound selectively by a ligand which has appropriate donor atoms, and which is able to accommodate the normal geometry of the metal ion. Other metal ions must be excluded by features in the ligand which do not accommodate their usual geometries, provided these are different from the stereochemical requirements of the analyte. Otherwise, a masking ligand may be used to remove competing metal ions.

1.6.2 Selective binding of metals in biology

Discrimination between the metals of biological importance results immediately from the nature of the groups presented in the binding site. If the site is made up of nitrogen or sulphur ligands, then the binding of the transition metals or zinc is favoured and the IA and IIA cations will be excluded. Each of these two sets of metals will now be considered further in turn.

(a) Selective binding of Na^+, K^+, Mg^{2+} and Ca^{2+}

A range of oxygen-donor ligands are available which distinguish effectively between Na^+ and K^+ and between Mg^{2+} and Ca^{2+} on a size basis. There are substantial differences in ionic radii between these cations (Na^+, 0.95×10^{-10}m; K^+, 1.33×10^{-10}m; Mg^{2+}, 0.65×10^{-10}m; Ca^{2+}, 0.99×10^{-10}m), a situation which contrasts with that found for the divalent transition metal ions. Figure 1.13 shows some illustrative examples of ligands which bind the IA and IIA cations selectively by providing a central cavity that matches the size of the cation, although this is an over-simplified view. The crown ethers 18-crown-6 and 15-crown-5 bind K^+ and Na^+ selectively, although the binding is still weak. The bicyclic cryptands bind the IA and IIA cation more strongly and more selectively (Lehn, 1978a,b). The cryptand 2.2.2 is selective for K^+, while the cryptand 2.2.1 (not shown) is selective for Na^+. Figure 1.13 also shows some naturally occurring ionophoric antibiotics which bind cations selectively on a similar size basis (Sigel, 1985). Valinomycin, a neutral cyclic peptide, binds K^+ selectively and can carry it across cell membranes. The resulting equalization of potassium concentrations across membranes accounts for its function as an antibiotic. The actins also provide a cavity appropriate for the binding of K^+. The smaller sodium cation misfits in these binding cavities, and interacts only weakly with the ligands.

The differentiation between Na^+ and Ca^{2+} must involve a more subtle distinction as these cations have similar radii. Selectivity necessarily has to exploit the different charges and polarizing power of the two cations. This has been achieved for crown and cryptand ligands with hydrophobic substituents, which show a selectivity for Na^+ and against Ca^{2+}. This results from the greater readiness of the aqua sodium ion to lose water from its coordination shell, which facilitates its approach to the hydrophobic binding cavity.

These simple concepts, established for model ligands, may well underly the selectivity of biological molecules towards the IA and IIA cations. When such discrimination is associated with enzymatic processes involving membrane-bound cation pumps, this will lead to the selective distribution of these cations inside and outside the cell, subject to due expenditure of energy.

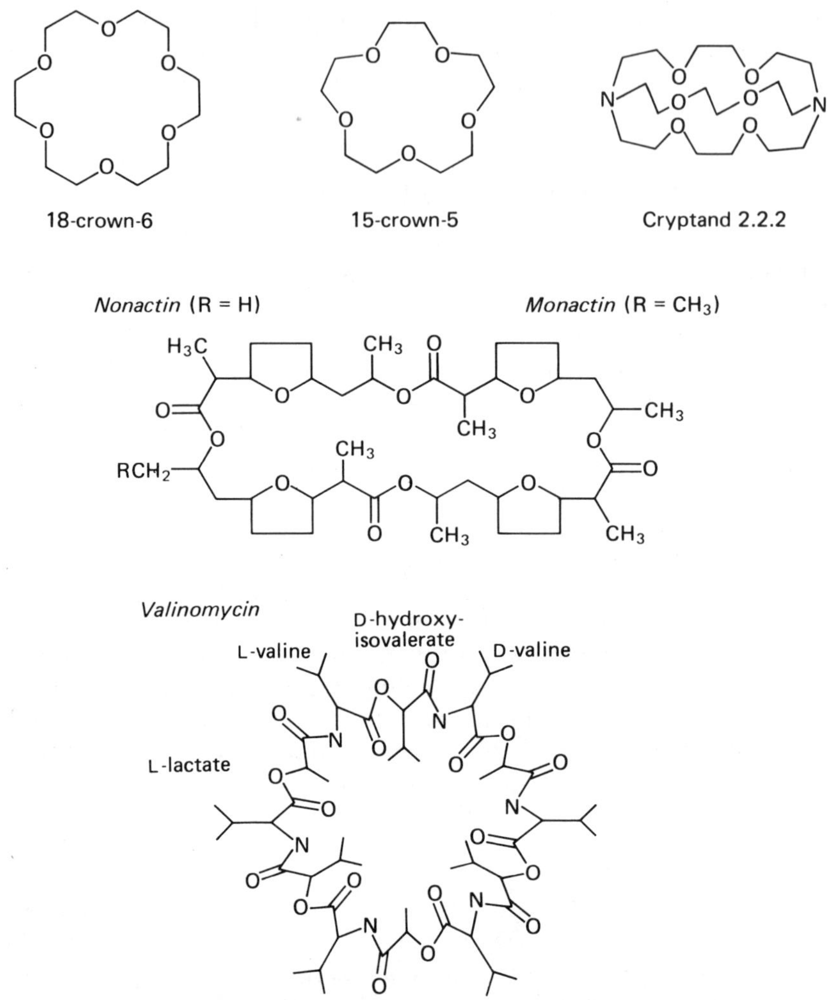

Figure 1.13 Cation-selective ligands: crowns, cryptands and potassium-selective antibiotics.

(b) Selective binding of the transition elements and zinc

The members of this group do not show extreme differences one from another in terms of their preference for particular binding groups. This is shown by a comparison of the formation constant data in Table 1.4, which also demonstrates the fact that Cu(II) is by far the strongest Lewis acid. In a situation where all these metals are in excess, every site will be occupied by Cu(II) to the exclusion of other metal ions.

It is clear that other cations will only be able to win the competition for sites if the concentration of copper in the cytoplasm is reduced dramatically. This could be achieved by excluding Cu(II) from the cell, which would not be straightforward. It is probably a simpler matter to allow Cu(II) to enter the cell and then to mask its presence, in an analogous fashion to that used in analytical chemistry, either by precipitation or by complexation to a protein, which will bind first to the strongest Lewis acid. The latter strategy appears to be used by cells, possibly through the synthesis of a metallothionein-like protein in response to the presence of copper(II) ions (Williams, 1984). Provided this protein is not in excess, other cations will then be able to compete for specific sites on intracellular molecules. However, Zn(II), as the second strongest Lewis acid, would probably dominate the new competition, and other devices would be needed to allow the other cations to be selected. Thus, provision of tetrahedral sites would discriminate against the Ni(II) cation, as this species prefers octahedral geometry. However, the importance of concentration must be emphasized, as this could have a marked effect on the outcome of any competition between metal cations for binding sites. Intracellular concentrations of cations are often reduced by precipitation or through uptake by a specific storage protein involved in homeostasis, such as iron in bacterial ferritin. The cell may thus be able to control site specificity by adjustment of the relative concentrations of metal ions in the localized environment.

1.7 SUMMARY

The subject of metal–microbe interaction is one of immense current interest and it encompasses a remarkable range of disciplines. Some metals are essential for microbial life while others are toxic: in both cases important questions about the underlying biochemistry and physiology have to be answered. Micro-organisms are now being used to help resolve industrial and environmental problems involving metals; these include the recovery of metals from low grade or refractory ores and the removal of toxic metals from aqueous effluent and process streams. The role of micro-organisms in geological processes, such as the laying down of mineral bodies, is only now being fully appreciated.

The essentiality of a metal may be demonstrated by setting up conditions where the supply of the metal is growth-limiting. Metals presently known to be essential for growth are the *s*-block metals sodium, potassium, magnesium and calcium, and the trace or ultra trace metals from the *d*-block of the Periodic Table (vanadium, chromium, manganese, iron, cobalt, nickel, copper, zinc and molybdenum). These metals have a variety of functions including the stabilization of structures, from protein conformations to cell walls, the catalysis of cellular reactions such as electron transfer, redox reactions and hydrolysis, and the triggering or moderating of a range of processes.

Other metals such as cadmium, mercury, tin, lead, arsenic, silver and aluminium are toxic towards micro-organisms, although the extent of the effect is dependent on the chemical form of the metal species in solution. The study of the mechanisms by which some micro-organisms resist these toxic effects is an area of intense current interest (discussed in Chapter 6).

Binding sites for metals are found in proteins, nucleic acids and polysaccharides, while a range of specialized ligands are also known. These ligands can bind metals selectively, in a way that can be understood in part by the study of basic chemical principles and simple model compounds. The factors that determine the magnitude and nature of metal–ligand interaction include the polarizing power and hard-soft character of the cation, together with special factors such as the chelate and synergic effects. The biochemistry of the transition elements and zinc is of particular interest. The electronic properties of the transition elements allow the exploitation of certain techniques that assist in the characterization of their local environment. Such environments around metals in active sites of enzymes are often distorted, and may be similar to the geometry required by the transition state of the reaction being catalysed. The transition metal ions are important redox catalysts, partly as a result of the way in which their redox potentials can be tuned finely to give particular values.

2

Metal nutrition: the supply and analysis of metals in microbial cultures

What does not poison, fattens

Italian proverb

2.1 INTRODUCTION

2.1.1 Historical aspects

It is axiomatic today that micro-organisms require metals for growth. However, the history of microbiology reveals that this requirement was appreciated rather late. The method of growing bacteria in transparent liquid media was employed by Pasteur, who made various attempts to obtain the most suitable nutritive mixture. Thus in 1861, he used 100 parts water, ten parts pure candy sugar, one part ammonium tartrate and one part of ash of yeast. This 'Pasteur's fluid' was improved by Mayer in 1869 and made suitable for a wide range of organisms by using solutions of the salts found in yeast ash. Cohn followed Mayer's idea to prepare a medium ('Cohn's fluid') which differed from that of Pasteur by omission of sugar and the inclusion of potassium phosphate, magnesium sulphate, calcium phosphate and ammonium tartrate. Simultaneously, Raulin developed a medium containing the ash of *Aspergillus* mycelium as a source of Zn and Fe. These appear to be the first conscious attempts to include inorganic compounds in microbial growth media and, incidentally, provided the earliest demonstration of a role for zinc in a living organism.

Interestingly, textbooks written half a century later pay scant attention to the requirements for salts in bacterial nutrition, and give the

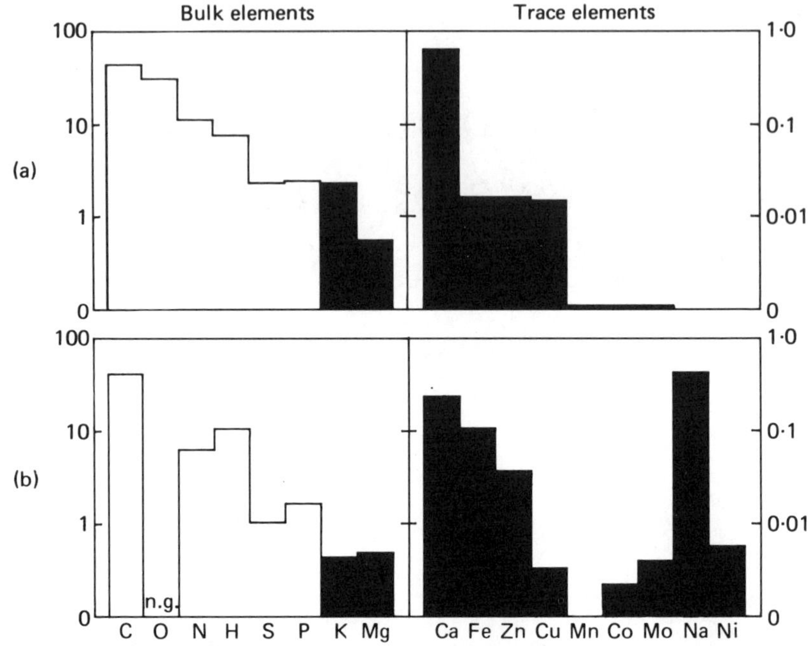

Figure 2.1 The elemental composition of bacteria: (a) shows common literature values for the concentration of elements as a percentage of dry weight; (b) shows the values (%, w/w of lyophilized cell material) determined for *Methanobacterium barkeri* strain 227 by Scherer, Lippert and Wolff (1983). Metals are indicated by solid columns, non-metals by open columns. Note the scale difference between left and right-hand sides. n.g. = not given.

impression that the phosphates of magnesium, calcium, sodium or potassium are necessary only because of the inclusion of phosphate.

2.1.2 Classification of metals as bulk or trace elements

Only a few metals, notably potassium and magnesium, are required as bulk or major elements; the rest are trace or minor elements. This somewhat arbitrary (and sometimes variable) distinction is based on the concentration at which metals are required for growth or at which they are detectable on chemical analysis of cells (Figure 2.1). The bulk metals are involved in maintaining osmotic balance and providing an appropriate cationic environment. These metals, from groups IA and IIA of the Periodic Table, are also the major cations in sea water (Table 2.1). The trace metals, on the other hand, have more specific physiological roles and are usually transition elements.

Table 2.1 The essential elements of life and their abundance in the ocean

Major elements (mM)		Minor elements (mM)			
C, H, N, O					
Cl	536	B	0.426	Se	0.000 051
Na	456	Si*	0.107	Cu	0.000 047
Mg	56	F	0.068	V	0.000 039
S (SO_4^{2-})	28	Cr	0.000 96	Mn	0.000 036
Ca	10	I	0.000 47	Ni*	0.000 036
K	9.7	Al	0.000 37	Sn	0.000 025
P(HPO_4^{2-})	0.001	Fe*	0.000 18	Co	0.000 008
		Zn	0.000 16	Pb(?)	0.000 000 1
		Mo	0.000 104		

* Values can vary between water masses by a factor of 1000.
Reproduced by courtesy of Harrison and Hoare (1980).

2.2 PROVISION OF METALS IN GROWTH MEDIA

2.2.1 Sources and concentrations of metal ions

In the complex or natural media still widely used, the necessary metals may be provided by extracts of plant or animal origin, e.g. corn steep liquor, peptose, tryptone or serum. Yeast extract, for example, contains all metals generally necessary for growth and has specially high levels of copper, iron, magnesium and zinc (Grant and Pramer, 1962). The iron contents of some common medium ingredients are given by Lankford (1973).

In a chemically defined medium, both bulk and trace nutrients are provided as organic or inorganic supplements of known composition. When only those nutrients essential for growth are provided, the medium is said to be minimal. Some metal requirements are only now beginning to be identified (nickel and vanadium being examples) and others may be revealed in the future, so that this definition should be used with caution. Although careful consideration is frequently given to the qualitative composition of growth media, quantification of cellular nutrient requirements is often ignored and medium composition is dictated by adherence to recipes that have been proved successful previously. It is possible, however, to design a chemically defined medium (Pirt, 1975) based on the elementary composition of biomass (Figure 2.1) and the calculated growth yields on each nutrient (Table 2.2).

However, there are pitfalls in this approach also. The concentrations of trace metals that permit optimal cell growth (judged by yield of biomass)

Table 2.2 Composition of a chemically defined medium to produce *Klebsiella aerogenes* at concentrations up to 10 g dry biomass/l in aerobic culture

Constituent	Major elements provided or other function of constituent	Theoretical growth yield (g dry biomass/ g element)	Mass of constituent (g)
Water (distilled)			1000
Glucose	C, energy	1.10	22.7
NH_4Cl	N	8.75	4.37
KH_2PO_4	P + K	39.1 for P 59.5 for K	1.13*
$MgSO_4.7H_2O$	S + Mg	333 for S 430 for Mg	0.232
$CaCl_2.2H_2O$	Ca	3.33×10^3	0.011
$FeSO_4.7H_2O$	Fe	6.7×10^3	0.007
$MnSO_4.4H_2O$	Mn	2.0×10^4	0.002
$ZnSO_4.7H_2O$	Zn	2.0×10^4	0.002
$CuSO_4.5H_2O$	Cu	10^5	0.000 4
$CoCl_2.6H_2O$	Co	10^5	0.000 4
EDTA, disodium salt dihydrate	Chelation		0.394†

* The potassium requirement is met by 0.586 g KH_2PO_4.
† 1 mol EDTA/mol Mg and other non-alkali metals. Reproduced by courtesy of Pirt (1975).

are frequently higher than the metal concentrations that result in optimal yield of metabolites. For example, zinc concentration is critical for secondary metabolic processes in fungi and actinomycetes, and manganese concentration is crucial in *Bacillus* species. Iron, copper and cadmium also exert important regulatory effects. A possible explanation is that since many secondary metabolites possess strong metal binding sites they could be involved in transport of specific metals. The yield of metabolites could also be depressed by concentrations of the metal that permit maximal vegetative growth (Weinberg, 1982). Siderophores (Chapter 3) are a good example of this class of metabolite. Conversely, the production of some bacterial secondary metabolites, such as pyocyanine and cyanide, requires higher iron concentrations than does growth of the cells so that, to maintain homeostatic control of metal metabolism, micro-organisms may need to form molecules to remove excess metals. Similarly, a concentration of manganese 10 to 100 times greater than that required for vegetative growth is required by various species of *Bacillus* for synthesis of antibiotics and toxins and development of bacteriophage and endospores. In a recent example (Scott, Jones and

Gaucher, 1984), supplementation of submerged cultures of *Penicillium urticae* with 152 μM manganese allowed accumulation of patulin (a polyketide antibiotic and mycotoxin) to high concentration. At 1.5 μM manganese, only patulin precursors accumulated, although the biomass achieved was routinely greater.

The stage of growth of a batch culture and environmental factors can also affect the apparent metal requirement. For example, the uptake of metals that are required for secondary metabolism in quantities higher than for vegetative growth is accelerated in the latter stages of batch growth and in the stationary phase. Copper and manganese, required in sporulation, are accumulated following exponential growth of *B. megaterium*, whereas zinc is lost, such that an elemental analysis of cells from a stationary phase culture would not reflect accurately the metal requirements for exponential growth. Temperature also regulates secondary metabolism and may do so by affecting trace metal uptake. Gram-negative bacteria, fungi and algae have elevated trace metal requirements at higher temperatures (Weinberg, 1977).

Calcium is required for DNA uptake during transformation of *E. coli* at a concentration that exceeds by tenfold the level (5 μM) required for optimal growth (Page and Doran, 1981). Even higher concentrations are required to render *Ps. aeruginosa* and some other bacteria competent (Mercer and Loutit, 1979).

2.2.2 Chelation of metal ions

One aspect of defined and minimal media that is of paramount importance, but frequently overlooked, is that of metal chelation. Hutner (1972) has bemoaned that 'inspection of catalogs of commercial media for cell cultures indicates a prevailing inadequacy in respect to trace elements, judging from the nutrition of free-living microbes . . . Trace elements besides Fe are seldom included and chelating agents, as such, are omitted. Hence the media are buffered, in respect to transitional metals, only by the weak chelation lent by histidine and other amino acids. Ascorbic acid or α-ketoglutarate in a medium fortified with Zn, Mn and Cu, besides Fe, may provide some additional chelation.' These observations are confirmed by our own experience of common growth media. As examples, the very widely used media 'M63' and 'M9' (Miller, 1972) for *E. coli* and a defined medium for yeast (Light and Garland, 1971) do not contain chelating agents. However, the absence of a chelating agent and the consequent formation of a precipitate that may contain a significant proportion of a metal (as phosphate) need not prevent an organism utilizing the metal (Gaunt, Trinci and Lynch, 1984).

A frequently misunderstood aspect of the use of chelators is their ability to control free metal concentration in the medium and thus widen the margin between adequacy of supply and toxicity in unchelated media. Hutner (1972) cites the example of potential copper toxicity from its presence in peptones, sea water and reagent-grade chemicals. Citrate prevents iron precipitation and in its presence bacteria may tolerate iron concentrations greater than 100 ppm (Waring and Werkman, 1942). Chelators can also be used to advantage when collecting aqueous samples for microbiological analysis by preventing injury to the sample from contaminating metals (Domek *et al.*, 1984).

Chelating agents themselves may cause growth inhibition, which can be relieved by addition of further metals. Thus the growth of *Legionella pneumophila* (the agent of Legionnaire's disease) in a complete defined medium is particularly prone to inhibition by EDTA, citrate or 2,2′-bipyridyl and inhibition is reversed by adding compounds of cobalt, copper, iron or other metals (Reeves *et al.*, 1981).

2.2.3 Metal–metal interactions in growth media

The presence or absence of metal may exert a profound influence on the effects of other metals due to competition between them. For example, in the design of a medium for *Legionella pneumophila*, concentrations of copper that were not toxic in a chemically defined medium (containing Ca^{2+}, Fe^{2+}, Mg^{2+}, Mn^{2+}, MoO_4^- and Zn^{2+}) had to be reduced 20-fold in a metal-deficient defined medium (Reeves *et al.*, 1981). Similarly, low concentrations of manganese enhanced the growth-stimulatory effects of iron, cobalt and other metals. Magnesium consistently stimulated a growth response to another metal.

2.3 REMOVAL OF METAL IONS FROM GROWTH MEDIA

2.3.1 Merits of metal-limitation studies

A classification of metal ions that is complementary to that of Figure 2.1, which is based on the abundance of the element in dry biomass, is one that reflects the ability or inability to demonstrate that the metal is required for growth. Thus, Pirt (1975) has defined three groups of trace elements which may be required in microbe and cell culture. The group A trace elements are generally essential for growth and comprise calcium, manganese, iron, cobalt, copper and zinc; omission of these from media or their removal affects growth and metabolism. Elements in group B can only rarely be shown to be growth-limiting and include sodium, nickel, molybdenum and selenium. In group C are those metals for which a requirement has not been unequivocably demonstrated: these include scandium, titanium and gallium.

Often, however, unequivocal assignment of a metal to one of these groups will require not only omission of the metal from the medium but also extraction of the medium to reduce the metal concentration to a level where a growth dependency can be observed. For this reason, and also because of the interest that attaches to the study of metal deficiency, many methods have been devised for removal of metals from solution, stemming from the work of the plant physiologist, Steinberg who, in the 1920s, developed fastidious techniques for removing trace elements. In batch cultures of micro-organisms, trace metal deficiency is generally reflected by a reduction of growth rate rather than by diminished biomass concentration (Pirt, 1975), but this may be an insensitive measure of metal deficiency (Archibald and DeVoe, 1978). In addition, the carry-over from the inoculum of the metal under study may mask an effect caused by addition or deletion of a metal, unless several sub-cultures in the metal-deficient medium are made. In chemostat culture, a trace metal deficiency is shown by a gradual decrease in the steady-state biomass as growth rate is increased (Pirt, 1975; Figure 2.2).

2.3.2 General methods

Trace elements may be removed by precipitation as insoluble hydroxides, phosphates, sulphides or carbonates (Nicholas, 1952; Hewitt, 1966; Lankford, 1973; Pirt, 1975) and as ferrocyanides (Choudhary and Pirt, 1966), by adsorption on alumina (Ratledge and Choudhary, 1971) and by Seitz filtration. In the case of iron, solubility depends so strongly on the nature of the anions present that the use of medium containing only potassium phosphates, and filtered to remove precipitated metals, can give iron-limited growth (Hartmann and Braun, 1981). More commonly, however, metals are extracted with an organic complexing agent in a non-aqueous phase or by using chelating resins. In some cases, simply adding a chelating agent to the growth medium results in metal limitation (Partridge and Yates, 1982). This approach is fraught with difficulties of interpretation, however, since more than one metal will be

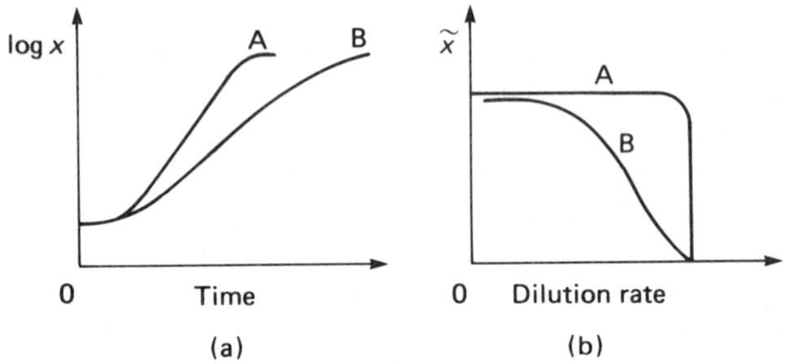

Figure 2.2 Effects of trace element deficiency on growth of micro-organisms in (a) batch culture and (b) chemostat culture: x = biomass in batch; x̃ = steady state biomass in chemostat. Curve A, optimum amount of trace element; curve B, deficiency of trace element (reproduced with permission from Pirt, 1975).

chelated and multiple metal deficiencies must be resolved. Thus, whilst EDTA has been used to limit nickel supply for hydrogenase (Partridge and Yates, 1982), the chelator also represses synthesis of MnSOD in *E. coli*. Other chelators, notably 2,2′-bipyridyl, increased MnSOD as a result of chelation of Fe^{2+} (Hassan and Moody, 1984).

(a) Extraction of iron with 8-hydroxyquinoline

This method was developed by Waring and Werkman (1942), who realized that demonstrating metal deficiencies (especially iron deficiency) in bacterial cultures required more stringent measures than those successful for culture of fungi. Extraction is achieved by shaking the medium (lacking metal salts other than those of potassium) with a solution of 8-hydroxyquinoline in chloroform (or some other solvent, e.g. 1,2-dichloroethane; Partridge and Yates, 1982). The organic phase containing the iron complex (and black as a result) is removed and the extraction repeated as necessary by adding 8-hydroxyquinoline directly to the chloroform-saturated aqueous layer. The extracted medium may contain 0.7–3.0 μg Fe l^{-1} and be tested for iron by adding diphenylthiocarbazone (dithizone). *Aerobacter* grown in this medium was cytochrome- and catalase-deficient (Waring and Werkman, 1944). Since 8-hydroxyquinoline also extracts other metal ions, particularly Cu^{2+}, Mn^{2+}, Mo^{n+} and Zn^{2+}, it is important to add these back as high purity reagents to the final medium when desired. A detailed procedure for extracting molybdenum in investigations of nitrogenase function and which may have achieved levels as low as 3.5 nM is described in detail by Eady and Robson (1984).

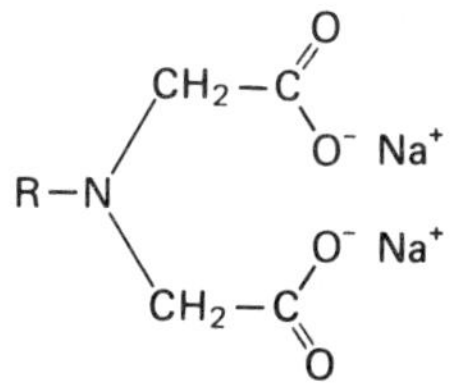

Figure 2.3 Structure and selectivity of Chelex 100, a chelating ion exchange resin. The structure of the sodium form is shown, with 'R' indicating the styrene-divinylbenzene copolymer matrix. The sodium ions, for which chelex exhibits the lowest selectivity, are readily replaced by metals. In nitrate or chloride-containing medium, for example, the order of selectivity for cations is $Cu^{2+} >> Pb^{2+} > Fe^{3+} > Al^{3+} > Cr^{3+} > Ni^{2+} > Zn^{2+} > Ag^{+} > Co^{2+} > Cd^{2+} > Fe^{2+} > Mn^{2+} > Ba^{2+} > Ca^{2+} >>> Na^{+}$ (Reproduced from Information Sheet 2020 with permission of Bio-Rad Laboratories).

(b) Extraction using ion-exchange resins

Chelex 100 (Figure 2.3), a polyaminecarboxylate-substituted styrene derivative (Bio-Rad Laboratories), is now widely used and is reported to reduce iron levels to 0.05 μM. A more conservative claim of 0.3 μM was made by Arnold, Evans and Denniston (1983). In a typical procedure (Davis, McCauley and Byers, 1971) three litres of medium, containing components at ten times the normal concentration, are passed through a column of washed Chelex 100. After adjusting the pH of the eluate, aliquots are frozen until needed, whereupon they are diluted 10-fold with distilled water and sterilized. The method can also be used to extract copper (Schwab, 1973) and other metal ions (Arnold, Evans and Denniston, 1983), including Mn^{2+}, Cu^{2+} and Ca^{2+} (Bio-Rad Laboratories, Caxton Way, Watford, Product Bulletin 2020). Careful use of ion-exchange columns in removal of metal ions from growth medium and the distilled water used in dilutions can reduce Mg^{2+} concentrations more than 200-fold to about 1 μM and Zn^{2+} more than 50-fold to 0.05 μM; vitamins are not removed from mediam by Chelex treatment (Arnold, Evans and Denniston, 1983).

(c) Miscellaneous methods for removing iron from medium

Lankford (1973) and Neilands (1983) survey alternative methods, including (i) the use of the siderophore, mycobactin (Chapter 3), and subsequent chloroform extraction of the water-insoluble ferric mycobactin complex, (ii) addition of excess, sterile, unheated egg white or the purified iron-binding constituent, conalbumin (ovotransferrin), and (iii) limiting the availability of iron by adding 2,2′-bipyridyl. The last is not entirely specific for iron, however, and may lead to cell lysis. The phenolic analogue of EDTA, ethylenediamine-di-(*o*-hydroxyphenyl)acetic acid (EDDA; Miles, Khimji and Maskell, 1979; Neilands, 1983) is an effective iron-removing agent.

A growth medium that imposes severe iron deficiency on enteric bacteria is Tris-succinate (Neilands, 1980). Tris is a weak chelator, has a pH near 8 and maintains the pH at a value where iron is maximally insoluble and also has permeability effects on membranes. Succinate as sole carbon source favours the synthesis of iron-containing cytochromes and FeS proteins. In addition, succinate utilization in a culture without pH control results in a rise in pH, further exacerbating the insolubility of the iron.

Autoclaving glucose with other medium components, especially nitrogenous ingredients, may lead to Schiff base decomposition products, some of which may be iron-complexing agents.

(d) Extraction of copper and cobalt

The chelator 1,5-diphenylthiocarbazone (dithizone) may be used to deplete a medium of copper (Iwasaki, Saigo and Matsubara, 1980; Matsubara, Frunzke and Zumft, 1982). The treated medium typically contains 0.1–0.01 μM copper (Iwasaki, Saigo and Matsubara, 1980). The simplest recommended procedure for the elimination of cobalt is co-precipitation with nitroso R salt at pH 6.5; this and other extraction techniques are described by Hewitt (1966).

(e) Culture vessels

Study of trace metal requirements demands not only reduction of the concentration or availability of the metal in the growth medium, and attention to subsequent contamination of the medium, but also the use of growth vessels from which metal ions cannot leach into the culture.

Various modifications to conventional chemostat designs have been used and reported. For use in studies of iron- and copper-limited *Candida utilis*, Ware, Thompson and Light (1970) modified a Porton-type vessel by fitting Tufnol and PTFE parts where possible and a hollow titanium stirring shaft, which also served to transmit aerating gases to the medium. Other less obvious practices were the use of compressed air generated from compressors fitted with unlubricated carbon piston rings and the use of waterproof lithium- and molybdenum-free grease. Archibald and DeVoe (1978) modified a commercial fermentor by (i) replacing all stainless steel components with glass, unpigmented silicone rubber or polyethylene, and (ii) coating the impeller with polyvinyl chloride. Teflon has also been used for constructing a chemostat and for media reservoirs (Strachan *et al.*, 1982). Although this design has the advantages of simplicity and low cost, the published description suggests that it may be difficult to attain oxygen-sufficient growth, even at only moderate biomass densities, by simple magnetic stirring and agitation. In our laboratory (Hubbard *et al.*, 1986a), we have been successful with small-scale (working volume less than 100 ml), all-glass chemostats aerated, using the vortex system, by a vaned magnetic stirrer with air sparging through the gas phase (Pirt, 1975). Larger chemostats (working volumes 800–1000 ml) have been modifications of a bench-top 'mini-fermentation' system (FT Scientific Instruments Ltd., Bredon, Tewkesbury, Glos., UK) using an all-glass Quickfit vessel. Temperature control (as in the small vessels) is by water circulation through the double-skinned growth vessel, and stirring by a glass impeller and shaft driven from above through nylon bearings.

Work of this kind requires the utmost care with respect to cleanliness of glassware and minimizing medium handling. For example, a typical procedure for glassware (Waring and Werkman, 1942) involves soap washing, followed by rinsing with water, alcoholic KOH, water, aqua regia and water (in that order) before filling with, or soaking in, water and autoclaving for 30 min. Hartmann and Braun (1981) recommend rinsing with 10 mM EDTA and water. The use of water purified by reverse osmosis and a charcoal demineralizer column system is recommended by Strachan *et al.* (1982). Cleaning glassware in 10% Decon containing 50 mM EDTA and extensive rinsing allowed achievement of molybdenum-limited growth of *Azotobacter vinelandii* (Eady and Robson, 1984).

2.4 EXAMPLES OF REQUIREMENTS FOR SPECIFIC METALS AND THE CONSEQUENCES OF DEFICIENCY

In this section, we survey the microbial requirements for selected metals and point to the effects of metal deficiency, metal-for-metal replacements, and metal–metal antagonisms. Such interactions can often provide clues to the biological function of a metal.

2.4.1 Iron

(a) Iron requirements

Virtually all living organisms have an absolute requirement for iron; deficiency causes anaemia in animals, chlorosis in plants, and poor growth in micro-organisms. Iron is required in the many redox reactions catalysed by haem proteins (e.g. cytochromes, catalases, peroxidases) and non-haem iron proteins such as iron-sulphur proteins (Chapter 5). The ability to acquire iron from the host via complex transport systems (Chapter 3) appears to be a virulence factor for many microbial pathogens. Some bacteria (Section 5.8) utilize Fe(III) as an electron acceptor. The ferric iron available for microbial reduction can be estimated from the hydroxylamine-reducible iron (Lovley and Phillips, 1987).

The amount of iron required for maximal growth is variable, but generally small (Lankford, 1973). Requirements of gram-negative bacteria range from 0.36 to 1.8 μM, although the apparent requirement can be influenced by residual traces of the material used for iron extraction. Higher iron requirements have been reported for nitrogen fixers and aerobes such as *Azotobacter* and *Pseudomonas* (Fekete, Spence and Emery, 1983) and for certain *Clostridium* species, possibly due to unusually high demands in synthesis of cytochromes and iron-sulphur proteins. With the exception of those relatively few organisms that require haem for growth (Section 2.4.1 (c)), iron is generally supplied as a ferric salt, although a *Bifidobacterium* appears to require Fe^{2+} and cannot utilize Fe^{3+} even in the presence of siderophores (Bezkorovainy and Topouzian, 1983).

Apparent requirements for iron may be influenced by the presence of toxic metals. Thus high iron concentrations or the addition of an iron chelator may be necessary to relieve inhibition by Al^{3+} or Cr^{3+} or an inhibitory imbalance of other metal ions. High temperatures and pH may also influence iron requirements (Lankford, 1973); the physiological basis of these effects is poorly understood.

In nature, the low solubility of iron generally demands that an organism synthesizes chelators. It has been suggested (Newsome and Wilhelm, 1983) that habitats in which the dissolved iron concentrations are relatively high (for example, the acidic waters of a eutrophic body of water), and the production of chelators is consequently repressed, are conducive to the growth of certain pathogens such as the amoeba *Naegleria*, which is normally inhibited by the microbial iron chelators, desferrioxamine B (Desferal) and rhodotorulic acid. Iron in concentrations of 0.1–1 mg ml^{-1} favours the growth of scum-forming blue algae at the expense of green algae, and iron in lakes has been implicated in the formation of troublesome algal 'blooms' (Morton and Lee, 1974) as a result of the ability of blue-greens to produce hydroxamate chelators (Murphy, Lean and Nalewajko, 1976).

The high affinities of microbial iron transport systems, especially of bacteria (Chapter 3) generally necessitate extreme measures (Section 2.3.2) for iron removal from growth medium if iron deficiency is to be demonstrated. Illustrative of such procedures are those used to demonstrate the apparent absence of an absolute iron requirement in *Lactobacillus plantarum* (Archibald, 1983). Medium (containing glucose-trypticaseyeast extract), which would contain high Fe levels, was passed over iron-complexing beads and, after re-addition of other trace elements, supplemented with a large molar excess (over Fe) of the powerful Fe(III) chelator EDDA ($K_{ML} = 10^{39.7}$). Such treatment had no significant effect on the rate or extent of growth of *L. plantarum*, which accumulated less than 0.1 μM intracellular iron during this growth. *E. coli* B in the same medium contained about 900 μM Fe. Increased sensitivity was obtained by growing cultures in ^{59}Fe-labelled low (Fe) medium and the use of a defined medium lacking haem. Fewer than two atoms of Fe per cell were detectable. It is suggested (Archibald, 1983) that the scrupulous avoidance of iron by this bacterium may result from the ability of traces of Fe to catalyse a variety of oxygen free radical reactions, which would aggravate the toxicity of hydrogen peroxide (H_2O_2) and superoxide ions (O_2^-) in the absence of iron-requiring catalase and superoxide dismutase.

(b) Consequences of iron deficiency

The consequences of iron deficiency on microbial growth, metabolism and ultrastructure are diverse (Table 2.3; Lankford, 1973; Macham, 1976) and include alterations of metabolism, particularly involving iron-containing enzymes, effects on toxin production (e.g., by *Corynebacterium diphtheriae*), enhancement of secondary metabolic production and restriction of growth of pathogenic micro-organisms in serum (Byers and Arceneaux, 1977).

There is much evidence for the preferential loss of iron-containing enzymes in micro-organisms provided with an inadequate iron supply. In aerobic and facultative bacteria, levels of catalase, peroxidase and cytochromes are reduced markedly (Waring and Werkman, 1944; Lankford, 1973; Hubbard *et al.*, 1986a). In addition, aerobic pathways are suppressed with a shift to increased anaerobic glycolysis to meet energy requirements. In some *Clostridium* spp. and enteric bacteria, gaseous

Table 2.3 Examples of the diverse effects elicited by restriction of iron supply

Organism	Approximate iron concentration resulting in deficiency (and method of growth)*	Effect(s) relative to iron-sufficient conditions	Reference
Bacteria			
Escherichia coli	0.013–0.054 μM (B)	Depression of aromatic amino acid biosynthesis	McCray and Herrmann (1976)
	? (B)	Decreased activity of poly-U-directed polypeptide synthesis *in vitro*	Iorio and Plocke (1982)
	? (B)†	Decreased levels of cytochromes, non-haem iron, oxidase activities and efficiency of oxidative phosphorylation	Rainnie and Bragg (1973)
	1 μM (C)	Diminished respiration and cytochrome content	Hubbard *et al*. (1986a)
Pseudomonas aeruginosa	? (B)	Inhibition of development of cyanide-resistant respiration	Trutko, Medentsev and Akimenko (1980)
Bradyrhizobium japonicum	0.25 μM (B)	Decreased cytochrome content and activities of δ-ALA synthetase and dehydrase; excretion of protoporphyrin	Roessler and Nadler (1982)
Clostridium formoaceticum	3.6 μM (B)	Diminished ferredoxin and flavodoxin, not rubredoxin	Ragsdale and Ljungdahl (1984)

Algae and Cyanobacteria			
Anacystis nidulans	0.62 μM (B)	Chlorosis, changes in chlorophyll c, photosynthetic action spectra and fluorescence. Decreased membranes, phycobilisomes and carboxysomes, increased glycogen storage granules	Öquist (1974); Sherman and Sherman (1983)
	0.25 μM (B)	Diminished H_2-dependent reduction of NO_3^- and NO_2^-; effects attributed to Fe–S clusters of hydrogenase	Peschek (1979)
Anabaena flos-aquae	? (B)	Diminished levels of nitrate reductase and glutamine synthetase	Verstreate, Storch and Dunham (1980)
Agmenellum quadruplicatum	? (B)	Loss of phycocyanin and chlorophyll a; degradation of ribosomes and thylakoid membranes; accumulation of intracellular glucose and polysaccharide	Hardie, Balkwill and Stevens (1983a, b)
Euglena gracilis	0.5 μM (B)	Changes in chloroplast polypeptides	Gaynor (1982)

Table 2.3 Continued

Organism	Approximate iron concentration resulting in deficiency (and method of growth)*	Effect(s) relative to iron-sufficient conditions	Reference
Yeasts			
Candida utilis	0.3 μM (C)	Loss of 'Site I' oxidative phosphorylation, lowered cytochrome and non-haem iron content, loss of sensitivity to piericidin A	Garland (1970);Clegg and Garland (1971); Light and Garland (1971)
	0.23–0.28 μM (C)‡	Lowered energy charge and adenylate pools; production of ethyl acetate	Thomas and Dawson (1977, 1978)
Candida lipopolytica	? (B)	Inhibition of cytochrome synthesis; development of cyanide-resistant respiration	Trutko, Medentsev and Akimenko, (1980)
Saccharomyces rouxii	0.3 μM (B)	Diminished growth and activity of Fe^{3+}-dependent acid phosphatase	Arnold, Evans and Denniston (1983)
Protozoa			
Acanthamoeba castellanii	4–6 μM (B)	Switch from cyanide-stimulated to cyanide-resistant respiration	Hryniewieck, Jenek and Michejda (1980)

Notes on methods
* (B) = batch culture; (C) = continuous culture.
† Iron levels reduced by successive sub-culture in iron-deficient medium (Section 2.3.2 (e)).
‡ 'Phased' culture technique; see references for details.

mixed acid fermentations show a marked shift towards homolactic fermentation in iron-deficient cultures due to the preferential loss of hydrogenase and formate dehydrogenase.

Iron deficiency may result in overproduction of porphyrins, predominantly coproporphyrin III, or δ-amino laevulinic acid (ALA), a precursor of porphyrin. This arises from the nature of regulation of tetrapyrrole synthesis in which haem, rather than protoporphyrin, acts as a feedback inhibitor and corepressor of ALA synthetase, the first enzyme in the biosynthetic sequence. Thus, iron deficiency and diminished haem synthesis result in failure of the regulatory control of ALA synthetase and continued overproduction of porphyrins. It is worth noting that the presence of cobalt in the medium exerts a similar effect, since it is inserted into protoporphyrin in place of Fe^{2+} (Porra and Ross, 1965).

In iron-deficient media, several bacteria (e.g. *E. coli*; Ratledge and Winder, 1964) grow as filaments which may result from defects in DNA synthesis (Lankford, 1973).

The production of haemolysin by *Listeria monocytogenes* occurs in response to low iron stress; iron is the only metal identified as thus influencing production (Cowart and Foster, 1981). The concentration of free iron in plasma is sufficiently low (<0.1 μM) to allow maximal haemolysin production.

In other instances, the presence of iron exerts repressive effects. For example, bioluminescence and the synthesis of luciferase in *Vibrio harveyi* are repressed (Makemson and Hastings, 1982) by the addition of iron to a HEPES-containing minimal medium, which would support iron-limited growth. In *E. coli*, iron limitation results in the synthesis of outer membrane proteins which are receptors for iron chelators (Chapter 3). A similar induction of proteins in the outer membranes of *Klebsiella aerogenes* and *Vibrio cholerae* has been described (Sigel and Payne, 1982). Lesion formation by *Colletotriatum musae* on banana fruits is reduced by the presence of iron and enhanced by anthranilic acid, 2,3-dihydroxybenzoate or a bacterial siderophore (Brown and Swinburne, 1981).

(c) Haem-requiring organisms

The iron requirements of most micro-organisms are met by inorganic iron, but in some instances haem is required as a growth factor, due to the inability of the organism to synthesize the porphyrin ring or insert Fe into it. *L. plantarum* has been classified as a facultative siderophobe (Archibald, 1983), indicating its (optional) ability to assimilate haem from haematin or heated blood and incorporate it into catalase. The fungus

Pilobolus lacks siderophores and alternative methods for Fe(III) uptake; haemin is commonly used in laboratory growth medium for the organism, although Fe(II) and ascorbate can support growth (Bourret, 1982).

Haemin (chlorohaemin, the usual source of haem) is frequently added to culture media (e.g. for the insect trypanosome *Crithidia fasciculata*) as a solution in triethanolamine or 2,2′,2′′,2′′′-(ethylenedinitrilo)tetraethanol. In bacterial growth media, haem is generally added as an aqueous solution, made alkaline with NaOH. The pathways for haem transport are poorly understood.

The respiratory systems of bacteria that require haem for cytochrome synthesis have been reviewed by Knowles (1980). There is an obligatory requirement for haem by certain *Bacteroides* spp. (strict anaerobes, which respire with fumarate as an electron acceptor) and *Haemophilus* sp. (facultative anaerobes that respire using fumarate, NO_3^- and O_2 as acceptors). Other organisms, such as *Streptococcus faecalis*, can grow anaerobically without haem but will utilize haem if provided. Haematin is provided for growth of *Bacteroides*, and protohaem for *Haemophilus*.

(d) Dissimilatory iron requirements

The above account has concentrated on the uptake of iron and its requirement as an essential trace element for assimilation. However, iron can also be oxidized ($Fe^{2+} \rightarrow Fe^{3+}$) and precipitated, or reduced ($Fe^{3+} \rightarrow Fe^{2+}$) and solubilized. These reactions are mentioned briefly here since (i) ferrous ion can serve as sole energy source for the growth of certain 'iron bacteria', and (ii) there may exist organisms that grow aerobically using ferric iron as sole electron acceptor. Thus, in these senses, as well as in the more familiar role as an assimilated nutrient, iron represents a requirement for growth. For a detailed discussion of iron oxidation and reduction, see Chapter 5.

2.4.2 Potassium

(a) Requirements

The requirement for potassium for growth of microbial biomass is higher than that for any other metal. A growth yield of about 60 g dry biomass/g potassium (Pirt, 1975) has been used in the design of culture media (Table 2.2). Because of the association of intracellular K^+ with RNA, factors that affect RNA content, such as growth rate, often influence K^+ requirements. Thus in K^+-limited steady-state cultures of *E. coli*, intracellular K^+ concentration increases with growth rate (Mulder *et al.*, 1988). Sodium chloride content and pH of the medium may also affect K^+ content and needs (Pirt, 1975). In *Candida utilis*, however, K^+ content is markedly dependent on the nature of the growth limitation, with only a weak correlation between cellular contents of Mg^{2+}, phosphate, RNA and K^+ (Aiking and Tempest, 1976). Instead, a correlation is evident between K^+ content and the yield values for O_2 and glucose. At low dilution rates (<0.2 h^{-1}), K^+-limited cells (irrespective of the carbon source) have depressed P/O ratios at phosphorylation 'sites' 2 and 3 and altered cytochrome composition (Aiking and Tempest, 1976; Aiking *et al.*, 1977a,b). There is other evidence that K^+ is directly involved in oxidative phosphorylation (Blond and Whittam, 1965; Gomez-Puyou *et al.*, 1972). Potassium is the predominant monovalent cation in the cytoplasm of most cells and has three important functions: the activation of enzymes, maintenance of osmotic pressure, and regulation of intracellular pH (Walderhaug, Dosch and Epstein, 1987). Enzyme activation is in part specific to K^+ (Bayley and Morton, 1978) but there is also non-specific activation of some enzymes by ionic strength. Internal cell concentrations of K^+ in bacteria range from about 18 mM (in a freshwater thermophile) to 4 M in *Halobacterium halobium*, which requires 2.5 M NaCl for growth and contains enzymes requiring >2M K^+ for optimal activity and stability. Generally, however, only about 19 mM K^+ is required for enzyme activation and so the much higher intracellular K^+ concentrations point to additional roles such as osmotic regulation and pH homeostasis (Rothstein, 1960; Harold and Altendorf, 1974; Booth, 1985, 1988).

Since K^+ transport is energy-dependent, the high intracellular concentration presumably makes a significant demand on energy-yielding reactions in the cell economy (Stouthamer, 1973), and may determine the maximum specific growth rate (μ_{max}). In K^+-limited chemostat cultures of *Paracoccus denitrificans*, μ_{max} depends on the input K^+ concentration. At low K^+ input values, high K^+ gradients may not be maintained (Stouthamer and Bettenhausen, 1980).

Potassium deficiency can generally be achieved by simply omitting the element from the medium. Although the dry weight of *Aerobacter aerogenes* has been shown to be linearly dependent on the entering K^+ concentration in chemostat culture, Tempest (1969) noted that extrapolation to zero [K^+] shows that the bacterial K^+ requirement can be met to a small extent by other ions (see below). In other respects, the response of the organism to K^+ limitation resembled the response to Mg^{2+}, both metals being related to RNA content. The molar ratio of cell-bound Mg^{2+}:K^+:RNA (as nucleotide):P was shown to be 1:4:5:8, regardless of growth rate and temperature. Tempest (1969) argued that although K^+

may be involved in osmoregulation, and possibly in association with anionic polymers such as teichoic and teichuronic acids in *B. subtilis*, there remains a pool or compartment of K^+ that is essential for ribosome configuration and function. In common with magnesium deficiency, K^+-limitation in *E. coli* leads to selective hydrolysis of ribosomal RNA (St. John and Goldberg, 1980), supporting a role for both metals in ribosome stability.

(b) Replacement by other metals

The only other alkali metal generally considered to replace K^+ is rubidium, an element not abundant in nature but of similar chemistry and ionic radius (1.48×10^{-10}m) to K^+ (1.33×10^{-10}m). In the yeasts *Candida utilis* and *Saccharomyces rouxii*, lithium, sodium, caesium and ammonium ions are all ineffective (Aiking and Tempest, 1977; Arnold, Evans and Denniston, 1983) in replacing K^+. Although addition of Na^+ to K^+-limited *C. utilis* gives a slight increase in cellular K^+ content, this may be due to increase in the osmotic pressure of the medium (Aiking, Van Holst and Tempest, 1977b). Replacement of K^+ by Rb^+ is often associated with a decrease in growth rate, probably as a result of the inability of Rb^+ to replace K^+ functionally rather than to poor Rb^+ transport (Aiking and Tempest, 1977). Sodium and rubidium have been reported to permit growth of *S. cerevisiae* at low K^+ concentrations that would otherwise result in cell death (Camacho, Ramos and Rodriguez-Navarro, 1981). In *E. coli*, Rb^+ cannot completely substitute for K^+ in the K^+ pool and the growth rate declines with the extent of replacement by Rb^+ (Nakajima, Yamato and Anraku, 1979). Very recently, the ammonium ion (NH_4^+) has been shown capable of replacing K^+ (which has similar ionic radius) in K^+-limited cultures of *Bacillus stearothermophilus* and other bacteria (Bunrman *et al.*, 1989).

2.4.3 Magnesium

(a) Requirements

Magnesium is essential to all cells and is the most abundant intracellular divalent cation. About 90% of intracellular Mg^{2+} is bound to ribosomes and polyanionic cell constituents; the remainder constitutes a relatively constant free concentration of 1–4 mM (Jasper and Silver, 1977). It has roles in macromolecular synthesis and phosphorylation reactions and may be considered a 'function-maintaining' element.

By the 1930s and 1940s, it had been demonstrated that the metal was a component of bacteriochlorophyll, was essential for nitrogen-fixing bacteria, and a co-factor for various enzymes, especially those involving phosphorylation, pyruvate decarboxylase and RNA polymerase (Chapter 4). As an integral part of ribosome structure, magnesium influences association and dissociation of 30S and 50S subunits in *E. coli* (Grunberg-Manago *et al.*, 1978). There is evidence from chemostat cultures for Mg^{2+} control of RNA synthesis (Tempest, 1969). The minimum intracellular magnesium requirement in these experiments varied substantially with growth rate, being greatest at high growth rates. The strikingly similar changes in RNA content, linked to growth rate, suggest an interdependence between RNA and magnesium. In addition, a growth rate-independent stoichiometry between RNA and Mg^{2+} has been observed, 4 mol of RNA nucleotide being synthesized per mole of culture Mg^{+} (Tempest, Hunter and Sykes, 1965).

It is commonly stated that gram-positive bacteria require much more magnesium than do gram-negative organisms (Brock, Smith and Madigan, 1984). The chemostat studies of Tempest, Dicks and Meers (1967) which compared *B. subtilis* and *Klebsiella aerogenes* at similar growth rates, however, revealed only minor differences in cellular magnesium contents. The gram-positive organism had a high capacity for Mg^{2+} adsorption but a lower affinity for the metal ion, such that it was outgrown by *K. aerogenes* in Mg^{2+}-limited mixed cultures. A similar competition has been demonstrated between *B. subtilis* and the yeast *Candida utilis* (Meers, 1971). Some *Halobacterium* spp. have remarkably high magnesium requirements (optimum 0.1–0.8 M), consistent with the extremely high magnesium concentrations (up to 1.8 M) of the Dead Sea where they are found (Cohen, Oren and Shilo, 1983).

(b) Magnesium limitation

Magnesium is a common contaminant of organic medium components such as peptone, casein, and malt and yeast extracts, but the relatively high concentration of magnesium required by most cells (Table 2.2) means that magnesium-limited growth can generally be achieved by omitting such organic components or magnesium salts from the medium (Tempest, Dicks and Meers, 1967; Light and Garland, 1971; Arancia *et al.*, 1980) without the need for laborious extraction procedures. Thus *Klebsiella aerogenes* in chemostat culture is Mg^{2+}-limited below about 1 mM Mg^{2+}; at a fixed dilution rate, the yield of cells is linearly dependent on added Mg^{2+} from 0–750 μM (Figure 2.4). The fact that a plot of yield versus culture Mg^{2+} concentration extrapolates to the origin indicates that other cations present cannot functionally replace Mg^{2+}. It is reported, however, that the magnesium requirements of *B. subtilis* and *B. megaterium* are reduced by 25 μM Mn^{2+}, presumably due to

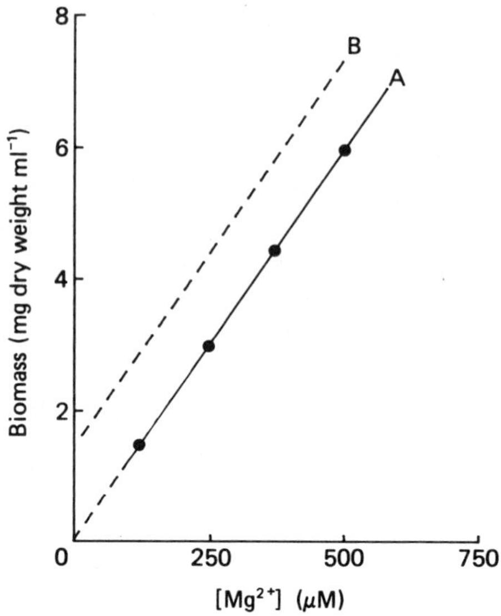

Figure 2.4 Plot of bacterial concentration versus culture magnesium concentration for a Mg^{2+}-limited chemostat culture of *Klebsiella aerogenes* (trace A; dilution rate, D = 0.4 h^{-1}; 37°C; pH 6.5). The upper trace (B) shows the relationship expected if other medium components had functionally replaced magnesium (with modification from Tempest, 1969). Reproduced with permission from *Microbial Growth*, Cambridge University Press.

enhancement of Mg^{2+} accumulation, whereas Cu^{2+}, Zn^{2+} and molybdate increase the Mg^{2+} requirements of both organisms (Webb, 1968).

Pleomorphic *Halobacterium* spp. from the Dead Sea require at least 75 mM Mg^{2+} for growth and lose their normal shape even at 50 mM Mg^{2+} (Cohen, Oren and Shilo, 1983).

Although the growth requirement of micro-organisms for magnesium cannot generally be met by other metals (Figure 2.4; Tempest, 1969), manganese can replace magnesium in some cellular processes, especially cell division (Utkilen, 1983). Similarly, the induction of an outer membrane protein (H1) in *Ps. aeruginosa*, believed to be involved in determining susceptibility to chelators, polymixin B and gentamycin, can be prevented by supplementation of low-Mg^{2+} medium with Mg^{2+}, Ca^{2+}, Mn^{2+} or Sr^{2+} but not certain other metals (Nicas and Hancock, 1983). Similar receptor specificity has been seen for the chemotaxis receptor for Mg^{2+} in bacteria (Koshland, 1979).

There are numerous examples of the effects of magnesium deficiency on bacteria (Jasper and Silver, 1977); macromolecular synthesis, especially of nucleic acids, seems most sensitive. Striking structural modifications have also been reported. Magnesium-limited growth of *Klebsiella aerogenes* or *B. subtilis* results in the synthesis of walls with an enhanced binding capacity and affinity for Mg^{2+}, suggesting changes in cell wall composition. Magnesium-depleted *E. coli* show pleiotropic membrane effects, notably permeability to lactose analogues that cannot be actively transported, and the formation of internal septate membranes and patches of opposing membranes where a regular particle pattern develops (Weiss, 1977).

A further effect on membranes is the increase in resistance of *Ps. aeruginosa* in Mg^{2+}-limited cultures to EDTA, EGTA, polymyxin and cold shock (Brown and Melling, 1969; Kenward, Brown and Flyer, 1979), correlated with large increases in the level of the outer membrane protein, H1. It is suggested that protein H1 acts to replace divalent cations at a site on the lipopolysaccharide and protects this site from attack by chelators of divalent cations and cationic antibiotics such as polymyxin B and gentamicin (Hancock, Raffle and Nicas, 1981). Indeed, sensitivity to such agents is retained by cultures grown in Mg^{2+}-depleted medium, supplemented with Ca^{2+} or Mn^{2+} (Kenward, Brown and Flyer, 1979).

The requirement for magnesium in chlorophyll is reflected in lowered chlorophyll levels in *Euglena gracilis*, but protein synthesis is more severely affected by Mg^{2+}-limitation (Zielinski and Price, 1978). In *Schizosaccharomyces pombe*, magnesium deficiency induces oxidative rather than fermentative metabolism and the production of highly cristate mitochondria (Walker *et al.*, 1982); the cause is unknown.

An unusual, unexplained result of magnesium deficiency is shown by chemostat-grown *Agrobacterium tumefaciens* (Kurowski, Fensom and Pirt, 1973). Large oscillations in biomass result, reflecting oscillations in growth rate.

(c) Magnesium and cell division

Like calcium, which has long been regarded as an ion involved in regulation, magnesium also has been implicated in the control of cell growth and division. Cell enlargement appears to be a general feature of Mg^{2+}-limited growth in prokaryotes (e.g. *Anacystis nidulans*) and eukaryotes (e.g. *Chlorella*, *Schizosaccharomyces pombe* and *Saccharomyces rouxii*) (Utkilen, 1982; Arnold, Evans and Denniston, 1983). The detailed studies by Tempest *et al.* on chemostat cultures of *Klebsiella aerogenes* demonstrated clearly that enlargement was not due merely to changes in growth rate, a factor well known to influence cell size in *E. coli* and other bacteria (Lloyd, Poole and Edwards, 1982).

Magnesium ions have been shown to be essential for cell division in yeast and protozoa. This requirement was demonstrated by exposing the fission yeast *S. pombe* to a variety of Mg^{2+}-chelating agents and transferring cells to Mg^{2+}-deficient media (Lloyd, Poole and Edwards, 1982). Under such conditions *S. pombe* continues to elongate, but nuclear division and cell plate formation are prevented, suggesting that Mg^{2+}-limitation inhibits events late in the cell cycle. Addition of Mg^{2+} (but not Ca^{2+}, Sr^{2+}, Be^{2+}, Mn^{2+} or Zn^{2+}) to Mg^{2+}-deficient cultures induces synchronous cell division of the arrested cells. There is evidence that intracellular Mg^{2+} levels may act as a transducer of cell size. Total cell Mg^{2+} concentrations in synchronous cultures of *S. pombe* and *Kluyveromyces fragilis* fall steadily, followed by a rapid influx of magnesium just before cell division, which brings about spindle formation. Some support for the proposal comes from the finding that a short (5 min) pulse of the divalent cation ionophore A23187 applied just before division accelerates cell division.

The requirement of magnesium for division is confirmed by other studies. In the protozoan *Tetrahymena pyriformis*, cell division is inhibited (and cyclic AMP levels are elevated) when cells are incubated in a medium with no added magnesium (Hilfiker and Higgins, 1981). In *Anacystis nidulans*, a shift up from growth with 5 μM Mg^{2+} to 1 mM Mg^{2+} elicits a parasynchronous division of the filamentous cells that result from metal-limited growth (Utkilen, 1982).

In higher eukaryotic cells also, protein synthesis and DNA synthesis are sensitive to small changes in intracellular Mg^{2+} concentration; such dependencies have also been taken as evidence for an involvement of this cation in cell cycle regulation (Lloyd, Poole and Edwards, 1982).

2.4.4 Manganese

(a) Requirements and deficiency

The most familiar examples of a manganese requirement are for growth of, and acid production by, lactobacilli, the growth and development of certain yeasts, and sporulation of *Bacillus* spp. In the first of these examples, the unusually high manganese requirement is probably related to use of the metal by *Lactobacillus* as a mechanism for oxygen detoxification (Section 5.12.2(b)). The medium used by Fridovich (1982) contains 710 μM Mn(II), 5000-fold higher than in many complex media. Enzymes requiring manganese include enolase, superoxide dismutase (Chapter 5), phosphoenolpyruvate carboxykinase and those involved in the early steps of fatty acid biosynthesis. Its requirement by plants is particularly high, probably because of the metal's involvement in the oxidation of water in photosynthesis (Barber, 1984).

As noted in the previous section, Mn^{2+} can lower, but not eliminate, the growth requirement for Mg^{2+}. The ability of other metals to replace Mn^{2+}, however, is variable and less well documented. Thus, the unusually high Mn^{2+} requirement of *Pediococcus cerevisiae* is not reduced by Zn^{2+} (Efthymiou and Joseph, 1972).

In many fungi (Barnett and Lilly, 1966) manganese is required for hyphal extension and branching, fruiting and asexual sporulation. These requirements are such that metal extraction from the medium is unnecessary for their demonstration. The difficulty in correlating the specific morphological effects observed with metal deficiency are well illustrated by the work of Zonneveld (1975a) who identified manganese as a requirement for cleistothecium development in *A. nidulans*. Manganese deficiency results in the virtual absence of α-1, 3 glucan, the main carbon and energy source for fructification, and the impairment of phosphoglucomutase activity (Zonneveld, 1975b). The enzyme is not known to be manganese-requiring or-containing, but glutamine synthetase is a manganese-requiring enzyme and loss of its activity could be responsible for a build-up of glutamate and increased NAD glutamate dehydrogenase activity.

Various effects of Mn^{2+} have been compiled by Silver and Jasper (1977) but in only a few cases can the physiological basis for the manganese requirement be explained. However, the manganese requirement (2 μM) of *Brevibacterium ammoniagenes*, which is used industrially to produce 5′-inosinic acid (a flavour additive), has been traced to a requirement in DNA synthesis (Oka, Udagawa and Kinoshita, 1968). The Mn^{2+} requirement of *Lactobacillus curvatus* late in batch growth arises from the role of the metal in an RNA polymerase; excess of Mg^{2+} could not substitute for 10 μM Mn^{2+} (Stetter and Kandler, 1973).

Although *Rhodopseudomonas* and other photosynthetic bacteria do not have Photosystem II and do not evolve O_2, they do accumulate Mn^{2+} during growth. A significant amount of Mn^{2+} is located in the chromatophores, but it does not appear to be present in porphyrins or to function in redox reactions. Retarded growth of *R. spheroides* has been demonstrated under conditions of Mn^{2+}-deficiency, but the level of Mn^{2+} required for normal growth was less than 1/100 of that required by green algae (Silver and Jasper, 1977).

It is well known that manganese deficiency is an essential condition for citric acid accumulation by *Aspergillus niger*. Manganese deficiency results in reduced activities of the hexose monophosphate pathway (HMP) and tricarboxylic acid cycle and an impairment of cellular anabolism (Kubicek and Röhr, 1977), reduced lipid levels (Orthofer, Kubicek and Röhr, 1979), impaired protein synthesis and turnover (Kubicek, Hampel and Röhr, 1979) and morphological and cell wall abnormalities (Kisser, Kubicek and Röhr, 1980). The complexity and diversity of these effects has prevented a full understanding of the consequences of manganese deficiency.

(b) Microbial oxidation and reduction

Many micro-organisms (fungi, algae and especially bacteria) oxidize or reduce manganese (Nealson, 1983a). In only a few cases has direct transformation of the manganese by a microbial enzyme been demonstrated, and it is probably inappropriate to regard this utilization as a growth requirement. Although manganese oxidation sometimes accompanies sporulation, it appears not to be obligatory (Vojak, Edwards and Jones, 1984). Neither a true manganese chemoautotroph nor a manganese mixotroph has been identified. Nevertheless, a few pertinent features are of interest.

Many biological activities alter the E_h or pH of the environment and enhance valence changes indirectly, sometimes at considerable distances from the cells. More direct effects usually involve the production of external or internal binding proteins, external acidic polysaccharides or other materials that bind, concentrate and enhance manganese oxidation (Nealson, 1983a). Some bacteria oxidize manganese enzymatically (Zapkin and Ehrlich, 1983) as do *Arthrobacter* extracts supplied with solid MnO_2. This cell-free activity has been partially purified. Like *Arthrobacter*, *Leptothrix* and *Metallogenium* are thought to oxidize manganese via a catalase reaction (Dubinina, 1978).

2.4.5 Zinc

Zinc has been clearly recognized as an essential element for growth of one organism (*Aspergillus niger*), since Raulin's experiments of 1869, but a requirement for bacterial growth was not demonstrated until almost 80 years later.

Chemically defined culture media prepared from reagent grade chemicals usually contain 0.5 to 1 μM zinc as impurities (Failla, 1977). In addition, its abundant release from borosilicate glass generally precludes the appearance of a deficiency in unchelated media (Hutner, 1972) and most micro-organisms achieve optimal yields without Zn^{2+} supplementation. Preparation of zinc-deficient media by solvent extraction or alumina treatment has proved difficult, although a zinc-deficient medium for *Rhizobium* allowed the demonstration (Wilson and Reisenauer, 1970) that 0.1 μM added zinc supported maximal growth and that addition of as little as 0.1 nM Zn^{2+} stimulated growth. *Mycobacterium* and *Nocardia* require greater quantities of zinc than do most other bacteria, and zinc-limiting media have been relatively easily made by alumina extraction (Webley, Duff and Anderson, 1962). A survey of the elemental composition of ten methanogens (Scherer, Lippert and Wolff, 1983; Fig. 2.1) revealed unusually high levels of the metal (0.005 to 0.063% (w/w)) in some organisms but strict growth dependencies have not been reported.

The numerous biocatalytic functions of zinc have been reviewed (Failla 1977; Sigel, 1983) and include roles in about 200 enzymes (Chapter 5), nucleic acid metabolism and cell division, and the synthesis of industrially and medically significant microbial secondary metabolites. More recently, the metal has been found to play an essential, but unknown, role in yeast morphogenesis during meiosis (Bilinski, Miller and Girvitz, 1983). The magnitude of the growth dependence of yeast on zinc is emphasized by a US patent assigned for the use of zinc as a nutrient additive for promoting yeast growth on hydrocarbons and increasing protein yields of the biomass (Cavallo and Hines, 1971).

A subsequent detailed study showed that zinc limitation of *Candida utilis* in a chemostat resulted in lowered growth yields ($Y_{sucrose}$ and Y_o), suggesting an altered efficiency of energy metabolism (Lawford *et al.*, 1980a, b). The level of cell-associated zinc in the steady state rose during carbon-limited growth when Fe^{3+} was omitted from the medium, implicating zinc in iron metabolism, specifically porphyrin and cytochrome synthesis (Failla, 1977). At the onset of zinc limitation in a chemostat, where the steady-state level of biomass was monitored as a function of zinc concentration, the zinc content of the cells was 0.8 nmol Zn^{2+}/mg dry weight, equivalent to the value predicted by Pirt (1975).

Other observations indicate interactions between zinc and its deficiency with iron metabolism. Thus, zinc deficiency results in a zinc-reversible inhibition of cytochrome synthesis in *Ustilago*; the presence of cobalt (a zinc antagonist) induces an iron deficiency, which is also reversible by zinc. Zinc seems to be required for the synthesis of δ-ALA dehydratase. The effect of zinc on cytochrome synthesis appears to be mediated by transcriptional control (Failla, 1977).

A striking example of zinc deficiency is its effects on growth of *Mycobacterium tuberculosis*. Omission of Zn^{2+} results in growth as pale, unfolded pellicles which become wet and sink. In this study the wetting was eliminated by addition of glycerol, which was later found to be contaminated with zinc (Drea, 1956). Zinc deficiency also results in liberation of polysaccharides,a slightly elevated content of lipids and phospholipids, suggesting alteration of the cell wall, and an increase in aldehyde accumulation (de Bruyn, Weckx and Beumer-Jochmans, 1981). Zinc may be needed for an NADP-dependent alcohol dehydrogenase.

Other examples of enzymes that have been reported to exhibit decreased specific activities in zinc-deficient organisms are (in fungi) alcohol and glutamate dehydrogenases, tryptophan synthetase and other enzymes of the tryptophan biosynthetic pathway, glucose-6-phosphate and 6-phosphogluconate dehydrogenase, and (in bacteria) glycerol and lactate dehydrogenases and alkaline phosphatase. However, as Failla (1977) points out, decreased activity alone does not distinguish between a requirement for synthesis, structure or function. Similarly, the activity of a zinc metalloenzyme will not necessarily be affected by zinc deficiency.

While many of the above effects are probably due to deficiencies in zinc metalloenzymes, there is now substantial evidence that zinc often acts as a 'stabilizer' of biological systems (Failla, 1977). The reduced numbers of ribosomes and mitochondria in zinc-deficient *Rhodotorula gracilis* (Cocucci and Rossi, 1972) are thought to reflect this role. However, only two zinc enzymes are proven to have a structural role for the metal aspartate transcarbamylase and α-amylase (Hughes, 1981).

Zinc deficiency of several organisms results in diminished levels of DNA, RNA or both; protein and especially polyphosphate levels may also be altered under these conditions. Supplementation of zinc-deficient cells with zinc increases RNA levels preferentially, suggesting that RNA rather than DNA metabolism is most sensitive to zinc deficiency (Failla, 1977).

The effects of zinc deficiency on *Euglena gracilis* have been extensively studied. Fortunately, severe zinc deficiency can be demonstrated in strain Z by decreasing the concentration in the medium to 0.1 μM, a level that supports optimum growth in other strains (Kempner and Miller, 1972). Zinc deficiency causes diverse effects, which are reversible by zinc addition. There are reductions in intracellular zinc levels, increases in cell size and density, accumulation of large paramylon granules, decreased RNA levels and greater intracellular accumulation of magnesium, calcium, manganese, nickel and chromium (Falchuk, Fawcett and Vallee, 1975a). Zinc concentration determines whether cells become blocked in G_1, S or G_2 phases of the cell cycle (Falchuk, Krishnan and Vallee, 1975b) and zinc deficiency profoundly alters histone complement. These findings suggest that zinc might affect gene expression through these chromatin constituents (Mazus, Falchuk and Vallee, 1984).

2.4.6 Calcium

Although, in typical growth media, calcium is an abundant trace element, its concentration in microbial cells is maintained at extremely low levels by highly specific active transport processes (Chapter 3).

Absolute requirements for calcium are rarely demonstrated. Calcium is a ubiquitous contaminant easily released from glassware and, being weakly chelated, is easily displaced by other metals. It is therefore available for growth even in highly chelated media. Because calcium-limited growth is difficult to achieve in a chemostat, Tempest (1969) measured the calcium content of bacteria grown under various conditions; calcium levels did not correlate with changes in RNA, Mg^{2+}, K^{+} or phosphorus but with total cell-wall content. Mg^{2+}-limited *K. aerogenes* contained less calcium than Mg^{2+}-sufficient cells, demonstrating that calcium did not replace magnesium.

The most striking role of calcium is in bacterial spores where, as the dipicolinate salt, it appears to be largely responsible for heat resistance. This role is discussed in Chapter 4 and the unusual calcium uptake systems involved are covered in Chapter 3.

Calcium deficiency frequently results in morphological abnormalities, as in *Saprolegnia diclina*, where inclusion bodies also accumulate (Fletcher, 1983). These vacuolar bodies appear to have sequestered various metals from the medium, perhaps aggravating the effect of calcium deficiency.

In the development of competence for transformation, calcium and magnesium are sometimes interchangeable and strontium can occasionally substitute for calcium (Page and Doran, 1981). Similarly, the finding that bacterial sensitivity to lysis by EGTA (which chelates Ca^{2+} effectively but Mg^{2+} poorly) is dependent on the amount of Ca^{2+} in the growth medium, suggests that either Ca^{2+} or Mg^{2+} may occupy the site attacked by the chelator (Boggis, Kenward and Brown, 1979).

Calcium (Ca^{2+}) is an important growth regulator (Lloyd, Poole and Edwards, 1982; Sisken, 1980), and there are numerous examples of the involvement of Ca^{2+} in growth and division of micro-organisms. In *Physarum polycephalum*, for example, cyclic uptake and release of Ca^{2+} occurs during mitosis and these fluctuations correlate with specific structural and kinetic events in the mitotic cycle. The isolation of calcium-dependent mutants (Ohya, Ohsumi and Anraku, 1984) should aid the identification of the mechanism of action of Ca^{2+} in the yeast cell cycle.

The relationship between calcium fluxes and cell division have been extensively studied in synchronous cultures of the ciliated protozoon

Tetrahymena (Charp and Whitson, 1982). Just before cell division there is a rapid increase in intracellular Ca^{2+} followed by an efflux to the medium. Both the influx and cell division are inhibited by verapamil, a calcium channel blocker. These studies suggest a requirement for calcium flux for cell division. In *Tetrahymena*, oscillations in the amounts of both Ca^{2+} and Mg^{2+} were found during synchronous growth (Walker and Zeuthen, 1980), with peaks coincident with cell division. Some doubt on an essential role for Ca^{2+} is cast by the finding that medium unsupplemented with calcium supports growth, division and DNA synthesis (Hilfiker and Higgins, 1981). However, traces of calcium must be present in non-extracted medium and may be sufficient to maintain 'trigger' functions.

2.4.7 Sodium

Sodium is generally regarded as an element whose requirement for growth can rarely be demonstrated, except for an occasional requirement in specialized functions.

One such exception to the rule is the requirement for an unusually high Na^+ level (0.1 M) by *K. aerogenes* growing on citrate, a requirement that reflects sodium involvement in oxaloacetate decarboxylase (Chapter 3). Sodium is now recognized as having a key role as a coupling ion in bioenergetics.

The requirement for sodium by marine bacteria is well known but poorly understood, although K^+ has been reported to substitute for two-thirds of the amount of Na^+ generally required by *Halobacterium* spp. (Brown and Gibbons, 1955), and the requirement for NaCl by *Micrococcus varians* var. *halophilus* can be partially fulfilled by LiCl, KCl or CsCl (Chan, Leung and Lee, 1979). Most halophilic bacteria specifically require NaCl but spontaneous mutants of *Micrococcus varians* var. *halophilus* have been isolated which will grow with 1.5 M LiCl or 2.5 M CsCl (Kamekura and Onishi, 1982). Clearly, even for halophiles, Na^+ is not always an absolute growth requirement.

The dependence of methanogenesis on sodium appears to be a general phenomenon (Perski *et al.*, 1981, 1982). Methane formation from CO_2 and acetate requires Na^+ in several organisms, including *Methanobacterium thermoautotrophicum*, *Methanosarcina* and *Methanospirillum* spp. The Na^+ ionophore monensin abolishes methanogenesis when the extracellular Na^+ concentration and pH are low. Methane formation may be coupled to an electrogenic extrusion of Na^+ which in turn re-enters the cell in exchange with a proton (Perski, Schöneit and Thauer, 1982). A recent elemental analysis of ten methanogenic bacteria has confirmed the importance of sodium in methanogens, some of which contained up to 4 g of sodium (100 g dry wt)$^{-1}$, although rather lower values were more typical (Figure 2.1; Scherer, Lippert and Wolff, 1983).

2.4.8 Copper

Copper is known to be the prosthetic group of a number of enzymes but its precise roles have been relatively neglected. Early studies were entirely descriptive. Forty years ago it was known that the presence or absence of copper affects pigmentation. Thus, *Azotobacter chroococcum* cultures do not blacken on ageing in the absence of copper and *Aspergillus niger* spores were known to be suitable for testing for the presence of copper in soils; as little as 0.06 μg Cu ml^{-1} permits normal spore colour.

Two copper-containing enzymes that have been relatively well studied in copper-sufficient and copper-limited growth conditions are cytochrome c oxidase (cytochrome aa_3) and nitrous oxide reductase, both of which are covered in detail in Chapter 5.

Copper-limited growth of yeast has been obtained in batch and continuous culture. Copper limitation of *C. utilis* in a chemostat (Light, 1972) caused a 10-fold reduction in cytochrome c oxidase levels. Partial recovery of the amount and activity of the oxidase ocurred, in a cycloheximide-sensitive fashion, on adding 40 μM $CuSO_4$ to an aerated cell suspension. Recovery experiments with cells grown in copper-limited batch culture (Keyhani and Chance, 1971) confirm the requirement for growth and protein synthesis for reconstitution of the oxidase. In the absence of copper, the protein is synthesized and inserted into the membrane and requires a cytoplasmically-synthesized protein for copper incorporation (Keyhani and Keyhani, 1975). Interestingly, copper limitation is manifest as a loss of the optical absorbance bands of the oxidase that arise from the two haems a, implying a role for Cu in haem insertion or a very close coupling between the insertion of the two types of redox centre. In *C. utilis* (Downie and Garland, 1973), *Neurospora crassa* (Schwab, 1973) and *Ps. aeruginosa* (Trutko, Medentsev and Akimenko, 1980), copper depletion gives rise to an alternative cyanide-insensitive pathway of electron transfer to O_2.

In bacteria, branched cytochrome-containing respiratory pathways sometimes provide alternative routes for electron transfer to oxygen when copper deficiency causes loss of an aa_3-type oxidase. In *Brevibacterium*, for example (Sugiyama, Kitano and Kanzaki, 1973), there is increased synthesis of the alternative oxidase cytochrome *d*, which does not require copper (Poole, 1983). In *Paracoccus denitrificans* (Hubbard, Poole and Hughes, 1986) cytochrome aa_3 synthesis is depressed but there is no clear evidence that an alternative oxidase functions.

The growth of methane oxidizers is particularly copper-dependent. For example, the optimal concentration of copper for growth of the obligate methanotroph '4025' is high at 1.2 μM (Vrdeljak, Maric and Johanides, 1984).

Growth of *Methanomonas margantol* on methane, but not methanol, is greatly diminished in the absence of added copper (Takeda *et al.*, 1976), and methane oxidation is inhibited by copper chelators such as thiourea, diethyldithiocarbamate and salicylaldoxime. Copper 'stress' appears to influence the intracellular location, inhibitor sensitivity and polypeptide composition of methane mono-oxygenase in the methylotroph *Methylococcus capsulatus* (Stanley *et al.*, 1983), even though copper is not known to be a component of the enzyme. Chemostat studies suggest that low oxygen tensions exacerbate copper deficiency and it may be that in these organisms growing on methane, copper is required for a terminal oxidase.

Nitrous oxide reductase, an enzyme involved in the anaerobic dissimilatory reduction of nitrate or nitrite to N_2, has been shown to be a copper-requiring enzyme in *Alcaligenes* spp. from experiments with copper-deficient media (Iwasaki, Saigo and Matsubara, 1980; Iwasaki and Terai, 1982). Similar findings have been reported for *Pseudomonas perfectomarinus* (Matsubara, Frunzke and Zunft, 1982), where reconstitution of nitrous oxide reduction by added copper, unlike the situation in cytochrome *c* oxidase, is independent of protein synthesis (Chapter 5).

2.4.9 Molybdenum

In many micro-organisms, molybdenum is an integral part of essential molybdoenzymes (Chapter 5), and is, therefore, an essential growth factor. Indeed growth of *Aspergillus niger* may be used as a biological assay for molybdenum (Hewitt, 1966). The concept of a molybdenum-containing cofactor, common to most molybdoenzymes, has received support from genetic and biochemical studies.

Growth of several bacteria and *Chlorella* in the absence of added molybdenum (with no special attention paid to extraction), or after including tungstate in the medium in large excess (e.g. 1 mM) with respect to molybdate (e.g. 1 μM), results in the synthesis of inactive enzymes containing empty or demolybdocofactors. The demolybdo-forms of several enzymes, including nitrate reductase, formate dehydrogenase and nitrogenase, can be reactivated by incubation of cells with molybdate, even in the presence of chloramphenicol (Scott, Sperl and DeMoss, 1979).

In *E. coli*, at least three genes (affected *chlA*, *chlB* and *chlD*) are required for the transport and incorporation of molybdenum into molybdoenzymes (Ingledew and Poole, 1984). Certain mutants (in the *chlA* and *chlB* genes) produce an inactive nitrate reductase, but do not synthesize an immunologically detectable formate dehydrogenase. Mutants affected in *chlD* will, however, grow on adding molybdate to the medium, and some such mutants accumulate an inactive nitrate reductase. Demolybdo forms of formate dehydrogenase, however, appear to be unstable and of unusual subunit composition (Giordano, Haddock and Boxer, 1980).

Molybdenum-limited steady-state growth of *Azotobacter vinelandii* in a chemostat was achieved by Eady and Robson (1984) after 8-hydroxyquinoline extraction (Section 2.3.2(a)). The Mo content of the cells was reduced 100-fold and they lacked a characteristic ESR feature of the MoFe protein of nitrogenase.

2.4.10 Cobalt

Cobalt is frequently essential for growth, being part of the vitamin B_{12} or cobalamin synthesized by prokaryotes, as well as having non-B_{12} functions in higher plants (Hutner, 1972). It is commonly leached from borosilicate glassware. The requirements for cobalt and magnesium are closely linked in some eukaryotes (Hutner, 1972) and in bacteria (Jasper and Silver, 1977; Section 3.4), since both metals may share uptake systems and intracellular binding sites.

2.4.11 Nickel

Nickel is now regarded as an essential element in several biological processes. Unitl recently, the most familiar requirement was for jackbean urease but its involvements in carbon monoxide dehydrogenases, factor F430 in methanogens, and in certain hydrogenases are now attracting great interest from chemists and microbiologists alike (Chapter 5).

Extraction of nickel from a medium is not always necessary to demonstrate nickel limitation, although Ni^{2+} commonly contaminates mineral salts at suficient levels for growth of Ni^{2+}-dependent autotrophic bacteria. Diekert and Ritter (1982) have used media that had previously supported the growth of a nickel-requiring methanogen to study nickel requirements in *Acetobacter woodii*. Enough nickel to support growth can readily be leached from stainless steel components of culture vessels; this indeed formed the basis for the discovery of a nickel requirement in bacteria (Chapter 5). Recently, Klucas *et al.* (1983), using media freed from nickel (by extraction with dithizone) and careful precautions to minimize metal contamination, have shown that free-living *Bradyrhizobium japonicum* requires Ni for growth and expression of hydrogenase activity, while the host soybean plant requires nickel for urease. Several chelating agents (*o*-phenanthroline $>$ NTA $\geqslant$ EDTA $>$ EDDA $\approx$ dimethylglyoxime) added to growth media resulted in depressed hydrogenase activity in *Azotobacter chroococcum* (Partridge and Yates, 1982; Pedrosa and Yates, 1983) and the effect was attributed to nickel deficiency.

Certain algae and cyanobacteria show a growth requirement for nickel (van Baalen and O'Donnell, 1978); although some seem to require the metal for urease (Rees and Bekheet, 1982), others, such as *Chlorella emersonii*, lack urease and the nature of the requirement is not known.

Studies of the effects of nickel-limited growth are in their infancy. Nickel-limited cells of *Alcaligenes eutrophus* contain catalytically inactive, but serologically active, hydrogenase. Some structural modification of the protein, however, results from nickel starvation (Friedrich, Schneider and Friedrich, 1982).

2.4.12 Vanadium

The biological roles of vanadium and its requirement for growth are still only poorly understood, but the metal has long attracted attention (Ramasarma and Crane, 1981). The early work that led to the identification of V as a growth factor for *Aspergillus niger*, *Candida stoofii* and mammalian cells is surveyed by Hutner (1972). Thirty years ago, vanadate was reported to stimulate growth and chlorophyll content of *Scenedesmus obliquus*, but only much later were traces of vanadate generally recommended for obtaining maximal biomass of *Chlorella* and *Scenedesmus* (Payer and Trultzsch, 1972). It is now well established that in such unicellular green algae, vanadium is essential for chlorophyll synthesis and in particular the formation of protoporphyrin IX (Meisch, Benzschawel and Bielig, 1977) and possibly δALA (Meisch and Bauer, 1978), intermediates in porphyrin biosynthesis, especially during iron deficiency. In other algae, the effects are less clear, for although vanadium increases growth yields and influences the morphology of the brown seaweed *Fucus* and the green alga *Enteromorpha compressa* (Fries, 1982), it appears to have no effect on certain red algae. As originally described by Warburg, vanadate (but not vanadyl) salts enhance photosynthetic O_2 production. It has been suggested that this effect is due to vanadium acting as a redox catalyst in electron transport from Photosystem II to Photosystem I (Meisch and Becker, 1981). At high concentrations, the VO^{2+}/VO_3^- couple could act as an artificial, i.e. non-physiological, redox catalyst.

Vanadium has also been reported to substitute for molybdenum as a trace element essential for nitrogen fixation in *Azotobacter vinelandii* (Bortels, 1936). This had been attributed to a 'sparing effect' in which vanadium stabilized the molybdenum nitrogenase (Benemann *et al.*, 1972), but it is now known that there is an alternative nitrogenase which contains vanadium (Robson *et al.*, 1986; Chapter 5).

Vanadium is the prosthetic group of a bromoperoxidase in the brown alga *Ascophyllum nodosum* (de Boer *et al.*, 1986).

2.4.13 Other metals

(a) Aluminium

Pirt (1975) classifies aluminium as an element which is rarely essential for growth. Aluminium, Sn, F and I are not recovered by wet ashing of microbes (Hutner, 1972). Aluminium can be bound tightly by Fe-transporting hydroxamates and ferrichrome (Byers and Arceneaux, 1977). Aluminium promotes growth of diatoms, possibly by adsorption into the silica frustules, thus impeding their dissolution, but no absolute growth requirement has been demonstrated (Stoffyn, 1979).

(b) Chromium

Chromium is an essential micronutrient in some biological systems, especially in insulin action, possibly glucose utilization (Haylock *et al.*, 1982) and stimulation of enzyme systems (Mertz, 1969). Little is known of microbial requirements; it is more frequently regarded as a toxic element to which some organisms have evolved resistance mechanisms (Chapter 6).

2.4.14 Selenium

Although the metalloid element selenium is not routinely included in defined growth media for microbes, its essentiality for higher animals is well established. At present four types of response to selenium are

Table 2.4 Requirements for selenium by microbial enzymes

Requirements	Organism	Reference
Formate dehydrogenase	*E. coli*	Ingledew and Poole (1984)
Formate dehydrogenase	*Clostridium thermoaceticum*	Andreeson and Ljungdahl (1973)
Formic hydrogen lyase	*E. coli*	Fukuyama and Ordal (1965)
Glycine reductase	*Clostridium stricklandii*	Stadtman (1974)

recognized. Organisms that (i) are inhibited by the element and (ii) can tolerate high levels, are discussed in Chapter 6. The few organisms that (iii) require selenium for induction of selenoproteins or (iv) the even fewer that require traces of selenium for normal growth are described briefly here (for recent surveys, see Stadtman, 1980, and Konetzka, 1977).

Table 2.4 lists a few selenoproteins identified in micro-organisms. In some cases the amounts of selenium found associated with purified enzymes is sub-stoichiometric with other metals. Thus, although the synthesis of xanthine dehydrogenase in *Cl. acidiurici* requires selenite compounds during growth, selenium is not involved in the FeS or Mo centres of the enzyme (Wagner, Cammack and Andreesen, 1984). Analysis of the purified native enzyme gives the molar ratios: 7.7 non-haem iron, 7.5 acid-labile sulphur, 1.7 FAD, 1.8 molybdenum, 0.12 tungsten and 0.13 selenium. Other examples cited by Wagner, Cammack and Andreesen (1984) are glycine reductase, the β-keto-thiolase of *C. kluyveri*, hydrogenase and formate dehydrogenase. In the case of xanthine dehydrogenase, the lability of the enzyme suggests that selenium is lost during purification and handling.

Two anaerobic bacteria have been described that not only require selenium for formate dehydrogenase and/or glycine reductase but require selenium for growth (Dürre and Andreesen, 1982; Jones and Stadtman, 1977). *Peridinium unatum* is a dinoflagellate that requires traces of the element for its growth (Lindström, 1980). Recently, an aerobic, gram-positive bacterium has been isolated from seeds of *Astragalus crotalariae*, a selenium-accumulating plant (Lindblow-Kull, Schrift and Gherna, 1982). Under certain conditions, the growth requirements of the bacterium are met only by concentrations of selenite or tellurate in the millimolar range. During growth, the culture becomes deep red (characteristic of elemental selenium) or black (due to elemental tellurium). Compounds of neither element are required for growth in brainheart infusion or Trypticase soy broth, however. An explanation for the exceptionally high levels of selenium required is awaited with interest.

2.5 THE QUANTITATIVE DETERMINATION OF METALS IN MICROBIOLOGICAL MATERIAL AND MEDIA

The determination of metals in biological material is not always a straightforward exercise. There are many pitfalls for the inexperienced worker. These are often due to complications arising from the biological matrix in which the metal is present. This section deals with the most widely used analytical techniques and outlines some common problems associated with sample preparation and analysis. An excellent account of these difficulties has been given by Veillon and Vallee (1978).

2.5.1 Techniques for the analysis of metals

A number of techniques for metal analysis are available, each having its own particular advantages and disadvantages (Willard *et al.*, 1982). The most commonly-met techniques are those of atomic absorption spectroscopy and inductively coupled plasma atomic emission spectroscopy (ICPAES), while under certain circumstances ion chromatography and stripping voltammetry (and other polarographic methods) are of value. Recently, the plasma source has been combined with mass spectroscopic MS detection to give a technique (ICPSMS) with significantly improved detection limits. Detection limits for various techniques are given in Table 2.5 for some metals of biological significance. These limits may be lowered further by a pre-analysis concentration stage in which the metal is extracted from the aqueous sample into a smaller volume of an organic solvent containing a chelating ligand.

The factors that are relevant to the choice of technique include detection limits, sensitivity, sample size and the number of samples to be analysed. In some cases, techniques that allow multi-element analysis are especially valuable. The ICPSMS technique has an advantge over ICPAES as the majority of elements show remarkably similar detection limits, while those for ICPAES are highly element-dependent, being spread over a wide concentration range (Table 2.5; Slavin, 1986).

Most techniques for metal analysis allow the determination of the total amount of the element in a sample, irrespective of its chemical form. The techniques of stripping voltammetry and ion chromatography may also

Table 2.5 Detection limits[1] ($ng\ ml^{-1}$, ppb) for analytical techniques

Metal	AAS[2] (flame)	AAS[3] (furnace)	AES (N_2O/acetylene)	ICPAES	ICPSMS	SV	Ion chromatography[4]
Na	0.4	0.004		7	0.1		50
K	1	0.004		30	1		50
Mg	0.3	0.0002	5	0.1	0.01		50
Ca	1	0.01	0.1	0.2	1		50
V	25	0.1	10	3	0.01		
Cr	3	0.004	5	2	0.1	<5	50, Cr (VI) 1000, Cr (III)
Mn	2	0.0005	5	0.3	0.1		
Fe	5	0.01	50	2	1		3, Fe (III) 10, Fe (II)
Co	4	0.008		3	0.03		3
Ni	5	0.05	30	4	0.01		25
Cu	2	0.005	10	1	0.01	<1	5
Zn	1	0.001		1	0.1	<1	10
Mo	600[5]	0.03	100	4	0.01		50
Cd	1	0.0002	2000	1.5	0.01	<1	10
Pb	9	0.007	200	20	0.01	<1	100
Hg	140	0.02		12	0.01		

1. Data from Johnson, Taylor and Skogerboe, 1979; Berman, 1980; Date and Gray, 1983; Slavin, 1986; Selby and Hieftje, 1987; Dionex specifications for ion chromatography.
2. Air-acetylene flame.
3. 100 μl samples.
4. 50 μl samples.
5. 10–25 with N_2O-acetylene flame

AAS: Atomic absorption spectroscopy. AES: Atomic emission spectroscopy. ICPAES: Inductively coupled plasma atomic emission spectroscopy. ICPSMS: Inductively coupled plasma source mass spectrometry. SV: Stripping voltammetry.

provide additional information on the oxidation state of the metal and its speciation in solution, but are probably too time-consuming for routine, high-throughput analyses.

Analysis for the metal content of a new biological material should be carried out by two independent techniques. Good agreement between the results provides reassuring evidence for the validity of the analytical methods. If it is not possible to do this then the analytical method in use should be tested on a standard reference material as similar as possible to the samples being analysed.

(a) Atomic absorption spectroscopy

This is probably the most widely used technique for trace metal analysis (Walsh, 1955; Kirkbright and Sargent, 1974; Ebdon, 1982). Detailed methodologies exist for many standard determinations (Berman, 1980, and manufacturers' handbooks). The technique employs a light source which emits the atomic spectrum of the element to be analysed, although some multi-element lamps are also available. The sample is atomized in the light path, absorbing the incident radiation and causing a decrease in the intensity of the light beam. Atomic absorption spectroscopy is, therefore, relatively free of spectral interferences, as the chance of an accidental overlap of an atomic absorption line of one element with an atomic emission line of another is almost negligible.

Several techniques for sample atomization are available, but only the two most common methods are discussed in this account. In the flame technique, the solution containing the analyte is sprayed into the flame to produce ground state atoms. An air-acetylene flame is usually hot enough to accomplish this (1300°C), but a nitrous oxide-acetylene flame may be necessary for refractory materials.

The use of electrically heated graphite atomizers gives detection limits two orders of magnitude lower than those obtained with the flame. The sample is placed in the graphite tube, which lies on the light path. The introduction of the sample is usually the main source of imprecision. The temperature of the graphite tube is then raised in stages to evaporate the solvent, ash the sample and finally to atomize the sample, with the temperature rising to nearly 3000°C in some cases. Optimization of the temperature and duration of these stages is an essential feature of the development of the analysis procedure. The heating rate must be fast enough to maintain a high density of atoms in the light path. Detection limits with the graphite furnace can be lowered by using a larger sample, while direct analysis of solid samples is also possible. An important advantage of the graphite furnace is the small sample size required.

Volumes of 10–50 µl are typical in graphite furnace atomization, while the flame technique usually involves an uptake rate of the sample into the flame of several ml min^{-1}, so that a sample volume of several millilitres is necessary.

The main problem in atomic absorption spectroscopy is that of interference from chemical and biological material in the sample which is not present in the standard solutions used to calibrate the instrument. Measurements in the u.v. range may be complicated by absorption of the radiation from the line source by molecular species, thus giving spuriously high concentrations of metal in the sample. Instruments usually have a 'background correction' facility to cope with this difficulty. The background, non-specific absorption is measured using a deuterium source and subtracted from the absorption determined using the line source. New methods for background correction have been developed recently (Knowles and Frary, 1988; Liddell and Brodie, 1980). Positive errors also arise from the presence in the samples of easily-ionizable elements such as the alkali metals. This may be overcome by the addition of ionization suppressors. Chemical interferences may result from the formation of compounds in the flame that differ in volatility from the original form of the metal. Absorbance readings will be depressed if less volatile compounds are formed. Thus, the presence of phosphate in a sample will depress a calcium signal as the formation of calcium atoms is hindered by the production of calcium phosphate. This type of interference may be reduced by the use of a hotter flame, or eliminated (in the phosphate example) by the addition of excess strontium or lanthanum compounds to bind the phosphate. Interferences involving the formation of volatile compounds are rare.

(b) *Atomic emission spectroscopy*

In this technique, a sample is converted to atoms, which are then excited to a higher electronic state. The emission of radiation due to the subsequent fall to the ground state is monitored, and will be linearly related to concentration over a wide range. Atomization and excitation of the sample may be accomplished by conventional flame techniques, but recent developments have centred on the use of plasma discharges involving temperatures as high as 10 000 °C (Fassel, 1978; Johnson, Taylor and Spogerboe, 1979).

Inductively coupled plasma atomic emission spectroscopy is a versatile tool for multi-element analysis, allowing the determination of about 80% of the elements in the Periodic Table with precision and having detection limits superior to flame atomic absorption spectroscopy (Table 2.5).

Graphite furnace atomic absorption spectroscopy, however, does involve lower detection limits than ICPAES, and is more convenient for micro-sample analysis. The advantages of ICPAES include the ability to analyse simultaneously for a range of elements, many of which would be considered difficult by atomic absorption methods. The high temperatures used in the plasmas means the ICPAES does not suffer from the problem of chemical interferences.

(c) Inductively coupled plasma source mass spectrometry

The ICP is an excellent source of inorganic ions for a mass spectrometer, as ionization efficiency is high but few ions become doubly ionized (Date and Gray, 1983). The main attraction in using mass spectroscopic rather than optical detection methods lies in the much higher sensitivity achieved by the former method, with detection limits in the range of 0.02–0.7 $\mu g\,l^{-1}$ (Table 2.5; Selby and Hieftje, 1987).

(d) Ion chromatography

This represents an important complementary technique to that of atomic spectroscopy. The metal species in a sample are separated using ion exchange columns and an eluant containing a complexing agent, and analysed as they come off the column by conductivity or optical methods. The metals are identified and quantified by comparison with a range of standards. Table 2.5 includes data on detection limits for ion chromatography based on the use of 50 µl samples. These values compare well in some cases with those found for the flame atomic absorption technique, and may be improved by preconcentration and/or the use of a larger sample.

The special advantage of ion chromatography lies in the ability to determine separately species with the metal in different oxidation states and to differentiate between metals with different coordination environments, for example between the aquo species and a complex of the metal with a biological molecule (Irgolic and Brinkman, 1986).

(e) Stripping voltammetry

This technique offers considerable advantages over that of flame atomic absorption spectroscopy for selected metals in terms of detection limits and interference problems. It also provides information on the speciation of a metal in solution. The method involves the reduction of the metallic species in the sample at a controlled potential using a hanging drop mercury electrode, followed by an analytical stage involving the electrochemical oxidation or stripping of the reduced metal.

(f) Colorimetric methods

These methods are now little used, but they do have detection limits as low as 10^{-8} M in some cases. They may be of use as confirmatory techniques when the validity of other methods is in question. Beinert (1978) has described the quantitative determination in biological material of iron and copper using 4,7-diphenyl-1,10-phenanthroline and 2,9-dimethyl-4,7-diphenyl-1,10-phenanthroline, respectively. This illustrates well how the selectivity of a ligand for a metal can be changed by alteration in steric factors, as noted in Chapter 1.

The indicator Arsenazo III may be used to determine calcium (Miledi, Parker and Schalow, 1977), as can the photoproteins aequorin and obelin, which emit light on binding calcium (Shimomura and Johnson, 1976). These photoproteins have detection limits for Ca^{2+} of about 5×10^{-8} M, and have novel applications in studying intracellular functions of calcium.

2.5.2 Preparation of samples for analysis

For all studies involving analysis of metal ions, particularly at trace levels, it is essential that contamination by adventitious metal be avoided. Unfortunately, metals such as iron and copper are ubiquitous contaminants of medium reagents: for example, a medium prepared from good quality reagents without added copper is likely to have copper present at about 0.3 μM.

(a) Water and chemicals

The water used in such experiments must be of consistently high quality. Doubly-distilled water from all-glass stills is probably adequate for most work, and should be stored in plastic tanks. Any air-intake points should have a filter to exclude dust. Reagent chemicals should be of the best grade available, and, if necessary, should be purified by recrystallization. If quality is in doubt the reagents should be analysed for metal content. Techniques for removing metals from buffer and media have been described in Section 2.3 (Thiers, 1957).

(b) Laboratory apparatus

Metals may be leached out of glass storage vessels and perturb expected concentration levels. The problem may be alleviated by washing glass containers in nitric acid or nitric acid/sulphuric acid mixtures, if necessary immersing the containers in acid for some days. The vessels should then be washed many times with best quality distilled water. Alternatively, equipment can be soaked in a solution containing a proprietary detergent and a chelating agent for a day, followed by thorough washing with distilled water. Plastic containers are much less likely to cause problems through leaching of metal into solution.

Glass and, to a lesser extent, plastic containers also take up metals from solution. At concentrations below micromolar, loss of metal ions to the wall of the container may be considerable within 30 minutes. Standard solutions at such concentration levels should not be stored but should be made up as required and used immediately.

(c) Airborne contamination

Contamination by metals in dust may be significant in trace metal work. At ultra-trace levels, preparations should be carried out in work areas shielded by a laminar flow of filtered air at a velocity of about 30 m min^{-1}.

(d) Treatment of biological samples

Organic or biological materials in a sample must be destroyed before analysis in order to minimize matrix effects. This can be achieved by dry-ashing or wet-digestion procedures. Dry-ashing involves heating the sample in a muffle furnace at 450°C for 24 hours or until ashing is complete. Complications arise for elements such as mercury, cadmium, iron, zinc and others which may be partly lost by volatilization. This may be overcome by using a low-temperature oxygen plasma system.

Wet digestion techniques are more widely used. The sample is heated or refluxed with mineral oxidizing acids. The relatively low temperatures used minimize losses by volatilization. Concentrated nitric acid is often effective but addition of concentrated sulphuric acid may also be necessary to achieve dissolution. Difficult samples may be treated with a mixture of nitric, sulphuric and perchloric acids or with a 1:1 mixture of hydrogen peroxide and sulphuric acid, but these methods should be used with great caution. Digestion should be continued until a clear solution is obtained, when the sample should be made up to a standard volume with distilled water.

2.5.3 Standard solutions for calibration

Standard solutions for calibration of instruments are usually prepared by dilution of stock solutions. Great care should be exercised in the preparation and maintenance of these standard solutions. As noted above, standards below $10\,\mu g\ ml^{-1}$ may be unstable due to loss of metal to the vessel wall, and should be freshly prepared as required if there is any doubt over their stability.

For the techniques of atomic spectroscopy, the concentration of the sample can either be taken from a calibration plot of emission signal or absorbance against concentration or obtained by the method of additions. In the latter case, known and increasing amounts of the analyte are added to aliquots of the sample. A plot of absorbance against concentration of the added analyte may then be extrapolated back to intercept the concentration axis, giving the concentration of the sample. A method of additions plot parallel to a conventional calibration plot suggests the absence of interferences.

The method of additions is often favoured for the analysis of biological material because it is thought that problems of interference are overcome by the standards being presented in the same matrix as the sample. However, the analyte may be bound specifically in a manner not available to the added metal. It is, therefore, essential that great care is exercised to detect any abnormal effects due to interferences. It cannot be stressed too strongly that the validity of the method should be verified using a second method.

2.6 INSTRUMENTAL TECHNIQUES IN THE STUDY OF METALS AND MICRO-ORGANISMS

The chapters that follow emphasize the importance of instrumental techniques in the study of the structure and reactivity of metal-containing macromolecules. Some of these techniques are dependent upon the electronic and nuclear properties of the metal centre, and give information on the symmetry of the metal-binding site, the identity of the ligands and the nature of the metal-ligand bond, and the redox state of the metal. The presence of a transition metal ion allows the application of *d–d* spectroscopy and, for paramagnetic species, ESR spectroscopy. The presence of the paramagnetic centre also allows distance measurements to specific groups from the effect of the paramagnetic centre on the relaxation rate of nuclei in these groups (Van Divender and Grisham, 1985). The technique of nuclear magnetic resonance spectroscopy has been used successfully in studying the dynamic state of inorganic ions and organic metabolites in whole cells, and promises to be of immense importance in the future (Fernandez and Clark, 1987).

2.6.1 The characterization of metal-binding sites

The coordination environments around metals in isolated biological molecules may be established by the use of X-ray diffraction techniques and extended X-ray Absorption Fine Structure (EXAFS) methods. The list of metalloenzymes and other metal-containing biomolecules whose structures have been determined is increasing rapidly. However, in this section, attention will be drawn to other methods for probing coordination environments, from both structural and dynamic view points (Brown, 1980).

(a) Electronic spectroscopy

The *d–d* spectra of transition metals give information on the geometry of the metal site, including deviation from regularity of structure, as found for many metalloenzymes. They may also give some indication, from the Spectrochemical Series, of the identity of the ligands in terms of their ability to cause splitting of the *d* orbitals (Hughes, 1981). Reflectance *d–d* spectra of transition metal ions bound to microbial cell surfaces have been interpreted in terms of the geometry and likely ligand character of the metal-binding sites (Savvaidis *et al.*, 1988).

The energies of charge-transfer bands (absorptions resulting from excitation of electrons between ligand orbitals and metal orbitals) will depend upon the ligand if the metal is fixed, and so help to define the ligand. The classic example of this approach is the work of Druyan and Vallee (1962) on metallothionein, where the role of –SH groups in cysteine residues as ligands was established. Zinc metallothionein has an intense band at 215 nm due to charge transfer, while substitution of zinc by cadmium gives a new band at 250 nm. The spectra of mercaptoethanol complexes of zinc and cadmium show charge-transfer bands at 215 and 250 nm respectively, providing good evidence for the role of cysteine residues as ligands in metallothionein, with ligand to metal charge transfer, Cys(S) $\rightarrow$ M^{2+}.

(b) Vibrational spectroscopy

A non-linear molecule with n atoms has a total of (3n−6) vibrations, so large biological molecules will have complex vibrational spectra with many fundamental vibrational frequencies plus overtone and combination bands. Nevertheless, much useful information may be obtained from Raman and infrared spectroscopy in terms of metal–ligand stretching frequencies and the properties of small molecules coordinated to metal centres in enzymes or cofactors.

The application of infrared techniques to biological systems has been transformed in the last decade by major developments in the design of spectrometers, which now allow spectra to be measured for samples in aqueous solution. Water absorbs very strongly in the infrared, except in certain limited regions, and so could not be used as a solvent in conventional spectrometers. Interferometers are now used to record spectra, and the resulting interferogram transformed by computer into a spectrum (the Fourier transform method). Multiple repeat scanning of the spectrum is also possible, increasing the signal to noise ratio. The background spectrum of water can be removed by subtracting a stored spectrum of water, so allowing the measurement of the infrared spectrum of a dilute sample in water. Special solution cells with multiple internal reflection are also available.

Raman spectroscopy is well suited for work in aqueous solution as the intensity of the band at 1640 cm^{-1} due to water is very weak. A laser light source is used to illuminate the sample and the scattered radiation collected at right angles to the incident beam. The scattered radiation includes the original laser frequency, and, on both sides of this frequency, a series of much weaker lines. The lines on the red (low energy) side of the incident frequency are the Stokes lines, and are shifted by amounts corresponding to the vibrational frequencies of the sample under study, that is the scattered Raman radiation has the frequency of the exciting line minus the vibration frequencies. The lines on the blue (high energy) side of zero are the anti-Stokes lines, and correspond to the frequency of the laser plus the vibrational frequencies. The Raman effect is weak and so benefits from increased laser power, provided this does not cause the sample to decompose. The intensity of certain Raman bands can be enhanced greatly if the molecule is irradiated with an exciting line whose frequency corresponds to an electronic transition of the molecule. The intensities of Raman bands associated with the vibration of the chromophore are enhanced, usually by a factor of 100–1000. The theory of this effect is complex. The electronic excitation is accompanied by some distortion of the molecule in the excited state, and it appears that only vibrations that support this distortion are enhanced in intensity. The effect is of great value in the study of metalloproteins since the metal is often involved both in the charge transfer electronic process and the active site of the enzyme. Resonance Raman spectroscopy thus allows specific study of the active sites of enzymes. The metal–ligand stretching vibrations will be enhanced in intensity if the electronic transition involves an increase in metal–ligand bond length, as observed for copper enzymes. In other cases, sometimes with octahedral Fe(III), the charge is transferred to a metal π orbital with little effect on the metal–ligand bond length. In such cases the intensity of the metal–ligand stretching vibration is not enhanced.

The following examples illustrate the use of vibrational spectroscopy. Rubredoxin is an iron-sulphur protein, in which the single iron is bound by four cysteine residues in an approximately tetrahedral stereochemistry, as shown by X-ray diffraction and EXAFS measurements. The tetrahedral geometry around the iron has been confirmed by the Raman spectrum of oxidized *Clostridium pasteurianum* rubredoxin in solution. This shows bands at 368 and 314 cm^{-1}, which may be assigned to ν_{Fe-S} modes of the $Fe(S-Cys)_4$ tetrahedron. Depolarization studies confirm that these are the ν_3 and ν_1 fundamentals of a tetrahedral complex (Long *et al.*, 1971). The spectrum of crystalline rubredoxin shows bands at 365 and 311 cm^{-1}, showing that the environment around the metal ion is the same in the crystalline state and in solution, an important observation. Adrenodoxin, a two-iron protein, involves a bitetrahedral structure with bridging sulphide and terminal cysteine groups. Replacement of sulphide by selenide allows the assignment of the Fe–S(Cys) stretching vibration at 350 cm^{-1} and the Fe–S vibrations at 397 and 297 cm^{-1}. This provides good evidence for the nature of the ligands and for the geometry of the sites.

The resonance Raman spectra of enzymes and electron transfer catalysts with haem groups are dominated by the vibrational modes of the porphyrin ring. The electronic spectra of porphyrins show a number of bands, so offering alternative possibilities for excitation. Excitation within the Soret band at about 400 nm results in the enhancement of the intensity of certain porphyrin vibrational modes, while excitation within the lower energy bands gives enhancement of a different set of vibrations. Furthermore, four of the ring modes are sensitive to the oxidation and spin states of the iron centre. These results allow interesting correlations to be made but detailed interpretations of these data are not yet possible. Certain anomalies have been explained in terms of the doming of the porphyrin, which will tend to reduce the extent of π-bonding in the porphyrin and hence produce a decrease in the Raman frequencies.

Resonance Raman spectroscopy is one of a variety of techniques that has been used to characterize the structures of intermediates formed during the action of peroxidase enzymes. Such work is usually carried out on horseradish peroxidase, but similar intermediates are found during catalysis by a range of peroxidases and catalases. The use of resonance Raman techniques has confirmed the presence of the oxoferryl structure in intermediates HRP-I and HRP-II, showing the characteristic frequency of the Fe=O group (Proniewicz, Bajdor and Nakomoto, 1986; Terner, Sitter and Reczek, 1985; Oertling and Babcock, 1988).

Most applications of infrared spectroscopy have centred on oxygenated haem intermediates, where the value of the O–O stretching frequency is indicative of the bond order of the bound 'O_2' groups. Assignment of the O–O stretching bands, confirmed unambiguously by isotopic substitution, shows whether the intermediate contains a superoxide or peroxide species. It is probable that there will be much greater interest in infrared spectroscopy in the future, and, hopefully, this work will be carried out in association with Raman studies. The two techniques are complementary and do not necessarily give the same information. Both techniques have been used to study the vibrational spectrum of the carbon monoxide adduct of haem proteins, which is a useful probe for monitoring the interactions of the haem group with protein residues in the haem pocket. The C–O stretching frequency is best measured from the infrared spectrum of the adduct, while the Soret-excited resonance Raman spectrum gives the Fe–C stretching vibration and sometimes the Fe–CO bending vibration. The assignments must be confirmed by the use of isotopically substituted adducts. Recently, these frequencies have been used to characterize the active site structures of a range of cytochrome *c* mutants prepared by site-specific mutagenesis techniques, particularly with respect to hydrogen bonding interactions with the coordinated carbon monoxide (Smulevich *et al.*, 1988). Various aspects of the infrared and Raman spectroscopy of biological systems are reviewed in the series edited by Clark and Hester (1988).

(c) ESR spectroscopy

Electron spin resonance gives a great deal of specific information on paramagnetic metal centres in metalloenzymes. The spectrum shows the absorption of energy from a continuous electromagnetic field plotted against the strength of the magnetic field. The parameter *g* specifies the magnetic field strength at which absorption of energy occurs. Transition metals may give broad, asymmetric ESR line shapes due to *g*-value anisotropy. The *g* values give information on the ligands around the paramagnetic ion and the geometry around the metal.

Metals with more than one unpaired electron may show resonances at several *g* values, but ESR spectra often show fine structure in cases where S = 0.5. Hyperfine splitting arises from the interaction of the electron spin with the nuclear spin of the metal (I). In general, 2I + 1 lines will be seen: for Cu(II), with I = 3/2, the spectrum shows four equally spaced lines. Superhyperfine splitting arises from the interaction of the electron spin with spins of nuclei other than the one on which the electron resides. Ligand donor atoms provide one example: an N donor ligand (I = 1) will give three lines in the superhyperfine structure.

The superhyperfine structure allows the specific identification of ligand donor atoms, although some donor atoms (e.g. oxygen) have a zero nuclear spin and so do not show this effect. This problem may be overcome by using isotopically labelled compounds or by growing microbes on labelled nutrients to produce isotopically substituted molecules where the isotope has a nuclear spin. Studies of this type played an essential role in the clarification of the structure of ferredoxin, when it was first isolated from nitrogen-fixing bacteria and was found to have an ESR spectrum with an unusual *g* value. The signal was established as originating from iron by growing the nitrogen-fixing organism on a ^{57}Fe source (I = 1/2) and showing the hyperfine splitting of the signal. Furthermore, similar experiments with ^{33}S showed that S was bound to the iron.

As noted above, ^{16}O (and ^{18}O) have zero nuclear spin, but ESR has been employed in the study of oxygen ligands by using the isotope ^{17}O, for which I = 5/2. Kalitzer (1987) has reviewed the use of ESR to determine the coordination mode of nucleotides in Mn-nucleotide-protein complexes. Complexes were prepared using ATP specifically labelled with ^{17}O at the α, or the β, or the γ phosphate groups. The ESR spectrum shows which of the three phosphate groups is coordinated to the Mn(II) as the ^{17}O interacts with the electron spin giving substantial line broadening (or six superhyperfine lines, if the resolution is sufficient).

Certain specialized developments in ESR spectroscopy focus on the characterization of coordination environments (Sigel, 1987b). Electron-nuclear double resonance (ENDOR) facilitates the study of hyperfine interactions in complex spectra, giving considerably enhanced resolution. Recent work has been reviewed by Huttermann and Kapple (1987). A newer development is that of pulsed ESR, the electron spin echo technique, and involves the analysis of the electron spin echo (ESE) envelope modulation. The decay of the ESE signal due to relaxation shows periodic falls and rises in the amplitude of the echo signal. These electron spin echo envelope modulation effects (ESEEM) result from the coupling of the unpaired electrons of the paramagnetic centre with the surrounding nuclei of non-zero spin, and allow the detection of nuclear spin couplings that remain unresolved in a conventional ESR spectrum due to the broad line widths often observed. The technique offers the possibility of exploring the second coordination shell around the metal (Tsverkov and Dikanov, 1987). These ESEEM effects have allowed the suggestion that the structure of the three-iron iron-sulphur protein discussed in Section 1.3.1 (e) does not involve coordinated water (Malikayil *et al.*, 1985), and that the copper at the active site of dopamine β-hydroxylase involves three, or more likely four, imidazole ligands (McCracken *et al.*, 1985).

In addition, pulsed ESR techniques can be applied to the study of rapid reactions involving paramagnetic intermediates, including radicals produced by photolysis or radiolysis.

(d) NMR spectroscopy

Nuclear magnetic resonance techniques have been applied extensively in structural studies on proteins, nucleic acids and biological membranes. The binding of metals by biological molecules has been elucidated in some cases by analysis of shifts in resonances in ^{1}H and ^{13}C spectra on addition of metal. Definitive conclusions may be reached if spin–spin coupling is observed in cases where the metal has a non-zero nuclear spin. Cass *et al.* (1978) have shown that ^{199}Hg(II) binds to the antibiotic bleomycin through a histidyl group by observation of splitting of the C–2(H) and C–4(H) resonances due to ^{199}Hg–^{1}H coupling. Structural studies have also been carried out using other nuclei. Thus, the use of ^{113}Cd NMR has shown the presence of seven non-equivalent CdS_4 sites in Cd-metallothionein (Sadler, Bakka and Beynon, 1978).

The radio frequency energy absorbed by a nuclear spin system is lost by transfer either to other spin systems (T_2, the spin–spin relaxation time) or to the local environment (T_1, the spin–lattice relaxation time). Paramagnetic centres increase the rate of nuclear relaxation by a 'through space' dipolar coupling and a 'through bond' contact hyperfine coupling, although the latter effect may be neglected in most cases (Sigel, 1987b). The measurement of relaxation rates ($1/T_1$) of substrate nuclei in the presence of paramagnetic probes has allowed the distances between various nuclei and the paramagnetic centre to be calculated. These distances compare well with values obtained by X–ray crystallography,

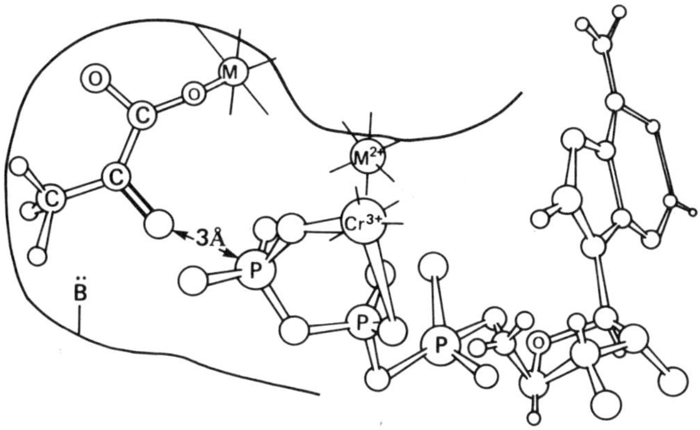

Figure 2.5 The active site of pyruvate kinase. This structure is based on distance measurements determined by NMR relaxation methods. Reprinted with permission from Dunaway–Mariano *et al.* (1979). Copyright 1989 American Chemical Society.

and so offer the possibility of obtaining three-dimensional structures of enzyme–substrate complexes in solution. Many studies of this type have been carried out by Mildvan and his coworkers, mainly on enzymes utilizing ATP. One example will be considered in detail. Pyruvate kinase catalyses the transfer of phosphate from ATP to pyruvate and requires Mg^{2+} and K^+ for activity. Substitution of these cations by Mn^{2+} and Tl^+, the use of Cr(III) ATP and relaxation measurements with ^{13}C, ^{31}P and ^{205}Tl NMR allowed nineteen distance measurements to be made. A model of the active site that accommodates these distances is shown in Figure 2.5. This model suggests there is contact between the phosphorus of the γ-phosphoryl group of ATP and the carbonyl oxygen of pyruvate, and, in turn, direct phosphoryl group transfer from ATP to substrate (Dunaway-Mariano *et al.*, 1979). A further example is that of glutamine synthetase from *Escherichia coli*, for which a model of the binding of Cr(III)ATP and Mn(II) to the active site has been proposed by Villafranca and Raushel (1982).

2.6.2 NMR Studies on intact cells

Nuclear magnetic resonance is a non-invasive technique and allows the study of ionic and metabolic processes in essentially unperturbed biological systems. Detailed information can now be obtained on structures, motion, reaction rates and binding sites *in vivo*. The nuclei used in this work have usually been ^{1}H, ^{18}F, ^{31}P, ^{13}C, ^{15}N and, to a lesser extent, ^{23}Na and ^{39}K (Fernandez and Clark, 1987; Gupta, 1987). These studies have been concerned with small, mobile metabolites.

Micro-organisms have been extensively studied using NMR techniques (Ezra *et al.*, 1984). Disadvantages associated with the rapid metabolism of microbial cells have been overcome by immobilization of cells in a diffusible matrix. New developments in NMR techniques have allowed high resolution spectra of solid samples to be obtained, including those of motionally-restricted compounds such as large biopolymers. High sensitivity and elimination of line-broadening are achieved using cross-polarization and magic angle spinning techniques.

Early experiments with micro-organisms involved the use of ^{31}P NMR to measure simultaneously various polyphosphates and inorganic phosphate for yeast (Salhany *et al.*, 1975) and *Escherichia coli* (Shulman *et al.*, 1979). Hayashi, Tompson and Agris (1981) used ^{13}C NMR to study the metabolism of methyl groups, while NMR techniques have also been used to determine intracellular pH (Ogawa *et al.*, 1978; Legerton *et al.*, 1983). Solid state studies have been reported on *N*-phosphonomethylglycine metabolism in a *Pseudomonas* sp. (Jacob *et al.*, 1985).

The uptake of sodium, potassium, magnesium and calcium ions has been investigated using ^{31}P NMR (Gupta, Gupta and Moore, 1984; Ogino, den Hollander and Shulman, 1983). Alkali metal ions have also been monitored directly by NMR. *Halobacterium* has high intracellular cation concentrations of about 4 M $[K^+]$ and 1.5 M $[Na^+]$ and is a good choice for NMR work. Bound and free intracellular cations have been identified by ^{23}Na, ^{87}Rb and ^{133}Cs NMR techniques (Goldberg and Gilboa, 1978). A distinction may be made between NMR signals from intracellular and extracellular metal ions by the use of lanthanide shift reagents. These interact with the extracellular cations and cause a shift in their resonances. The dysprosium tripolyphosphate anion, $Dy(PPP_i)_2{}^{7-}$, at about 3 mM concentration, shifts the extracellular ^{23}Na resonance by about 7 ppm. The efflux of Na^+ from Na^+-rich yeast cells has been monitored by the use of other anionic dysprosium shift reagents, present only outside the cells, to induce a splitting of the ^{23}Na resonance into components due to intra- and extracellular sodium (Balschi, Cirillo and Springer, 1982). The time course of efflux of Na^+ agrees well with values obtained by the use of conventional analytical techniques for sodium. However, NMR techniques have the major advantage of allowing direct study of the transport process.

2.7 SUMMARY

Metals play roles as both bulk and trace requirements for micro-organisms. Potassium and magnesium are regarded as bulk elements, with typical theoretical growth yields of 60 and 430 g dry biomass/g element, respectively. The requirements for calcium, iron, manganese, zinc, copper, cobalt and other metals are generally at trace levels, although some organisms require abnormally high levels for specialized roles. Metal requirements for growth are frequently not similar to optimal metal levels for other processes, such as secondary product formation. Careful design of defined growth media with due regard to metal chelation, metal–metal interactions and the removal of specific metals from media and growth vessels can provide valuable information on the roles of metals in micro-organisms and on substitution of a metal by another for a particular cellular function. However, the avidity of most micro-organisms for certain trace metals, notably iron, may necessitate extreme measures to demonstrate metal-limited growth. Iron limitation affects the synthesis and function of haem-containing molecules and may affect toxin and secondary metabolite production. Potassium and magnesium requirements are strongly influenced by growth rate and the demands for these metals in association with RNA and ribosomes. Extraordinarily high K^+ requirements are displayed by certain halophiles. Both magnesium and calcium are important in the control of cell growth and division and limitation of the former, a bulk element, leads to cell enlargement and inhibition of division. As is the case for most trace elements, the quantitative requirement for manganese and zinc by micro-organisms can only be estimated but the amounts are probably an order of magnitude lower, on a weight basis, than the Fe requirement. Nevertheless, manganese deficiencies can be demonstrated and the metal implicated in oxygen detoxification, photosynthesis and fungal growth. Zinc is important in numerous metalloenzymes and nucleic acids but growth requirements are frequently met by contaminating levels of the metal. Although calcium is an abundant trace element in the environment and laboratory media, intracellular concentrations are kept low by cellular export systems. Calcium fluxes play important signal roles in the cell division cycle and are crucial in bacterial sporogenesis. Sodium is also expelled by many cells and growth requirements are rarely demonstrated, except in the case of obligate alkalophiles, methanogens and other organisms that utilize Na^+ as a bioenergetic coupling ion. Copper, molybdenum and nickel limitation may all be demonstrated and used to study the roles of these metals in redox proteins.

The quantitative determination of metals in micro-organisms and microbial material is central to many of the studies described above, and often involves analysis at very low concentrations of metals. Unfortunately, the presence of the biological matrix represents a major complication in the analysis, and careful acid-digestion of the sample is necessary. A range of analytical techniques are available, but the most commonly used ones are atomic absorption spectroscopy and, more recently, inductively coupled plasma emission spectroscopy. The latter technique has the advantage of allowing the determination of all the metals in solution, many of which would be regarded as difficult by atomic absorption spectroscopy, while the high temperature of the plasma means that there is less chemical interference in analyses. Neither of these techniques gives information on the chemical form or oxidation state of the metal, but this may be obtained by the use of ion chromatography or stripping voltammetry.

The presence of a transition metal often allows the application of instrumental techniques such as electronic and electron spin resonance spectroscopy. These provide information about the geometry of the metal site and the nature of the ligands binding the metal. The use of multi-nuclear magnetic resonance techniques may also allow the characterization of site geometry through the measurement of distances from a paramagnetic centre to a particular group. Furthermore, NMR techniques may be applied to the study of whole cells and offer the exciting prospect of following intracellular processes directly.

The techniques of vibrational spectroscopy are finding increasing application to biological systems, notably for Fourier transform infra-red spectroscopy and resonance Raman spectroscopy.

3
Metal Transport

It is easy to get everything you want,
provided you first learn to do without
the things you can not get

Elbert Hubbard
The Note Book (1927)

3.1 INTRODUCTION

3.1.1 Pathways: organization of the cell's limiting membranes

For every metal required by a cell, there exist membrane transport systems with some degree of specificity to assure the cell of an adequate supply of the metal in the face of variable or limited amounts in the external environment. These transport systems for inorganic cations thus constitute an important proportion of the 60 or so carriers estimated to be present in the cytoplasmic membrane of, say, *E. coli*.

The boundary layers (walls and membranes) of eukaryotic microbes are diverse in structure and function but in general it is the lipid bilayer of the cell membrane adjacent to the cytoplasm that provides the major barrier to the influx and efflux of metal ions and other solutes (Rogers, Perkins and Ward, 1980).

The cell envelopes of bacteria are multi-layered. The inner layer is the cytoplasmic membrane, which is in direct contact with the cytoplasm and consists of a phospholipid bilayer in which proteins are embedded. Exterior to this are layers whose identity depends on the bacterium. The inner membrane is in contact with a layer of peptidoglycan (a network of polysaccharides, cross-linked by short peptides) which, in gram-positive organisms, is considerably thickened and forms the outermost boundary of the cell envelope. In gram-negative bacteria, the peptidoglycan is covered by an outer membrane consisting of phospholipids and lipopolysaccharides in which proteins are also embedded. The space between the outer and inner membrane is the periplasmic space. The outer membrane of gram-negative bacteria also forms a barrier for solutes but differs from the cytoplasmic membrane in having proteins (porins) that form non-specific hydrophilic pores through which all metal ions (and other solutes up to 600 M_r) can pass (Nikaido and Nakae, 1979). Other layers, most peripheral to the cell, namely capsular and slime material, play an important role in metal binding to the cell (Chapter 7) but are not relevant to specific membrane transport processes. In this chapter, the mechanisms of metal ion transport across the cytoplasmic membrane will be the main concern, since this layer constitutes the major diffusion barrier between the cytoplasm and the external material: it is provided with translocation systems for solutes and the necessary energy-transducing systems, namely electron transfer systems, ATPases and light-driven pumps (Rosen and Silver, 1987).

3.1.2 Driving forces: mechanisms of translocation

The molecular mechanisms of transport processes in general are outside the scope of this volume but some introductory comments are in order. Metal ions may diffuse through aqueous channels or through the lipid phase of the membrane, resulting in equilibration of solute concentrations in the extra- and intracellular compartments. Most metal ions, however, are imported or exported from cells against concentration gradients of considerable magnitude, so that the translocation process must be coupled to an energy-transducing machinery. The mechanism of coupling that we will adopt in describing these phenomena is that proposed in Mitchell's chemiosmotic hypothesis (Nicholls, 1982). It is proposed that the initial 'intermediate' is an electrochemical potential gradient of protons or proton-motive force, $\triangle p$. This potential is produced by one or more energy-generating processes such as aerobic or anaerobic respiratory electron transfer, ATP hydrolysis or photosynthetic electron transport. The $\triangle p$ is consumed, on the other hand, by energy-demanding processes, which include 'reversed' electron transport, flagellar movements and ATP synthesis and of which the most important, for present purposes, is solute transport. Such transport in Mitchell's terminology is 'secondary' in that a concurrent chemical transformation is not involved; transport may use thermodynamically passive pathways for the diffusion of ions in accordance with the concentration gradient, the electrical potential, or both. Some such secondary transport processes are electroneutral, others translocate one or more charges (electrogenic). In some cases, the ion moves singly (uniport); in others, its movement is coupled to that of another metabolite in the same direction (symport) or in the opposite one (antiport). The solute is translocated in an unmodified form.

There are a few metal ion transport systems that, again using Mitchell's terminology, are primary. These are integrally linked to some enzymatic reaction; many, but not all, are electrogenic. Examples include the Na^+ and K^+-stimulated ATPase of mammalian cell membranes and ATP-driven Na^+ efflux from certain bacteria, frequently referred to as 'ion pumps' (Harold, 1986).

The mechanism of action of membrane-associated primary and secondary translocators is not well understood. For this reason, much attention has been given to the ionophore antibiotics that exemplify many features of secondary transport, and some examples of these will be dealt with first.

3.1.3 Ionophores

A number of antibiotics of microbial origin exhibit ion-selective membrane channel activity. These ionophores possess antimicrobial activity in nature and are widely used in laboratory studies of membrane phenomena. Many of these facilitate the transport of metal cations in both natural and artificial membranes. Among the more familiar are the neutral cyclic peptide valinomycin (Figure 1.13; Section 1.6.2(a)), which forms charged complexes with monovalent cations such as K^+, and the negatively charged carboxylic acid polyether ionophores such as nigericin and monensin which form neutral metal–antibiotic species. The carboxylic acid X537A binds alkali metal cations, divalent cations and lanthanides. Another carboxylic acid, A23187, binds predominantly divalent cations, especially Ca^{2+} and Mg^{2+} (Nicholls, 1982; Painter and Pressman, 1985). The structure of these ionophores is also shown in Figure 1.13. Whereas the neutral ionophores such as valinomycin effect electrogenic movement of monovalent cations, the carboxylic acid types catalyse electroneutral exchanges of alkali metal cations for protons. A recently discovered antibiotic produced by *Streptomyces coelicor* appears to be unique in forming, in a calcium-dependent fashion, channels for monovalent cations (Lakey *et al.*, 1983). (For a review of antibiotics and their complexes see Sigel, 1985.)

3.1.4 Terminology

The terms 'uptake', 'transport', 'accumulation' and 'binding' have all been used to describe various stages in solute transfer from the medium to the interior of the cell, and intracellular binding. This transfer generally comprises two phases: an apparently instantaneous metabolism-independent binding of the metal to the microbial surface and, in some cases, a metabolism-dependent translocation of the solute through the surface to the interior. This distinction is particularly important in studies of metal removal in a biotechnological context where the extent and mechanism of solute movements are unclear. The terminology appropriate to such cases is discussed in Chapter 7. Here, we are concerned with the mechanism by which metal ions are translocated (or transported) across microbial cell membranes and, in the case of iron, are first sequestered by soluble ligands. 'Translocation' has been taken by some to be synonymous with 'uptake' (Failla, 1977) but we prefer to use translocation and reserve the latter, somewhat vague, term for use in situations where the mechanism and extent of metal movement is also poorly-defined (Chapter 7).

3.2 IRON TRANSPORT IN PROKARYOTIC AND EUKARYOTIC MICRO-ORGANISMS

3.2.1 Introduction

With the notable exception of some lactobacilli, iron is essential for all cells (Archibald, 1983). However, the solubility product of $Fe(OH)_3$ or Fe(O)(OH) is near 10^{-38} M and the maximum concentration of free iron(III) at pH 7 is about 10^{-17} M (Hill, 1982). Iron is generally in the oxidized iron(III) form in most growth media. Even under anaerobic, reducing conditions, the metal appears to be taken up as Fe(III) (Lodge and Emery, 1984), although certain *E. coli* strains transport predominantly Fe(II) (Hantke, 1987).

Thus, free-living, pathogenic and commensal bacteria require specific and effective ferric iron transport systems to cope with the low availability of the metal, a situation which is exacerbated for microbes growing within higher organisms by the presence of animal transferrins and other host compounds that compete strongly for iron. The microbial solution to this problem is the production of specific ligands, now generally called siderophores (Greek: 'iron carriers'; Lankford, 1973) and previously called siderochromes, which compete for the limited supplies of iron. An enormous literature, chemical and microbiological, has grown up on this subject, so that this section in particular must be illustrative rather than comprehensive. Extensive reviews include those by Lankford (1973) and Neilands (1980, 1981, 1982, 1983, 1984), Hider (1984) and Crichton and Charloteaux-Wauters (1987).

Siderophores are produced by diverse bacteria, including mycobacteria and cyanobacteria (Reeves *et al.*, 1983), but not by strict anaerobes, lactobacilli, nor, apparently, *Legionella*, even though these bacteria do require iron for growth. Certain lactobacilli may not require iron (Sections 2.4.1(a) and 2.4.1(c)). Other bacteria, e.g. *E. coli* and *Rhodopseudomonas sphaeroides* (Moody and Dailey, 1984), can utilize siderophores (ferrichromes and parabactin, respectively) produced by other organisms. Probably all *Penicillium* spp. produce ferrichromes, as do species of *Aspergillus* and *Neurospora*, but not *Saccharomyces* spp. Several other fungi are thought not to produce siderophores and may use a less specific transport system such as a citrate-mediated mechanism (Winkelmann, 1979). *Pilobolus* lacks both siderophores and alternative pathways and requires either iron-containing compounds such as haem or Fe(II)-ascorbate (Bourret, 1982). The marine dinoflagellate *Protocentrum minimum* is probably the most highly developed micro-organism to produce a siderophore (prorocentrin; Trick *et al.*, 1983), although a range of diamino-carboxylic acids, called phytosiderophores, occur in higher plants. These compounds, however, may bind various metals and their presumed function in iron transport is largely inferential. An impression of the diversity of microbial siderophores can be obtained from Table 3.1.

3.2.2 Diversity of iron transport mechanisms

Not explicit in Table 3.1 is the fact that many micro-organisms possess multiple mechanisms for iron binding and transport, thus ensuring a continuous supply of this essential metal. *E. coli*, for example, has at least five independent iron transport systems. Firstly, iron(III) is bound by excreted enterochelin and the complex taken up by the cell, whereupon the Fe(III) is reduced to Fe(II) to facilitate release, possibly accompanied by degradation of the enterochelin. Secondly, during growth in the presence of citrate, a separate system is induced which is unusual in that the inducer (citrate) need not enter the cytoplasm. *E. coli* can also take up ferrichrome, a siderophore produced by certain fungi. Strains that possess the *colV* (virulence) plasmid synthesize aerobactin and,

Table 3.1 A survey of siderophores from micro-organisms

Family	Examples	Chemical characteristics	Distribution of siderophore family	log K_f
Cyclic tri-catechols	Enterobactin (enterochelin)	See Figure 3.1	All enteric bacteria	52
Linear catechols	Agrobactin parabactin vibriobactin	Based on spermidine, amino acid moiety cyclized in oxazolidine ring which provides one ligand (*N*) to iron	*Agrobacterium tumefaciens' Paracoccus denitrificans*	≈ enterobactin
Ferrichrome	Many	See Figure 3.1 (prototype of hydroxamate class)	Used by many bacteria, but synthesized by various fungi	
Rhodotorulic acid	Rhodotorulic acid, dimercimic acid, coprogen, triornicin	Derived from N^5-hydroxyornithine	*Rhodotorula* and related yeasts; various fungi	31.2
Citrate hydroxamate	Schizokinin arthrobactin aerobactin	See Figure 3.1	*Bacillus* spp., *Arthrobacter*, *Klebsiella aerogenes*, *E. coli* and *Shigella flexneri*; *Anabaena* spp.	22.9
Mycobactin	P, S, etc. (designating species)	Two hydroxamic acids	*Mycobacterium phlei*, *M. smegmatis*,	
Exochelins	Many types	Not known, low M_r	Mycobacteria	
Fusarinine		linear and cyclic forms of N^8-acyl-N^8-hydroxy-ornithine	Various fungi, including *Fusarium* spp.	
Ferrioxamine	Ferrioxamine B (Desferal)	unit of 1-amino-ω-*N*-hydroxyaminoalkane, succinic and acetic acids	*Streptomyces*, *Actinomyces*, other bacteria	30.6
Pseudobactin*		Three iron-binding functional groups: hydroxamic acid, catechol, α-hydroxy acid	*Pseudomonas spp.*	

Data largely from Neilands (1983, 1984).
* Azotobactin, from *Azotobacter vinelandii*, has a very similar set of iron-binding groups (Demange *et al.*, 1988).

there is also a low-affinity process whose energy-dependence has not been established. Such a system is useful only during growth in an iron-sufficient environment and in the absence of a ligand such as nitrilotriacetate (NTA) which may bind iron in a form unavailable to the cell.

The existence of a multiple system is also well illustrated by mutants of *Salmonella typhimurium* LT2 that are defective in enterochelin, the single siderophore synthesized by this organism. Such mutants are, however, able to grow at normal rates even in minimal media when a low-affinity system is used. Such systems are probably widely distributed but are poorly understood. Certain strains of *Shigella flexneri* thought to synthesize only aerobactin, have been shown to synthesize also phenolate (enterobactin) siderophores (Payne *et al.*, 1983).

Iron uptake by *Mycobacterium smegmatis* is also more complex than hitherto believed (Macham, Stephenson and Ratledge, 1977; Ratledge, 1987). Several mycobacteria produce three iron-chelating agents – mycobactin, salicylic acid and exochelin. Mycobactin occurs exclusively in the cell, close to or in the cytoplasmic membrane (Ratledge, Patel and Mundy, 1982). It is a low molecular weight, lipid-soluble, iron-binding compound, whose synthesis is enhanced in iron-deficient conditions. It takes up both Fe(II) and Fe(III) chelated to salicylic acid and may also act as an iron store as well as being involved in iron transport. Salicylic acid, or 6-methylsalicylic acid, is secreted into the medium by mycobacteria. The latter compound forms part of the larger mycobactin molecule. It seems, however, that the ability of salicylates to solubilize iron is limited (Ratledge *et al.*, 1974).

Exochelin occurs extracellularly and accumulates under iron-deficient conditions. There are four main classes, distinguished by their water solubility in the iron-bound form. They solubilize iron and participate in an energy-coupled uptake by mycobacteria; exochelin-mediated iron uptake does not necessarily involve the participation of mycobactin (Stephenson and Ratledge, 1979). They are low molecular weight compounds (<1000) and are probably substituted peptides (Barclay and Ratledge, 1983).

3.2.3 High-affinity, siderophore-mediated iron transport

Siderophores are relatively low molecular weight, iron(III)-specific ligands induced under conditions of low iron availability (Hantke, 1981) for assimilation of Fe^{3+}. They constitute the first of three stages in the uptake of iron by many prokaryotes. In the second, a membrane receptor for the iron-loaded form (ferri-siderophore) transports the chelate across

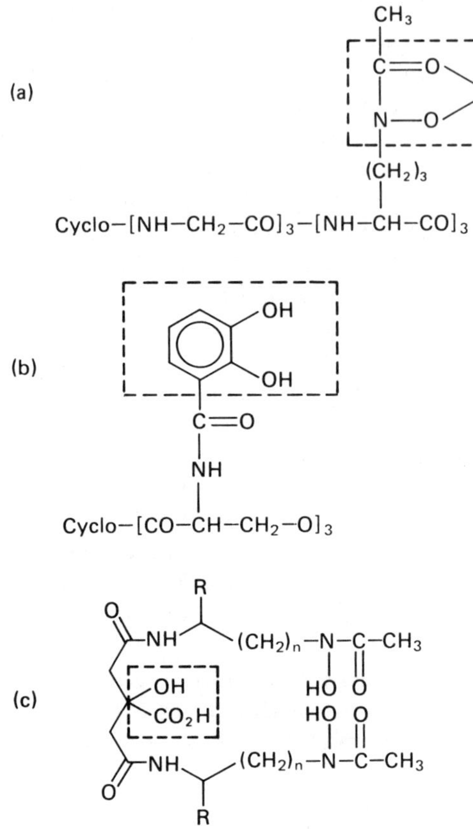

Figure 3.1 Structures of representative siderophores. (a) shows ferrichrome, prototypical of the hydroxamate siderophores. The bidentate hydroxamate liganded to iron is enclosed in the box. (b) shows ferric enterobactin, prototypical of the catechol siderophores. The bidentate catechol is enclosed. Fe(III) is bound by the two hydroxyls. (c) shows the structure of the siderophores of the citrate family. The α-hydroxylic acid responsible for iron binding is boxed. For schizokinen, R=COOH, n=2; arthrobactin R=H, n=4 and for aerobactin, R=COOH, n=4. Modified from, and reproduced by courtesy of, Neilands (1984).

the membrane, and in the third the 'iron' is released within the cell. Many have been characterized from bacteria, yeasts and fungi, and the structures of more than a dozen have been determined by X-ray diffraction techniques. Most are either catechols (e.g. enterobactin) or hydroxamates (e.g. ferrichrome) (Table 3.1; Figure 3.1). About 200 natural siderophores are known. The iron-free molecules are used as chelators in the treatment of iron overload in man.

(a) Assay of siderophores

The detection of siderophores in culture supernatants is described by Neilands (1983). The accumulation of brown oxidized and polymerized products in several-day-old cultures of iron-starved organisms may be taken as prima facie evidence for the presence of catechol-type siderophores. Chromogenic chemical assays exist for these and the enterobactin-type ligands. Microbiological assays offer greatly increased sensitivity (down to $100–200 \times 10^{-12}$ moles). The use of wild-type cells requires addition of a synthetic chelator (e.g. EDDA) to invoke iron deficiency. Mutants defective in a siderophore may be used for assay if the low-affinity uptake systems are suppressed using, for example, iron-binding ligands such as citrate or, in the case of *E. coli* (which has an inducible system for transport of ferric citrate), synthetic (α, α-dipyridyl) or natural (deferriferrichrome A) chelators.

(b) Structure and chemistry of siderophores

The iron is bound by the ligand, predominantly by oxygen atoms most often provided by bidentate hydroxamic acid or catechol moieties. Partial structures of ferrichrome (a hydroxamate-type siderophore and the first to be isolated) and enterobactin (a catechol-type) are shown in Figure 3.1 (Crichton and Charloteaux-Wauters, 1987). In these siderophores, three such bidentate groups are disposed within the molecule to yield a six-coordinate system with stability constants for Fe^{3+} ranging from 10^{52} for enterobactin to 10^{23} for aerobactin (Table 3.1). In siderophores of the citrate family, such as schizokinen, arthrobactin and aerobactin, and in the pseudobactins, the α-hydroxycarboxylic acid is linked to the iron (Figure 3.1). The distribution and chemical characteristics of other siderophores are given in Table 3.1. Rhodotorulic acid, ferrichrome and aerobactin have been synthesized in the laboratory (Maurer and Miller, 1982).

The remarkable affinity and thus specificity of the siderophores for iron (also shown in Table 3.1) arises from pairing of the 'hard' acid cation with a correspondingly 'hard' base such as oxygen to form a stable, high spin d^5 compound. The ligands also form very stable compounds with gallium(III), which is predictable in view of its similarity in charge and radius to iron(III). Aluminium (Al^{3+}) is a much more abundant metal, but the affinity for this and for all divalent cations is several orders of magnitude lower than for iron(III). Plutonium(IV) is strongly bound and synthetic hydroxamic acids have been used to extract plutonium from mixed fission products. Chromium (Cr^{3+}) can also be inserted into siderophores but only by refluxing under drastic conditions. Such chromium(III) siderophore complexes are kinetically inert (i.e. ligand exchange is slow) and can be used as probes of iron transport. Iron(III) enterobactin has an appreciable affinity for Ca^{2+}; indeed, its affinity and specificity compare favourably with recognized calcium ionophores (Section 3.1.3).

(c) Receptors for siderophores

The requirement for a receptor arises from the inability of the iron-laden siderophore (molecular mass 500 to 1000 Da) to permeate the small water-filled pores of the outer membrane of gram-negative bacteria (Neilands, 1982). Low levels of available iron derepress the synthesis not only of siderophores but also of membrane receptors (McIntosh and Earhart, 1977). For example, under iron-limited conditions, *E. coli* synthesizes seven new outer membrane proteins (74–83 kDa), five of which are siderophore receptors. These *E. coli* receptors will be described in turn.

The receptor for ferrichrome was the first to be identified. This receptor was originally called TonA in *E. coli* and regarded only as the attachment site for certain bacteriophages, colicine and albomycin (Neilands, 1982). However the identification in *Salmonella typhimurium* of mutants defective in ferrichrome transport and the mapping of the mutations to a locus equivalent to the *E. coli* tonA locus led to the realization that TonA was also a ferrichrome receptor. Similar reasoning led to the realization that the colicine B outer membrane receptor was also the ferric enterochelin receptor. This is an outer membrane protein of molecular mass 81 kDa, designated FepA. The gene is part of a cluster of nine genes involved in ferric enterobactin synthesis and transport extending across 26 kb of DNA. Loss of potency of cells or membranes in binding ferric enterobactin is associated with processing of the Fep protein to give a slightly smaller protein. The modification activity has been ascribed to protein *a*, the product of a gene mapping at 13 minutes on the *E. coli* chromosome. The receptor protein for ferric aerobactin, product of *nitA*, present in the outer membrane, has been sequenced (Krone *et al.*, 1985). The receptor for hydroxamate siderophores from fungi (such as coprogen) and ferrioxamine B (from *Streptomyces* spp.) is the *fhuE* gene product. *E. coli* takes up iron from ferric citrate through the receptor protein FecA. Figure 3.2(a) illustrates the outer membrane receptors and the inner membrane transport systems they serve in *E. coli pcolV*.

(d) Mechanism of iron transport and release

The mechanisms of removal of iron from the siderophores are only poorly understood even in *E. coli*, but three general schemes have been proposed (Figure 3.2(b)).

In the first, the metal complex is taken up intact and the iron released intracellularly, probably by reduction. Iron(II), unlike iron(III), is only weakly bound to the ligands (Kf $\approx 10^8$ L mol^{-1}) and so reduction provides a convenient mechanism for iron release. This leaves the ligand intact, which is re-excreted for further shuttling of iron. Ferri-siderophore reductase activity (NADH or NADPH:oxidoreductase) has been detected for a number of siderophores in an assay that uses ferrozine (Section 5.8.1) to chelate the iron(II) product. The number of enzymes involved is unclear; one such enzyme, catalysing the reductive release of iron from ferric-2,3 dihydrobenzoate, has been characterized (Gaines *et al.*, 1981) as a flavin reductase.

In the second (Figure 3.2), an intact metal–ligand complex is taken up, but the release of the metal is accompanied by modification or destruction of the ligand; enterobactin and the fusarinine-type siderophores may be examples. It has been suggested that the very low redox potential of the ferric enterobactin could preclude reduction as a means of liberating the metal except after cleavage of ester bonds. This degradation may involve an esterase encoded by the *fes* gene in *E. coli* (Langman *et al.*, 1972).

In the third mechanism (Figure 3.2), the complex or ligand does not penetrate the membrane, but donates the metal to a second membrane-bound chelator followed by reduction. Examples of such 'iron relays' are provided in the uptake of iron by rhodotorulinic acid (in *Rhodotorula*), the exochelin/mycobactin system in *Mycobacterium* (Section 3.2.2) and of ferrichrome by *Ustilago* (Ecker *et al.*, 1982). Here, the uptake of ferrichrome by the fungus is conformation and structure-dependent in that desferriferrichrome (i.e. lacking iron) is not taken up, but other metal-substituted ferrichromes such as the Al^{3+} and Ga^{3+} complexes are (Emery and Hoffer, 1980). Chromium(III) desferriferrichrome, from which the metal cannot be released by reduction is taken up at the same rate as the iron complex and so the transport system cannot rely on partial dissociation of the complex (Raymond and Carrano, 1979). Uptake of iron(III) aerobactin by *K. aerogenes* resembles the *Ustilago*

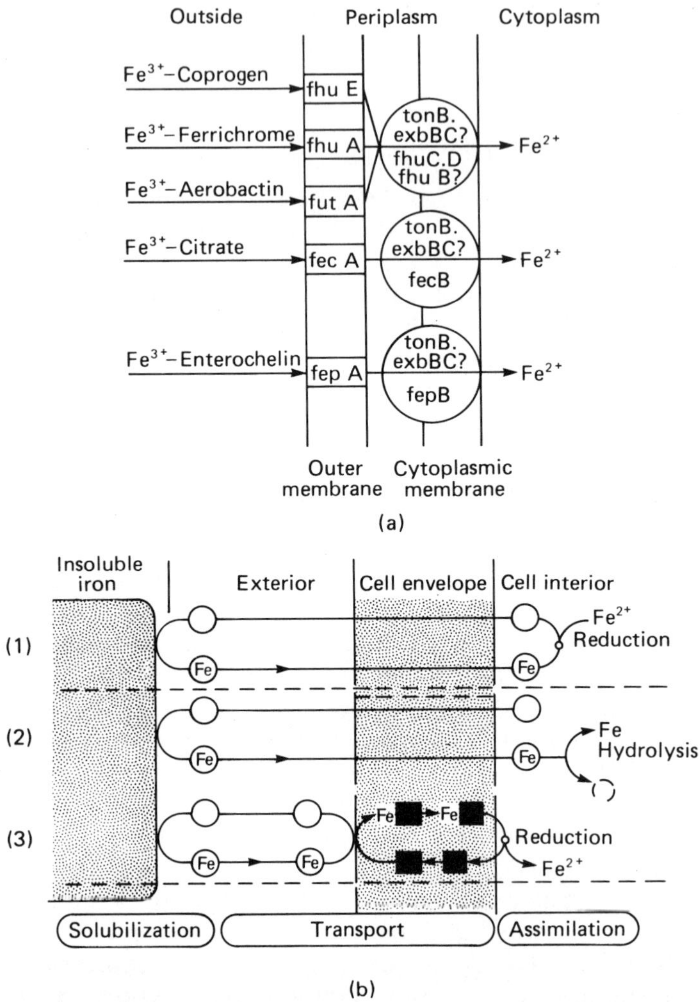

Figure 3.2 Mechanisms of iron uptake through bacterial membranes. In (a), the ferrisiderophore uptake systems of *E. coli* K12 *pcolV* are shown (from Hantke, 1985). Nomenclature of the gene products is described in the text. In (b) (Raymond and Carrano, 1979) an open circle O indicates a ligand. The diagram shows three general mechanisms for siderophore-mediated iron uptake. In (1), the metal–ligand complex is taken up intact, the iron released, largely as a result of reduction to Fe(II), and the ligand re-excreted. In (2) the metal is taken up as in (1) but iron release is accompanied by destruction of the ligand. In (3), the iron-binding ligand donates the metal ion to an envelope-bound carrier. (Adapted with permission from *Accounts of Chemical Research* **12,** 183–90. Copyright 1979 American Chemical Society).

system. In *Streptomyces pilosus* (Müller and Raymond, 1984), both chromic and gallium desferriferrioxamine B are transported at rates similar to the Fe complexes, implying that no decomplexation or metal ion reduction is required for transport.

The mechanism of ferrichrome transport across the cytoplasmic membrane of *E. coli* has been studied with membrane vesicles from an enterobactin-deficient strain (Leong and Neilands, 1976). The K_m for ferrichrome was 0.2 μM and the transport was dependent on membrane energization, being sensitive to uncoupling agents and valinomycin. A symport mechanism has been proposed in which a divalent cation, probably Mg^{2+}, is complexed to the ferrichrome cyclohexapeptide. There is evidence for the direct involvement of ATP in the uptake of iron(III) schizokinen by *Anabaena* spp. (Lammers and Sanders-Loehr, 1982).

(e) Other components of the iron transport systems in E. coli

As might be expected, genetic studies have been pursued most intensively with *E. coli* and have led to the identification of other components (Braun *et al.*, 1982; Neilands, 1984; Hantke, 1985). The genes encoding outer membrane receptors for siderophores are given in Table 3.2 and the products of others are shown in Figure 3.2(a). Thus, the uptake of iron from ferrichromes requires, in addition to *fhuA*, at least five more genes: *tonB* and *exbB* (essential for all ferrisiderophore uptake systems) and *fhuC*, *fhuD* and *fhuB*. The last three map close to *fhuA*; the gene products of *fhuC* and *fhuD* are plasma membrane proteins. Interestingly, translocation of iron compounds across the cytoplasmic membrane is less specific than through the outer membrane, in marked contrast to transport systems for other metal ions, sugars, etc. (Braun, 1985).

The genes for synthesis and transport of aerobactin are organized on *colV* plasmids in an operon. Like other siderophore-mediated mechanisms in *E. coli*, the aerobactin operon is regulated by the 17kDa product of the *fur* gene (ferric uptake regulation); *fur*, in conjunction with Fe(II), acts as a classical negative repressor (Bagg and Neilands, 1987).

A gene involved in the transport of ferrous iron (*feo*) has recently been described (Hantke, 1987) and interestingly is regulated by *fur* as are the siderophore transport systems for Fe(III). The large (90 kb) plasmid, *pcol V*, present in clinical isolates of *E. coli*, encodes a bacteriocin (colicin V) as well as bearing the recently-cloned aerobactin gene complex responsible for biosynthetic and transport functions. Five proteins have been identified as products of the cloned sequence: the largest is the outer

Table 3.2 Siderophore receptors detected in the outer membrane of *Escherichia coli* K12

Receptor	Gene mnemonic	Map locus (min)	Molecular weight (SDS)	Secondary substrates	Properties
Ferric enterobactin	*fepA*	13	81 000	Colicin B	Strongly induced at low iron
Ferric coprogen	*fhuE*	16			
Ferric citrate	*fecA*	7	80 500		Induced by iron citrate
Ferrichrome	*fhuA* (*tonA*)	7	78 000	Phages T1, T5, 80, colicin M, albomycin	Moderately induced at low iron
Ferric aerobactin	*iutA*	(plasmid *co1V*)	74 000	Cloacin	

Reproduced by courtesy of Neilands (1984) with minor changes.

membrane receptor for ferric aerobactin and the smallest is the synthetase or a component thereof.

(f) Implications of microbial iron transport in animal and plant infections and their treatment

The remarkable binding constants of bacterial siderophores allow micro-organisms invading the bloodstream to compete for iron with transferrin, the blood iron-binding protein (Hider, 1984). Similar protection to that afforded by transferrin is also displayed by lactoferrin (in milk) and conalbumin (in eggs), and has been called nutritional immunity. The soil bacteria *Ps. fluorescens* and *Ps. putida* produce a siderophore, pseudobactin and other iron chelators which stimulate the growth of plant pathogens such as *Fusarium*, *Gaeumannomyces* and *Erwinia* (Kloepper *et al.*, 1980; Vanderbergh *et al.*, 1983).

Neilands (1980) has surveyed the role of iron transport systems in bacterial pathogens. The derepressible, high-affinity transport systems enhance iron-scavenging capacity *in vivo* and may be regarded as a virulence factor. In *E. coli*, the *colV* plasmid, bearing the aerobactin system, enhances growth in calf serum (Braun, Brazel-Faisst and Schneider, 1984). The host may respond by attempting to diminish the efficacy of iron-scavenging by the bacterium. During infection, the mammalian host reduces the already small amount of iron in the blood plasma and increases its stores of iron in the liver. It appears that the small temperature increase accompanying fever may serve to depress bacterial siderophore synthesis.

The production of siderophore-like substances, such as agrobactin, by phytopathogenic bacteria and fungi is widespread (Leong and Neilands, 1982), but the growth of these pathogens may be limited by bacterial siderophores. The effect has been reproduced using partially purified siderophores and a synthetic iron-chelating agent (Davison, 1988).

The membrane-bound transport pathways for iron in bacteria will translocate certain antibiotics as well as the ferrisiderophores (Hider, 1984). Albomycin and ferrimycin are two iron-containing antibiotics which function this way. This has been successfully exploited in the preparation of semi-synthetic substances such as the sulphonamide derivatives of ferricrocin, a siderophore produced by *Aspergillus* and *Penicillium* spp. In principle, the enzymes involved in siderophore synthesis should represent good targets for future chemotherapy, since there is no evidence for analogous siderophores in higher plants and animals.

Another chemotherapeutic approach is to use as antimetabolites complexes of enterochelin with ions other than Fe^{3+}. Thus, indium(III) or scandium(III) complexes of enterochelin act as antimetabolites to iron(III) enterochelin in *K. pneumoniae*, *E. coli* and *Ps. aeruginosa*, exerting therapeutic effects on infections in mice (Rogers *et al.*, 1980, 1982, 1984).

The synthesis of a number of toxins (e.g. diphtheria, tetanus) occurs only under conditions of limited iron availability. Whether toxin production serves to exploit iron resources such as haemoglobin by cell lysis remains to be established (Braun, 1985).

(g) Ecological factors

The synthesis and export of siderophores and the ensuing effects on the availability of iron in the environment may play important roles in determining the stability and composition of ecosystems. The ability of a species to survive a period of low iron availability may depend both on its ability to take up iron and its capacity to antagonize competing species, for example, by producing antibiotics that are structural homologues of a competitor's siderophores.

Ecological phenomena that may be explained by iron availability factors include the fact that filtered water from blue-green algal blooms inhibits the growth of other algae. This is due to synthesis of hydroxamate chelators by the blue-greens (Murphy, Lean and Nalewajko, 1976). Growth of the pathogenic amoeba *Naegleria fowleri* (which causes primary meningoencephalitis) in freshwater lakes is stimulated by iron and depressed *in vitro* by added desferrioxamine B and rhodotorulic acid. Thus, eutrophication of a lake (e.g. by discharge of a thermal or organic effluent) would increase acidity and hence solubility of Fe^{3+}, which in turn would suppress the synthesis of microbial iron chelators and favour growth of *Naegleria* (Newsome and Wilhelm, 1983).

3.3 POTASSIUM TRANSPORT IN BACTERIA

3.3.1 Introduction

Potassium intake seems to be a characteristic feature of all cells. The high cellular content plays a role in the maintenance of turgor pressure, in enzyme activation and perhaps pH homeostasis. The intracellular K^+ concentration generally corresponds to about 0.25 M, but may be as great as 3 M in *Halobacterium halobium* (Wagner, Hartmann and Oesterhelt, 1978), where it is suggested that the huge pool may act as a long-term store of the light energy required to accumulate K^+. Transmembrane concentration gradients as high as 10^6:1 may be achieved. Such gradients are not obligatory for growth and biosynthetic functions, however, as shown by the ability of *Streptococcus faecalis* to grow in the presence of gramicidin and other ionophores (Section 3.1.3) such that gradients of K^+ and of Na^+ and H^+ are all short-circuited (Harold and Van Brunt, 1977).

Much progress has been made since the review of Harold and Altendorf (1974) and this is addressed in Sections 3.3.2–3.3.4.

3.3.2 *Escherichia coli*

E. coli possesses four (or possibly five) systems for the transport of potassium (Figure 3.3(a)). The high intracellular concentration is maintained by the impermeability of the membrane to K^+; this impermeability is breached in certain K^+-leaky mutants. Further details of the K^+

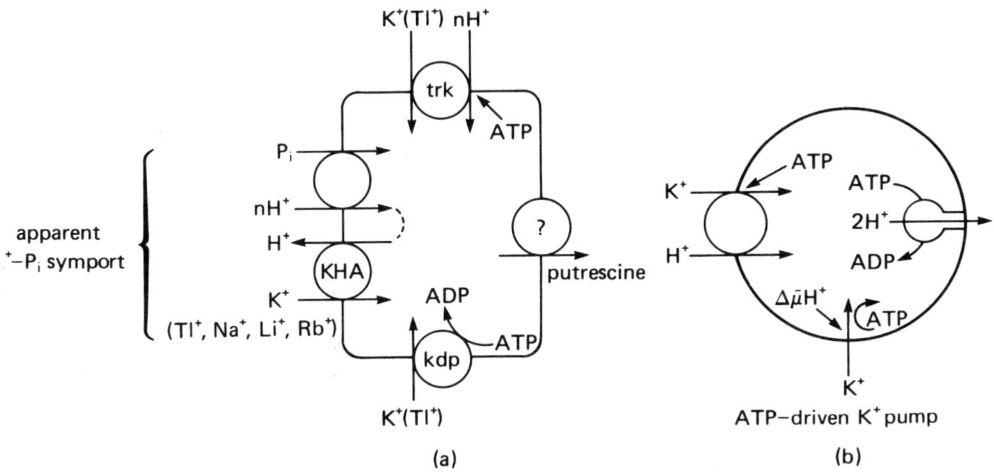

Figure 3.3 Potassium uptake systems and their interactions with other ion movements in *E. coli* (a) and *Streptococcus* (b). For *E. coli*, three of the major systems are shown. *trk*, of which there may be several systems, is a K^+-H^+ symport, *kdp* is a K^+-dependent ATPase and KHA is a K^+-H^+ antiport, loosely associated with a phosphate transport system. Putrescine is lost via an unidentified system during K^+ uptake in some cases. Other substrates for the three uptake systems are shown. In *Streptococcus* (b), two systems are shown. An ATP-driven pump is regulated by $\Delta\tilde{\mu}H^+$. Alternatively, K^+ uptake occurs on a K^+/H^+ symport. The proton-motive force derives from a proton-translocating ATPase. (b) is modified and redrawn from Bakker and Harold (1980).

retention mechanisms have been provided by the finding that the thiol reagent, *N*-ethylmaleimide, also causes K^+ leakage (Meury, Lebail and Kepes, 1980), probably by reaction with a non-protein molecule. Glutathione-deficient mutants of *E. coli* require high K^+ concentrations for growth and for retention of intracellular K^+ (Meury and Kepes, 1982), suggesting that this molecule might be glutathione.

(a) Trk

Under most conditions of growth, including in media of high [K^+], a constitutive system, encoded by *trk*$^+$ genes (for *tr*ansport of *K*$^+$), regulates and maintains a constant intracellular K^+ concentration. This high K_m system (about 1 mM K^+) is only moderately specific and also accepts Rb^+ and Tl^+ in addition to K^+. The *trk* system may function as a K^+/H^+ symport (Epstein and Laimins, 1980). However, a direct role for ATP in energy coupling is also indicated (Rhoads and Epstein, 1977). Several unlinked *trk* genes have been identified (Epstein and Laimins, 1980; Hanann, Bassemeyer and Bakker, 1987) and cloned. The loci seem to encode interacting components that contribute to Trk activity. The product of *trkA* is a 53 k-Da inner membrane protein, perhaps directly binding K^+ (Walderhaug *et al.*, 1987).

(b) Kdp

Under conditions of osmotic stress or low extracellular [K^+], a high-affinity (K_m about 2 μM K^+) K^+-dependent ATPase is induced, the *kdp* system (for *K*$^+$-*dep*endent growth) (Epstein, Whitelaw and Hesse, 1978; Wieczorek and Altendorf, 1979; Laimins, Rhoads and Epstein, 1981). The ATPase activity is quite distinct from the (Ca^{2+}, Mg^{2+})-stimulated ATPase of *E. coli* in being DCCD-insensitive, but is functionally similar to the ATP-driven transport processes of animal cells. Unlike these, however, the system does not require Na^+ and is insensitive to ouabain and *N*-ethylmaleimide. It is unique in having a narrow cation specificity that does not include Rb^+ or Cs^+. A divalent cation (Mg^{2+}, Mn^{2+} or Co^{2+}) is required for activity.

Although other ATP-driven transport systems in gram-negative bacteria have been shown to contain a periplasmic 'binding' protein component, no soluble component of the *kdp* system has been firmly identified. The molecular weights of the three inner membrane proteins *kdpA*, *kdpB* and *kdpC* are 47 000, 90 000 and 22 000, respectively. Since a mutation in the *kdpA* gene causes a change in the K_m value for K^+ binding it has been suggested (Wieczorek and Altendorf, 1979) that the *kdpA* protein is responsible for K^+ binding and possibly translocation, with KdpB functioning in energy transduction. Turgor pressure control requires expression of *kdpD* and *kdpE*, located adjacent to the *kdpABC* operon. Both genes encode positive regulatory proteins. Models for turgor control of operon expression are discussed by Walderhaug *et al.* (1987).

(c) Efflux pathways

Multiple pathways for K^+ efflux are indicated. K^+ efflux can be elicited by a variety of cell treatments such as alkalinization of the cytoplasm, increased turgor pressure or *N*-ethylmaleimide (NEM). This compound may titrate the glutathione pool. Two genes encode the NEM-stimulated efflux, *trkB* and *trkC*. However, since these genes are no longer thought to affect K^+ uptake *per se* they have been renamed *kefB* and *kefC* (for *K^+efflux*) (Bakker *et al.*, 1987). The presence of a third system is deduced from studies of mutants lacking all the other systems; this may be a K^+/H^+ antiporter (KHA).

(d) Electroneutral K^+/H^+ antiporter

This has also been demonstrated in everted membrane vesicles (Brey, Rosen and Sorensen, 1980; Plack and Rosen, 1980; Slonczewski *et al.*, 1981). In EDTA-treated cells (in which the outer membrane is disrupted), *N*-ethylmaleimide induces a similar antiporter (Bakker and Mangerich, 1982). This antiporter is thought to function in maintenance of intracellular pH during growth at alkaline pH or during a sudden increase in medium pH (Yamasaki *et al.*, 1980), net K^+ uptake leading to an increase in internal pH. In support of this, the antiport has an alkaline pH optimum and loss of the antiport leads to inability to grow at alkaline pH.

The transport of K^+ may be linked to that of inorganic phosphate (P_i); thus, P_i entry is absolutely dependent on the presence of K^+ and vice versa but there is no obvious preference by any of the P_i-transport systems (Pit or Pst) for a particular K^+ transport system (Russell and Rosenberg, 1979), indicating that P_i and K^+ do not use the same porter. In fact, the use of a series of K^+ transport mutants carrying mutations in either of the two phosphate transport systems (Russell and Rosenberg, 1980) shows that there is no stringent association of the K^+ and P_i-uptake systems and that the K^+/H^+ antiport and proton symport with phosphate or other anions are responsible for the apparent K^+ symport with these anions.

A sudden elevation in osmotic pressure of the medium causes uptake of K^+ and loss of the polyamine putrescine (1,4-diaminobutane). A degree of coupling between these fluxes is demonstrated by the diminished rate of putrescene excretion in various mutants defective in K^+ transport (Munro and Sauerbier, 1973). Linkage to one specific K^+ transport system has not been firmly established.

(e) Transport of other cations by potassium transport systems

Rubidium substitutes only partially for K^+. There appears to be a stable K^+ pool of about 25 mM (in a total intracellular K^+ concentration of about 240 mM) that is not replaceable by rubidium (Nakajima, Yamato and Anraku, 1979). Rubidium appears to be transported by the K^+/H^+ antiporter (Section 3.3.2(d)) which also transports Na^+, Li^+ and Tl^+. The high affinity *kdp* transport system is insensitive to Rb^+ (Rhoads, Woo and Epstein, 1977; Nakajima, Yamato and Anraku, 1979) and Rb^+ does not stimulate the associated ATPase activity. The isotope $^{86}Rb^+$ is not a satisfactory tracer for net K^+ influx by any of the K^+ transport systems in *E. coli*.

Bakker (1978) proposed that thallium(I), even though having similar crystal and hydrated radii to K^+, was passively distributed across the cytoplasmic membrane in response to the membrane potential ($\Delta\psi$). It was further suggested that $^{204}Tl^+$ could be used to quantify $\Delta\psi$ in untreated intact cells with the attendant advantages of avoiding potentially perturbing, permeabilizing treatments. Earlier work, however, had suggested that Tl^+ is accumulated by K^+ transport systems (Norris *et al.*, 1976) and it has now been confirmed that Tl^+ is a substrate for one or more K^+ pathways in *E. coli* (Damper *et al.*, 1979). The participation of the various pathways is complex. Damper *et al.* (1979) reported accumulation by Kdp and TrkA, whereas Brey, Rosen and Sorensen (1980) have shown that Tl^+ is also a substrate for the K^+/H^+ antiporter which could serve to excrete Tl^+ in intact cells or be responsible for uptake in everted membrane vesicles (Figure 3.3(a)).

3.3.3 *Streptococcus*

In this fermentative, gram-positive bacterium, which is conveniently sensitive to ionophores and antibiotics, the primary energy-linked process is the electrogenic extrusion of protons by the membrane-bound ATPase. Potassium is thought to accumulate in response to the electrical potential, interior-negative, but the concentration gradient is too large to be explicable solely by such a mechanism and, as in the TrkA system of *E. coli*, an additional ATP requirement has been identified (Bakker and Harold, 1980). In one proposal (Figure 3.3(b)), ATP serves as energy donor for a primary electrogenic, inwardly directed K^+ pump, which is regulated by the pmf. A second transport system for K^+ has been proposed that does not require intervention of a pmf, and in this respect resembles the Kdp system of *E. coli*. The system is derepressed by lowering the intracellular concentration of K^+ with monactin (Kobayashi, 1982). One or both of these pathways is also capable of transporting Tl^+, invalidating for this organism, as well as for *E. coli*, the use of Tl^+ as a probe of $\Delta\psi$ (Kashket, 1979).

3.3.4 Other bacteria

Table 3.3 lists the properties of some K^+ transport systems in several bacteria. Accumulation of intracellular K^+ to levels that cannot be explained solely by invoking uptake in response to a membrane potential is frequently observed. A general picture is emerging in which at least two distinct systems can be identified, the appearance and properties of each being influenced by the prevailing extracellular K^+. Some pathways require energy input in the form of ATP or a high energy compound derived therefrom. Further studies are required on the ions involved in maintaining charge balance and on the uptake by these systems of Rb^+ which, despite (i) a paucity of detailed study and (ii) the remarkable discrimination against Rb^+ by *E. coli*, continues to be used as a probe of K^+ uptake.

3.4 MAGNESIUM TRANSPORT IN BACTERIA

3.4.1 General features

Magnesium is the most abundant intracellular divalent cation. The requirements for magnesium are met by selective, active transport processes under genetic and physiological control. Magnesium transport in bacteria was intensively studied between 1976 and 1978 by Silver and his collaborators. Early work in this area was reviewed by Jasper and Silver (1977) and Silver (1978) and a recent survey of the limited, subsequent studies is that of Silver and Lusk (1987).

The magnesium transport systems described to date are energy-linked, being strongly temperature-dependent and inhibited by uncouplers, respiratory chain inhibitors and sulphydryl compounds. Many are non-specific in being competitively inhibited by cobalt, manganese, nickel or other divalent metal ions. Although such metals can enter via the Mg^{2+}

Table 3.3 Examples of potassium transport systems in various prokaryotes

Organism	Transport pathways	Properties	Reference
Alteromonas haloplanktis	$2H^+/K^+$ antiport	Na^+-dependent, ATP-driven	Sedgwick and Macleod (1980)
Anabaena variabilis	Two uptake mechanisms	(a) $K_m = 0.04$ mM (b) $K_m = 4.5$ mM Varied discrimination against Rb^+	Reed, Rowell and Stewart (1981)
Paracoccus denitrificans	Two systems	(a) $K_m = 1–2$ mM (b) $K_m = 0.1$ μM (a) transports Rb	Erecinska, Deutsch and Davis (1981)
Bacillus subtilis	Single system	K_m 50 μM transports Rb^+	Kay and Ghei (1981)
Halobacterium halobium	Charge-balanced by Na^+	Light-driven; ATP-independent. Similar uniport for Rb^+	Wagner, Hartmann and Oesterhelt (1978) Garty and Caplan (1977)
Mycosplasma mycoides	(i) ATP-consuming Na^+/K^+ exchange (ii) K^+ uniport ($\Delta\psi$-driven)	Sensitive to DCCD and NBD Insensitive to vanadate, ouabain	Benyoucef, Rigand and Leblanc (1982)
Rhodobacter capsulata		$K_m = 0.2$ mM	Jasper (1978)
Methanospirillum hungatei	NH_3/K^+ exchange (at alkaline pH, inward diffusion of NH_3 plus H^+, and operation of K^+/H^+ antiporter).		Sprott, Shaw and Jarrell (1984)

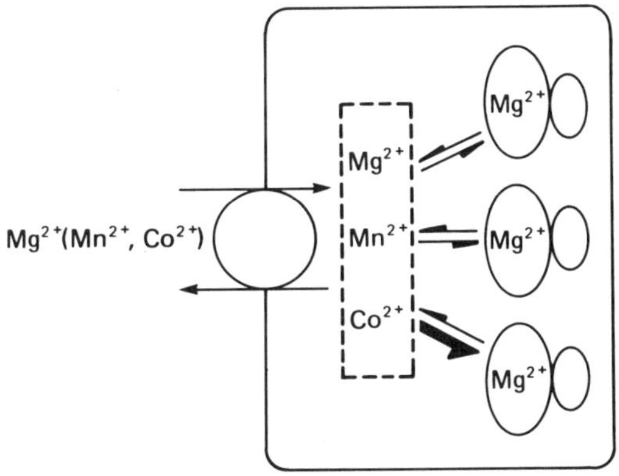

Figure 3.4 Magnesium transport in *E. coli*. The carrier also transports Mn^{2+} and Co^{2+} which, from an intracellular pool (dashed), can displace Mg^{2+} bound to ribosomes (oval structures).

transport pathway(s), there are separate systems for some. Cations that affect Mg^{2+} influx also affect the efflux, indicating that both entry and exit of Mg^{2+} are carrier-mediated.

Accumulated Mg^{2+} (or Mn^{2+} or Co^{2+}) constitutes a pool in equilibrium with ribosome-bound metal ions (Fig. 3.4). Thus, ribosome-associated Mg^{2+} (90% of the total cell magnesium) exchanges with non-bound intracellular Mg^{2+}, with Mn^{2+} (which also binds reversibly to ribosomes) or with Co^{2+} (which binds irreversibly). Exchanges with the last two metals result in an accelerated loss of Mg^{2+} through its transport system and explain in part the growth-inhibiting effect of these metals. However, some of the displaced Mg^{2+} may be recaptured in the periplasmic space by the high-affinity transport system of the cytoplasmic membrane. The use of the radioisotope ^{28}Mg and the availability of mutants lacking either or both of the two transport systems, have allowed a partial description of these systems in *E. coli* and other bacteria (Table 3.4).

3.4.2 Magnesium transport in *Escherichia coli* and *Salmonella typhimurium*

E. coli possesses two transport systems for Mg^{2+}. System I is less specific and constitutive; the other, System II, is inducible, repressible by growth on high Mg^{2+} and discriminates better against other metal ions (Table 3.4).

Cobalt is toxic to *E. coli*, probably due to the irreversible binding of the

Table 3.4 Properties of magnesium transport systems in bacteria

Organism	K_m (μM)	V_{max} (nmol/min/mg)	Alternative substrates					
			Co^{2+}	Mn^{2+}	Fe^{2+}	Ca^{2+}	Ni^{2+}	Zn^{2+}
Intact cells								
E. coli System I (constitutive)	30*	11	+	+			+	+(?)
E. coli System II (repressible)	30*	8	−	(±)				
Rhodobacter capsulata	52–55	1.8	+	+	+	+†		
B. subtilis (log phase)	250	4.4	+	+	(±)			
B. subtilis (sporulating phase)	<10	−						
B. subtilis (citrate system)‡	450	123		+				
S. aureus	70	9.6	?	+				+
Membrane vesicles								
E. coli	13	0.6	+	+			+	

* Values dependent on medium used in measurement (Park, Wong and Lusk, 1976).
† Nature of inhibition unknown (Jasper and Silver, 1978).
‡ Citrate and Mg^{2+} accumulated simultaneously.
Data from Jasper and Silver (1977) and Silver and Lusk (1987).

metal to ribosomes or intracellular polyanions in place of magnesium. On the assumption that cobalt resistance may be due to altered magnesium transport, Nelson and Kennedy (1972) isolated mutants that could survive exposure to cobalt (Cor^-), as well as a mutant (Mgt^-) able to grow with high (0.2M) magnesium concentrations. Jasper and Silver (1977) survey this and other early work on mutants with altered magnesium transport. Two types of System I Cor^- mutants were isolated, *corA* and *corB*. Both lack Co^{2+} transport when grown in low Mg^{2+} medium and rely on System II for Mg^{2+} transport under these conditions. The *corB* mutation seems to affect the inducibility of System I. Growth of *corA* strains is inhibited by Co^{2+} but the rate of Mg^{2+} transport is not. The *corA*$^+$ gene product has been identified as an M_r 3700 polypeptide (Silver and Lusk, 1987).

Salmonella typhimurium closely resembles *E. coli* in having two Mg^{2+} transport systems. One is repressible and transports only Mg^{2+}; the other is constitutive and transports Co^{2+} or Mg^{2+} (Hmiel *et al.*, 1986). Furthermore, *corA* and *corB* mutants have been isolated. A plasmid complementing the *corA* mutation conferred enhanced sensitivity to Co^{2+} and increased V_{max} values for both Mg^{2+} and Co^{2+}; this plasmid may bear a structural gene for magnesium transport.

3.5 MANGANESE TRANSPORT IN BACTERIA

Highly specific manganese transport systems seem to be a universal feature of cells and enable them to concentrate Mn^{2+} even in the presence of high levels of calcium and magnesium.

3.5.1 *Escherichia coli*

In this organism (whose growth cannot be demonstrated to be dependent on Mn^{2+}), the metal ion is transported by low-affinity magnesium transport systems (Section 3.4), the study of which led to the discovery of a specific manganese transport system (Silver and Jasper, 1977).

This specific transport system for Mn^{2+} exhibits a high affinity but low rate of accumulation (Table 3.5), consistent with the presence of adequate supplies of manganese in most microbial environments. Silver *et al.* (1972) have calculated that below 40 μM Mn^{2+}, the specific, high-affinity system will predominate, whereas above this level the Mg^{2+} system (with a K_m for Mn^{2+} of 2 mM) would be the principal source of cellular Mn^{2+}.

Table 3.5 Properties of manganese transport systems in bacteria

Organism	K_m (μM)	V_{max} (nmol/min/g dry wt)	Specificity*	Alternative substrates	Other properties
E. coli	0.2	4–16	10^5	Fe^{2+}, Co^{2+}	Temperature-dependent; sensitive to inhibitors of energy metabolism; only partially cyanide-sensitive
Bacillus subtilis	1–2.5	100–5000	10^5		Temperature-dependent; cyanide-and uncoupler-sensitive
Rhodobacter capsulata	0.5	20	10^5		Uncoupler-sensitive. Sensitive (aerobic cells) or insensitive (anaerobic cells) to cyanide
Lactobacillus plantarum	0.2	24 000†		Cd^{2+}	Proton gradient-driven
Brevibacterium ammoniagenes	0.65	120‡	2×10^4	Cd^{2+}, Zn^{2+}	Active zinc uptake not observed. Mn^{2+} uptake is CCCP-sensitive

* Molar excess of Mg^{2+} and Ca^{2+} not affecting system.
† Rate expressed per g protein.
‡ At pH 7.2 (optimum is 9.5).
Data from Silver and Jasper (1977), Archibald and Duong (1984), Silver and Lusk (1987), and Schmid and Auling (1987).

Cadmium is not accumulated by the Mn^{2+} transport system of *E. coli* (Silver and Perry, 1983), accounting for the relative resistance to Cd^{2+} of this organism.

3.5.2 *Bacillus subtilis*

Bacillus subtilis has a well-established, absolute requirement for Mn^{2+} in sporulation. The rate of manganese accumulation by the *Bacillus* transport system is considerably greater than that in *E. coli* although the K_m is higher too (Table 3.5). The presence of a high-affinity, specific system and a second system of lower affinity and specificity has been shown by Kay and Ghei (1981). During sporulation, intracellular and extracellular Mn^{2+} is quantitatively sequestered in a form not released by treatment with toluene or lysozyme, in synchrony with the packaging of Ca^{2+} into the developing spore. Energy-dependent Mn^{2+} uptake has also been demonstrated in membrane vesicles, but the observed K_m was 13-fold higher than that measured in intact cells.

Manganese uptake is derepressed during Mn^{2+}-starvation and reduced (by inactivation of existing transport function) upon addition of high (1 μM or more) Mn^{2+} to derepressed cells. Derepression requires *de novo* protein synthesis. Dramatic changes in V_{max} occur with no alteration in the K_m for uptake and may, therefore, be explained by an increase in the number of 'gates' per cell.

A low M_r stimulator of Mn^{2+} transport has been identified which is excreted by derepressed cells (i.e. during Mn^{2+} starvation) and which facilitates uptake by such cells. It is assayed *in vitro* by its ability to carry ^{54}Mn through a gel filtration column. The molecule is highly specific for Mn^{2+} and has a M_r of about 1500 but its chemical nature is unknown; it is not identical, however, to bacitracin, the polypeptide produced by *B. licheniformis*.

3.5.3 Other bacteria

Manganese transport in *Lactobacillus plantarum* is of special interest (Table 3.5) because of the ability of this organism to accumulate over 30 mM Mn(II) intracellularly (Archibald and Duong, 1984), where it acts as a scavenger of toxic oxygen-derived species (Chapter 5). The Mn^{2+} uptake mechanism in *L. plantarum* resembles similar pathways in other bacteria in having high affinity for the metal ion, but the rates of uptake are greater than those of most bacteria studied. Interestingly, the rate of uptake is increased at pH 5.5 by citrate and other tricarboxylic acids, which could serve a functional role in metal transport from the fermenting plant material from which the bacterium is commonly isolated.

In *Rhodobacter*, the kinetic parameters are very similar to those for *E. coli* (Table 3.5). *S. aureus* also has a highly specific, energy-dependent transport system for Mn^{2+}.

3.6 ZINC TRANSPORT IN BACTERIA

Failla (1977), in a comprehensive review, suggests several reasons for our current state of ignorance regarding the mechanisms and regulation of zinc transport, although the comments are of general relevance to studies of trace metal transport: (i) most micro-organisms achieve optimal growth yields in media containing only 0.5–1.0 μM zinc; (ii) extraction of zinc from media using solvents or alumina has generally been unsuccessful; (iii) binding to anionic sites on the microbial surface is not readily distinguishable from transport (Section 3.12.6; Chapter 7).

The speciation of zinc in natural environments is variable and complex. For example, Zn^{2+}, $ZnCl^{+}$, $ZnCO_3$, $Zn(OH)_2$ and zinc complexes with organic ligands have all been proposed as the major species in seawater. Zinc availability is decreased at higher pH where the metal tends to be present in insoluble precipitates or adsorbed to colloids. As yet, there is no direct evidence for the secretion by microbes of zinc-chelating organic ligands.

Zinc transport has been demonstrated in a few bacteria (*Serratia marcescens*, *B. megaterium*, *E. coli* and *Salmonella*), generally radioisotopically. Bucheder and Broda (1974) demonstrated an energy-dependent zinc transport system in *E. coli*, which exhibited saturation kinetics (K_m 20 μM; V_{max} 2.67 nmol/min/mg dry wt.) and was competitively inhibited by Cd^{2+}. In selection-synchronized cultures, uptake of zinc (but not of K^{+}, Co^{2+} or Mg^{2+}) showed a stepwise increase early in the cell cycle (Kung, Raymond and Glaser, 1976). A sudden influx of zinc could conceivably be related to periodicities in the activities of zinc metalloenzymes (Chapter 5).

3.7 CALCIUM TRANSPORT IN BACTERIA

3.7.1 Transport by intact cells

Despite the relative abundance of calcium in the Earth's crust, natural microbial habitats and laboratory growth media, a low intracellular concentration of calcium is maintained in microbial and other cells. In unstarved *E. coli* cells, and several eukaryotic cells, the intracellular concentration of free Ca^{2+} is as low as 90 nM (Gangola and Rosen, 1987). The mechanisms by which microbial cells actively secrete calcium by highly specific active transport processes have been reviewed by Silver (1977, 1978), Devés and Brodie (1981) and Rosen (1987). There are relatively few studies on calcium transport by intact cells, due in part to the complications arising from the binding of calcium by cell walls. Nevertheless, the first evidence that bacteria possessed active extrusion mechanisms for calcium was provided by the demonstration that extremes of temperature (<15°C, >50°C), inclusion of metabolic inhibitors, or killing of *E. coli* and *B. subtilis* cells resulted in increased levels of intracellular calcium. Active efflux of Ca^{2+} has been demonstrated more recently in intact cells of *B. megaterium, Rhodobacter capsulata*, and *E. coli* (Tsujibo and Rosen, 1983).

In *S. faecalis*, Ca^{2+} is thought to enter electrophoretically in response to the membrane potential generated by the proton pump and to be extruded by an ATP-requiring transport system (Kobayashi, Van Brunt and Harold, 1978). There is no evidence that this Ca^{2+} circulation serves any physiological function; instead, Ca^{2+} entry is unavoidable and detrimental and the function of the ATP-driven pump is to prevent accumulation of Ca^{2+} in the system.

3.7.2 Membrane vesicles from *E. coli*

Studies with isolated cytoplasmic vesicles have the advantage of eliminating cell wall binding of calcium and, by the ability to prepare at will either 'inside-out' or 'right-side-out' vesicles, allow experimental determination of the polarity of cation movements. An elegant exploitation of this approach is the work of Rosen and others on *E. coli* (Rosen and Brey, 1979). Membranes prepared by osmotic lysis of lysozyme-treated cells retain the right-side-out orientation of intact cells and have been used extensively to study uptake of amino acids, sugars and other nutrients, but do not accumulate calcium. Disruption of *E. coli* in a French pressure cell, however, yields everted vesicles which are able to accumulate calcium, an event that is technically easier to measure than is efflux. Uptake by such everted vesicles can be driven by provision of (i) oxidizable substrates (e.g. NADH, succinate, D-lactate) when uptake is cyanide-sensitive, or (ii) by ATP when uptake is sensitive to DCCD or antiserum raised to purified ATPase. These findings, and the uncoupler-sensitivity of transport, suggest that calcium transport is

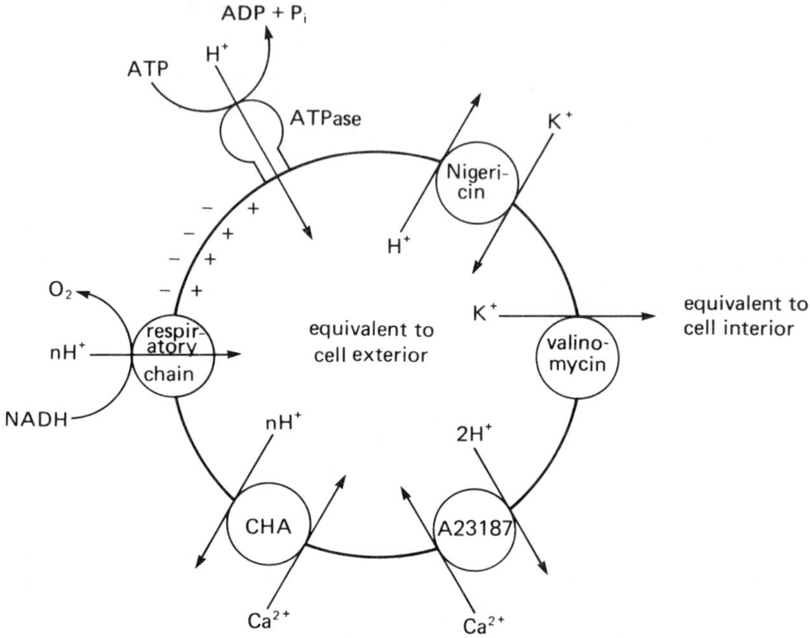

Figure 3.5 Model of the Ca^{2+}/H^+ antiport and the effects of ionophores in everted membrane vesicles of *E. coli*. The orientation of the proton-motive force is reversed from that *in vivo*. Primary proton pumps (ATPase or respiratory chain) form a proton-motive force (Δp), positive and acid interior. The antiport (CHA) couples to the ΔpH portion of Δp resulting in Ca^{2+} uptake. The Ca^{2+} : H^+ stoichiometry is not known. The ionophore A23187 mimics the operation of the natural antiporter. Nigericin ($+K^+$) and valinomycin ($+K^+$) are shown; the former, which dissipates ΔpH, inhibits Ca^{2+} accumulation whereas the latter is without effect. (Adapted from Rosen and Brey, 1979, with permission).

a secondary porter system driven by the proton motive force (Δp). Neither La^{3+} nor the dye Ruthenium Red inhibit ^{45}Ca accumulation by *E. coli* membranes (Silver, 1977; Gibson *et al.*, 1982), although these agents specifically and competitively inhibit calcium accumulation by mitochondria. A dimeric, mixed-valence compound of ruthenium is, however, a potent inhibitor of respiration-driven calcium transport by *E. coli* vesicles (Gibson *et al.*, 1982); the compound is not a respiration inhibitor *per se* and its mode of action remains to be established.

Experiments with other ionophores have suggested that the uptake of Ca^{2+} involves exchange with H^+. Thus, treatment with nigericin plus K^+ (which dissipates the ΔpH portion of Δp) inhibits calcium uptake, whereas valinomycin plus K^+ (which dissipates $\Delta\psi$ but not ΔpH) does not inhibit. The ionophore A23187, which catalyses the exchange of Ca^{2+} and H^+, is an effective stimulator of calcium uptake, suggesting also that the calcium transport system is coupled to exchange with H^+. The antiport (CHA) and these ion movements are shown diagrammatically in Figure 3.5. The stoichiometry of H^+ and Ca^{2+} movements is uncertain. An electrogenic $1H^+/1Ca^{2+}$ antiport is favoured by Tsuchiya and Takeda (1979a), but the data of Brey and Rosen (1979) suggest a $H^+:Ca^{2+}$ stoichiometry of >2. Contradictory results have been obtained when cation transport has been followed using radioisotopes or by following the effect of divalent cations on quenching of aminoacridine fluorescence (an indirect measure of ΔpH). Thus, the stoichiometry problem has still to be resolved.

The specificity of the antiporter is, in order of affinity, $Ca^{2+} \approx Mn^{2+} > Sr^{2+} > Ba^{2+}$ (Brey and Rosen, 1979).

A second system for calcium extrusion, namely a calcium phosphate symport/proton antiport has also been identified (Ambudkar, Zlotnick and Rosen, 1984).

A genetic study of the number and function(s) of calcium transport systems has been initiated by isolating mutants sensitive to calcium; wild-type strains do not require calcium for growth, and growth is unaffected at 50 mM $CaCl_2$. Membrane vesicles from a mutant sensitive to 25 mM $CaCl_2$ accumulate more Ca^{2+} than vesicles from the wild-type. Two mutations have been described: one is in the *corA* locus which is involved in Mg^{2+} transport (Section 3.4.2). It is thought that $CaCl_2$ may suppress the remaining Mg^{2+} transport system, resulting in Mg^{2+} starvation. The other mutation, *calA*, is probably responsible for the elevated Ca^{2+} levels in vesicles.

3.7.3 Membranes from other bacteria

Because most bacteria excrete Ca^{2+}, everted membranes (e.g. from *Bacillus megaterium*) prepared in a French pressure cell take up calcium but do not take up molecules (e.g. glutamate) normally transported inwards by whole cells. The surprising observation that right-side-out membrane vesicles prepared by the method of Kaback from this organism accumulate Ca^{2+} in an energy-dependent fashion is attributable to the presence of significant and variable populations of everted vesicles (Coakley, Bater and Lloyd, 1977).

Azotobacter vinelandii appears unique in requiring calcium for vegetative growth; the membrane contains, in addition to an electroneutral $Ca^{2+}/2H^+$ antiporter, an electrophoretic Ca^{2+} uniporter (Zimniak and

Barnes, 1980) responsible for Ca^{2+} uptake. The two carriers constitute a Ca^{2+} circulation similar to that found in mitochondria and it is interesting that Ruthenium Red and lanthanides exhibit a similar pattern of inhibition in the two systems.

In *Mycobacterium phlei*, also, the mechanism for calcium efflux probably involves a Ca^{2+}/H^+ antiport, though the system (studied in vesicles prepared by sonication) differs in detail from that in *E. coli* (Devés and Brodie, 1981). Calcium transport by everted membrane vesicles is sensitive to vanadate when driven by the ATPase (Yoshimura and Brodie, 1981), and to trivalent lanthanide cations and uranyl ions (UO_2^{2+}) (Agarwal and Kalra, 1983) which are substitutes for Ca^{2+} in other biological processes.

In *Halobacterium halobium*, the movement of Ca^{2+} appears to be directly coupled to Na^+ circulation rather than H^+ circulation (Belliveau and Lanyi, 1978), similar to the situation in certain eukaryotic cells. Although the primary gradient generated is one of H^+, an H^+/Na^+ antiporter converts this to a gradient of Na^+, which is then used to extrude Ca^{2+}. Transport is not inhibited by proton-conducting uncouplers, but is sensitive to monensin, an ionophore that catalyses H^+/Na^+ exchange.

In *Streptococcus faecalis* vesicles, prepared using a French press, Ca^{2+} accumulation is not inhibited by protein conductors or by DCCD and is perhaps energized by ATP directly rather than by proton circulation (Kobayashi, Van Brunt and Harold, 1978).

A further example of outwardly directed calcium transport in bacteria is provided by studies on chromatophore membranes (Silver, 1977). These intracellular membranes of photosynthetic bacteria such as *Rhodobacter capsulata* and of nitrogen-fixing bacteria such as *Az. vinelandii* are formed by invagination of the cytoplasmic membrane. Such structures are believed to be everted with respect to their orientation *in vivo*, and accumulate protons and, significantly, ^{45}Ca in an energy-dependent fashion.

A summary of the three major mechanisms for Ca^{2+} exclusion by bacterial cells, as deduced from studies with subcellular membranes, is presented in Figure 3.6. A useful compendium of relevant data is given in Devés and Brodie (1981).

3.7.4 Calcium accumulation during sporulation

During sporulation, *Bacillus* species accumulate large amounts of calcium (2–4% on a dry weight basis) and other divalent cations. Current evidence favours a two-step incorporation of calcium into the spore,

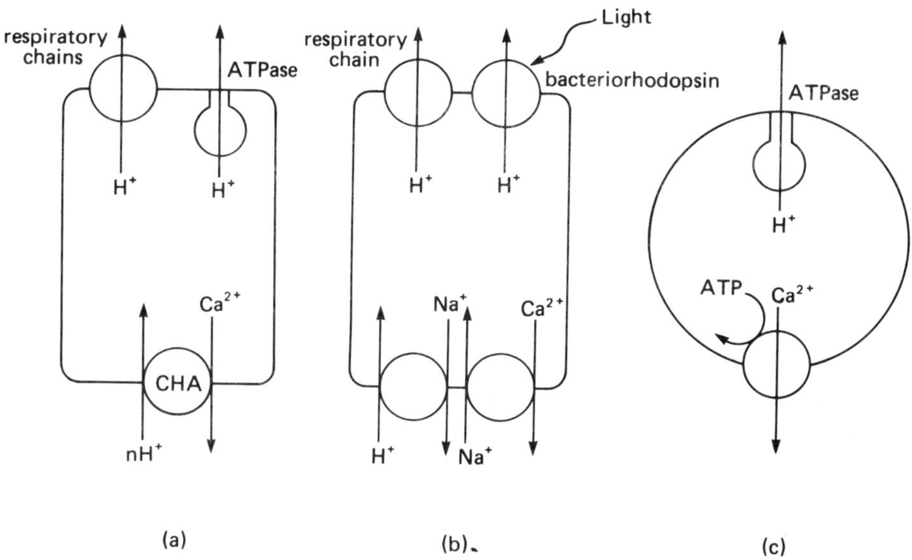

Figure 3.6 A summary of three different mechanisms for Ca^{2+} extrusion by bacterial cells, as inferred from studies of everted membrane vesicles. (a) shows the active efflux of Ca^{2+} according to the Ca^{2+}/H^+ antiport mechanism proposed for *E. coli*, *Az. vinelandii* and *M. phlei*. The $\Delta\bar{\mu}_{H^+}$ can be established by substrate oxidation by the respiratory chains or by ATP hydrolysis. Note that the directions of ion movements are the reverse of those shown in Figure 3.5, which illustrates the situation extant in everted vesicles. (b) shows the Ca^{2+}/H^+ antiport described in *H. halobium*. The outwardly directed Na^+ gradient that energizes Ca^{2+} transport is generated at the expense of the $\Delta\bar{\mu}_{H^+}$ which results from respiratory electron transport or illumination of bacteriorhodopsin. (c) shows the direct energization of Ca^{2+} transport by ATP, despite the generation of $\Delta\bar{\mu}_{H^+}$ by ATPase activity. Such a scheme has been proposed for *S. faecalis*. Redrawn and modified from Devés and Brodie (1981).

namely transport of calcium across the mother cell membrane followed by uptake into the forespore by another accumulation system. The first step may be a calcium-proton antiport, like that of *E. coli*, but operating in the reverse direction and driven by $\Delta\psi$ (Seto-Young and Ellar, 1981). It is not known whether this system is separate from that which normally expels Ca^{2+}. The second step involves calcium translocation across the outer and inner forespore membranes. The electroneutrality of this transport may be assured by co-transport of newly-synthesized dipicolinic acid, with which Ca^{2+} is associated in an approximately 1:1 complex in the mature spore. The problems of membrane sidedness and directionality of Ca^{2+} movement are particularly interesting in this system. The inner forespore membrane has a right-side out orientation since it originates from the cytoplasmic membrane of the small protospore cell which is enveloped by the mother cell as spore development proceeds. The enveloping membrane, however, has an inside-out orientation, being derived by invagination of the mother cell membrane. Nevertheless, calcium translocation is inward across both membranes. The mother cell plasma membrane thus differs markedly from its progenitor vegetative cell membrane which effects Ca^{2+} efflux; the development of the Ca^{2+} uptake system is not understood.

Calcium uptake by *B. megaterium* protoplasts is distinct from that in isolated forespores (Hogarth and Ellar, 1978, 1979). The protoplast transport mechanism exhibits a much lower K_m and operates in an energy-dependent fashion against a concentration gradient. In contrast, Ca^{2+} uptake by isolated forespores is probably by facilitated diffusion, in which both the rate and extent of uptake are influenced by external $[Ca^{2+}]$. Calcium uptake into the forespore is insensitive to uncouplers and respiratory inhibitors. It is proposed that Ca^{2+} is actively transported into the mother cell cytoplasm until a concentration of 3–9 mM is attained, sufficient to allow facilitated diffusion into the forespore if a low concentration of free Ca^{2+} were maintained by complexation with dipicolinic acid.

3.7.5 Properties of isolated calcium carriers

A carrier protein has been solubilized from *M. phlei* membranes and reconstituted both into artificial liposomes and Triton-treated membrane vesicles from which the carrier had been removed. Uncoupler-sensitive Ca^{2+}-transport has been demonstrated in each case (Lee, Kalra and Brodie, 1979; Devés and Brodie, 1981).

3.8 SODIUM TRANSPORT IN BACTERIA

Sodium, like calcium, is actively expelled by bacteria to attain $[Na^+]_{in}/[Na^+]_{out}$ gradients of between 1:3 to 1:50 (Silver, 1978; Lanyi, 1979; Skulachev, 1984). Instead of, or in addition to the Na^+, K^+ ATPase utilized in higher animal cells, bacteria exhibit several mechanisms for Na^+ extrusion, suggesting the considerable importance, still not entirely understood, of this process in bacterial metabolism.

At least four different functions of Na^+ in bacteria are known, three of which directly relate to membrane transport phenomena. These are (i)

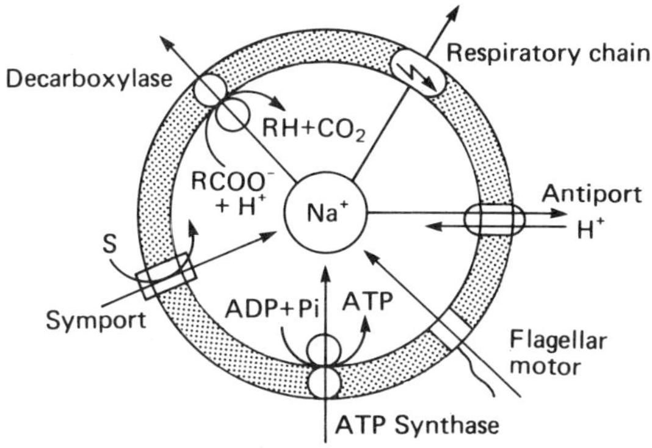

Figure 3.7 Systems performing energy coupling by Na^+ circulation in bacteria. (Reproduced from Dimroth, 1987 with permission of the American Society for Microbiology and the author).

Na^+-solute transport systems, (ii) pH homeostasis, and (iii) Na^+-coupled energy conservation and transmission mechanisms. Indeed, Na^+ is second only to the proton as a major coupling ion (Figure 3.7) and is especially important in marine, methanogenic, halophilic and alkalophilic organisms and in rumen bacteria that live in Na^+-rich environments (Krulwich, 1983; Skulachev, 1984, 1987; Dimroth, 1987).

3.8.1 Formation of an Na^+ gradient by primary sodium pumps

(a) Sodium transport decarboxylases

Several examples of decarboxylations that drive active Na^+-transport are now recognized. The facultative anaerobes *Klebsiella pneumoniae* and *Salmonella typhimurium* utilize oxaloacetate (generated by the action of citrate lyase) by decarboxylation to pyruvate. Subsequently pyruvate is split to give acetyl-CoA and formate, the former yielding ATP by substrate level phosphorylation. Only one mole of ATP is synthesized per mole of citrate but concomitantly a transmembrane Na^+-gradient is generated by oxaloacetate decarboxylation (Dimroth, 1980, 1987). Sodium pumping is primary and not mediated by a Na^+/H^+ antiporter. Other Na^+-transport decarboxylases are now recognized in strict anaerobes, namely methylmalonyl-CoA decarboxylase leading to propionate formation (e.g. in *Veillonella* and *Propigenium* spp.) and glutaconyl-CoA decarboxylase leading to butyrate formation in *Peptostreptococcus* spp.

(Dimroth, 1987). These decarboxylases have been purified and shown to consist of three or four different subunits. All contain the prosthetic group, biotin. The *K. pneumoniae* gene encoding oxaloacetate decarboxylase has been cloned in *E. coli* (Schwarz and Oesterhelt, 1985).

(b) Respiratory Na^+ pumps

The respiratory Na^+ pump of *Vibrio alginolyticus* (Tokuda and Unemoto, 1982, 1984) is another example of a primary Na^+ pump. At pH 8.5, the bacteria generate CCCP-insensitive $\Delta\psi$ and ΔpH. CCCP-sensitive mutants lacking the pump have shown that the NADH:quinone oxidoreductase pumps Na^+. Similar Na^+-motive NADH oxidases have also been found in other marine bacteria and may be a general system for bacteria living in Na^+-rich environments (Dimroth, 1987).

3.8.2 Na^+-coupled transport

(a) Na^+-solute symports

Anaerobic citrate transport by *Klebsiella pneumoniae* exhibits an absolute requirement for Na^+ or Li^+. Several other bacterial transport systems exhibit metal requirements. Citrate, for example, may be transported with Na^+ and K^+ (*S. typhimurium*), Mg^{2+} (*B. subtilis*) or Na^+ and H^+ (*K. pneumoniae*); Na^+-dependent citrate transport is probably restricted to organisms operating a decarboxylase Na^+ pump (Dimroth, 1987). A similar system exists in *E. coli* where glutamate uptake is accompanied by Na^+ and H^+ symport (Figure 3.8). Other transport systems, e.g. for melibiose in *E. coli* or various permeases in alkalophiles can be coupled to either Na^+ or H^+, whilst a third group uses Na^+ exclusively as a coupling ion and is exemplified by the proline transport systems of *E. coli* and *S. typhimurium*. It has been suggested that the Na^+ coupling of prokaryotes is a primitive form of coupling which evolved into the Na^+ energetics of animal plasma membranes.

(b) Na^+/H^+ antiports

A Na^+/H^+ antiport was first reported in *E. coli* by West and Mitchell (1974) and its properties were further described using membrane vesicles (Schuldiner and Fishkes, 1978; Tsuchiya and Takeda, 1979b). Similar systems are now being reported in an increasing number of bacteria and membrane systems (Lanyi, 1979; Krulwich, 1983; Dimroth, 1987).

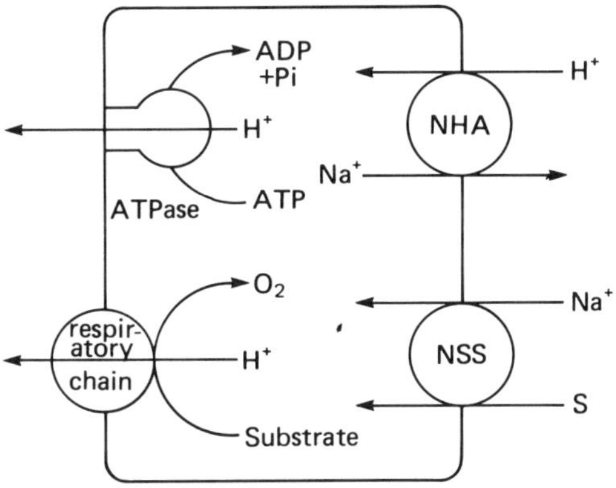

Figure 3.8 Coupling of primary proton translocation to secondary Na^+-driven solute transport in *E. coli*. The primary proton translocators are the outwardly directed ATPase and respiratory chain. The Na^+/H^+ antiport (NHA) converts the H^+ gradient thus formed to a Na^+ gradient that can be utilized to drive solute (S) transport by, for example, a Na^+-solute symport (NSS), where the solute might be glutamate or melibiose.

A major function of such antiports is to generate Na^+ gradients for the Na^+-coupled inward translocation of solutes (Section 3.8.2(a)). In *E. coli*, for example, methyl-β-D-thiogalactoside transport (via the melibiose carrier) and glutamate transport are energized by an electrochemical gradient of Na^+. The primary ion pumps, however, are the H^+-translocating respiratory chain and ATPase, both of which establish a H^+ gradient; this in turn is converted to a Na^+ gradient by the Na^+/H^+ antiport (Figure 3.8). A Na^+/H^+ antiporter is also well documented in *H. halobium* which uses the $\Delta\tilde{\mu}H^+$ established by bacteriorhodopsin to drive Na^+ extrusion.

A second function of Na^+/H^+ antiporters is to regulate intracellular pH especially in alkalophiles. Obligately alkalophilic bacteria have to maintain a cytoplasmic pH near 7 even when external values are 10.5 or higher. Homeostasis is achieved by a Na^+/H^+ antiporter which, together with respiration-driven proton translocation, achieves $[H^+]_{in} > [H^+]_{out}$ and $[Na^+]_{out} > [Na^+]_{in}$.

3.8.3 Na^+-driven flagellar motors

Motility in certain alkalophilic bacteria, like many solute uptake systems in these bacteria, is dependent on Na^+, being coupled to the electrochemical Na^+ gradient. The energy for flagellar rotation is provided both by membrane potential and the Na^+ concentration gradient (Dimroth, 1987).

3.8.4 Na^+-coupled ATPases and ATP synthesis

Heefner and Harold (1982) have described a Na^+ efflux pathway in membrane vesicles from *S. faecalis* in which an ATP-driven primary transport system mediates Na^+/H^+ exchange. The enzyme involved is thought to be a Na^+-stimulated ATPase that is distinguishable from the F_1F_0 ATPase by its differential response to mutation and inhibitors. This mechanism is thus analagous to the ATP-driven pump of animal cells and may be necessitated in *S. faecalis* by the limited capacity of this anaerobe to generate a proton motive force. More recently (Kakinuma and Harold, 1985), it has been suggested that a potassium transport system of this organism (Section 3.3.3) is identical to the above ATPase which might therefore provide a Na^+/K^+ exchange system, similar to that proposed in *Mycoplasma mycoides* (Benyoucef, Rigaud and Leblanc, 1982) and the plasma membrane ATPase of higher organisms.

Sodium-coupled ATP synthesis has been reported in the strictly anaerobic bacterium *Propionigenium modestum.* Although the Na^+ gradient generated in this organism by decarboxylation of methylmalonyl-CoA could in principle drive ATP synthesis via a Na^+/H^+ antiport to convert it into a proton gradient, Hilpert, Schink and Dimroth (1984) have demonstrated the presence of an ATPase specifically stimulated by Na^+ and monensin-sensitive ATP synthesis. This first demonstration of ATP synthesis linked to a Na^+ gradient has been followed by a similar claim for *Vibrio alginolyticus* (Dibrov *et al.*, 1986).

3.9 NICKEL TRANSPORT IN BACTERIA

In recent years, nickel has proved to be an essential component of certain bacterial enzymes (Chapter 5). Diekert, Graf and Thauer (1979) demonstrated the uptake of nickel by *Clostridium pasteurianum*, the subsequent stimulation of CO dehydrogenase activity and the inhibition of nickel transport by the metal chelator, NTA. Nickel uptake has been more extensively studied in *Methanobacterium bryantii* (Jarrell and Sprott, 1982) and involves active transport with a high affinity for nickel (K_m = 3.1 μM) but low rates of accumulation (V_{max} = 24 μmol min/(mg cells dry wt). Nickel transport appears not to be dependent on the membrane potential or intracellular ATP but is coupled to proton movement. The system is highly specific for nickel and unaffected by all cations tested (including Mg^{2+}), except for Ca^{2+}. A similar energy-dependent system ($K_m = 17\,\mu M$) has been identified as the major nickel transport pathway in *Alcaligenes eutrophus* (Lohmeyer and Friedrich, 1987). However, in the other few reported cases of nickel transport (e.g. in *E. coli*, Jasper and Silver, 1977), the metal is transported as an alternative substrate by the Mg^{2+} transport system (Jarrell and Sprott, 1982), not surprising in view of the similar ionic radii of Mg^{2+} and Ni^{2+}. In *Azotobacter chroococcum* (Partridge and Yates, 1982) $^{63}Ni^{2+}$ uptake is biphasic; a rapid binding is followed by a slower linear increase in nickel taken up. The transport is temperature-, but not energy-dependent, being insensitive to respiratory inhibitors, starvation and 2, 4-dinitrophenol. Although cyanide inhibits uptake, this appears attributable to the formation of a nickel-cyanide complex.

3.10 MOLYBDENUM TRANSPORT IN BACTERIA

Azotobacter spp. have long been known to accumulate high intracellular levels of Mo, their high affinity for the metal being exemplified by the use of *Az. paspae* in a biological assay for Mo in soil (Franco, Peres and Nery, 1978). However, the mechanism of molybdenum transport has been described in more detail in other nitrogen-fixers, which require molybdate for nitrogenase. In *Clostridium pasteurianum*, an energy-dependent system has been described (Elliott and Mortensen, 1975) which accumulates molybdate against a concentration gradient in a markedly temperature-dependent fashion and with a K_m of $48\,\mu M$. Uptake is competitively inhibited by sulphate and tungstate, but not by $S_2O_3^{2-}$. Molybdenum accumulation by *K. pneumoniae*, but not by *C. pasteurianum* or *Az. vinelandii*, is tightly regulated, probably by the *nif* operon, such that repression of nitrogenase synthesis (e.g. by O_2 or NH_4^+) also inhibits Mo transport (Pienkos and Brill, 1981).

A molybdenum-binding small peptide (M_r 550), termed chelin, has been identified in the culture medium of *Bacillus thuringiensis* grown under iron-deficient conditions (Ketchum and Somerville-Owens, 1975). The metal is bound as molybdate and the oxidation state is unchanged during binding. A similar siderophore-type system has recently been proposed to result from molybdenum limitation of *Az. vinelandii* (Page and von Tigerstrom, 1982). Interactions between such soluble compounds and the membrane-associated molydenum transport systems remain to be established. Similarly, the intracellular fate of accumulated molybdenum is poorly understood. In *Azotobacter*, the accumulated molybdenum is bound to a storage protein that may act as a reserve supply of the metal for nitrogenase synthesis. The protein binds 15 molybdenum atoms per tetramer; a tungsten-containing analogue is synthesized during growth in the presence of tungstate.

3.11 BACTERIAL TRANSPORT OF OTHER METALS AND METALLOIDS

Very little information is available on other trace metal ions. Copper(II) has a high affinity for the hydroxamate and catechol ligands of siderophores and forms stable complexes with schizokinen and desferrioxamine as well as with synthetic enterobactin analogues (Hider, 1984). Thus although copper(II) might employ siderophore-mediated transport, its higher solubility (as the hydroxide) and (unlike iron(III)) its high affinity for amino acids and peptides makes it unlikely that this would be the only means of transport. There is evidence from cyanobacteria that copper uptake involves a proton electrochemical gradient (Singh, 1985). In *Ps. stutzeri*, a 61 000 M_r protein has been identified which inserts copper into nitrous oxide reductase (Mokhele *et al.*, 1987) but is probably not directly involved in copper transport across the membrane.

Selenate, an analogue of sulphate, is transported by *E. coli* independently of sulphate (and sulphite) by a specific transport pathway. Uptake appears to be regulated in response to the synthesis of formate dehydrogenase (Brown and Shrift, 1982), a selenium-containing enzyme.

The efflux transport systems for toxic metal ions such as Cd^{2+} and metalloidal anions (e.g. AsO_4^{3-}) are described in Chapter 6.

3.12 METAL ION TRANSPORT IN EUKARYOTIC MICRO-ORGANISMS

3.12.1 Introduction

The most conspicuous feature of eukaryotic cells is their extensive compartmentation. The membranes of nuclei, vacuoles, mitochondria and chloroplasts surround and define these organelles, which have distinctive functions and requirements for metal ions. It is likely that, in principle, mitochondria and chloroplasts retain the central mechanism found in bacteria for transport of most metals, namely by exploiting transmembrane circulations of protons, secondarily coupled to metal transport. Here, we explore the mechanisms of transport both across organelle membranes and also the eukaryotic cytoplasmic membrane, emphasizing where possible comparisons with well-studied bacterial systems. Iron transport in eukaryotes, which is inseparable from the bacterial siderophore systems, was included in Section 3.2.

3.12.2 Potassium transport in fungi and other eukaryotic micro-organisms

Like bacteria, yeasts take up potassium very rapidly, possibly in exchange for H^+ (Jennings, 1976). The metal subsequently accumulates in the cytoplasm, nucleus (Roomans and Seveus, 1976) and intracellular vacuoles (Okorokov, Lichko and Kulaev, 1980). The pathway of uptake, although complex, is well-defined in terms of its kinetics and specificity (Rothstein, 1974). In general, it follows Michaelis–Menten kinetics with specificity in the order $H^+ > K^+ > Rb^+ > Cs^+ > Na^+ > Li^+$. Thus H^+, taken up with high affinity, acts as a competitive inhibitor. Divalent cations such as Mg^{2+} may also be taken up by the system, even though separate systems exist.

Evidence for a second K^+-transport system of much higher affinity (K_m = 24 μM) has come from studies of yeast grown at very low concentrations of K^+ (Rodriguez-Navarro and Ramos, 1984). Under certain conditions, the situation can be more complex. Apart from the substrate-binding site, responsible for K^+ translocation through the membrane, the carrier protein possesses a 'modifier' site. This has no transport function *per se*, but its occupation by cations other than the substrate leads to partial non-competitive inhibition of K^+ uptake. At low substrate concentrations, a third site is involved. Occupation of this 'activation site' by K^+, Cs^+ or Na^+ leads to an increase in the rate of Rb^+ uptake (Derks and Borst-Pauwels, 1980).

The K^+ efflux pathways are less well understood. Potassium efflux is generally slow but an electrogenic efflux of K^+ down a concentration gradient can be elicited by the antibiotic Dio-9 (Foury, Boutry and Goffeau, 1977). There are also efflux pathways for Li^+ and H^+. The latter ion is ultimately derived from metabolism and the pathway allows a stoichiometric efflux of H^+ to balance K^+ uptake. Unlike the K^+ (and H^+) uptake pathways the discrimination of the efflux pathway for H^+ over K^+ may be as high as 10^4 to 1. The Li^+ efflux pathway allows control of the intracellular concentration of this cation, to which yeast cells are permeable via the K^+ carrier.

Yeasts normally contain very little Na^+ but can be loaded by exposure to high concentrations of the cation in the absence of K^+. Suspension of

Na^+-loaded cells in a K^+-containing medium leads to Na^+ efflux, tightly coupled to K^+ uptake. The number of carriers involved has been the subject of some controversy (Rothstein, 1974).

The transmembrane K^+ gradient may provide the driving force for transport of divalent cations (Ca^{2+}, Mn^{2+}), amino acids (leucine, glycine) and glucose (Lichko, Okarokov and Kulaev, 1980) as well as cationic dyes (Peña, Mora and Carrasco, 1979).

In *Neurospora*, where transmembrane electrical currents and potentials can be measured directly with microelectrodes, K^+ influx has been interpreted in chemiosmotic terms by Slayman (1977). A large proportion, perhaps 25%, of the total cellular turnover of ATP is used to pump H^+ outwards via an ATPase and generate a membrane potential of about 200 mV (interior negative). Proton re-entry is predominantly via systems in which K^+ and H^+ entry is coupled to Na^+ efflux, while a smaller fraction of the proton flux is accompanied by efflux of organic acid anions or by H^+-coupled uptake of organic solutes. Thus, as in bacteria, but unlike animal cells, a gradient of H^+ and not Na^+ provides the primary driving force for K^+ transport.

However, more recent reports on higher plants, algae and yeasts suggest that potassium uptake, unlike that of many sugars, cannot be explained solely by invoking the membrane potential and may be directly coupled to ATP hydrolysis (Tromballa, 1978, 1980). Consistent with this are results on a purified plasma membrane ATPase from yeast, which has been reconstituted into proteoliposomes. Sensitivity to ionophores of the observed $^{32}P_i$-ATP exchange reaction suggests two types of proton transport, i.e. electrogenic proton transport and electroneutral H^+/K^+ exchange (Malpartida and Serrano, 1981).

In *Chlorella*, transport of alkali cations is energy-dependent and appears to involve a membrane-bound antiporter that takes up mainly K^+ (and possibly Na^+ and H^+), extruding Na^+, H^+ and also K^+. Net K^+ movement, however, is probably usually balanced by H^+ since, as in bacteria, the intracellular Na^+ concentration is very low. Such a system could operate to regulate internal pH (Tromballa, 1978). Indirect measurements of membrane potential (using a lipid-soluble cation) suggest that (as in yeast) K^+ uptake is unlikely to depend entirely upon a potential generated by an electrogenic H^+ pump, and may be mediated by an electroneutral or only slightly electrogenic K^+/H^+-exchanging ATPase.

At the tonoplast membrane, K^+ appears to be passively distributed in most algae, although in certain genera active K^+ transport from cytoplasm to vacuole may be important in cytoplasmic volume regulation (Raven, 1976).

A role of unique importance for K^+ has been proposed in the chytrid (water mould) *Blastocladiella emersonii* (van Brunt, Caldwell and Harold, 1982). Here, the membrane potential arises, not from H^+ extrusion, but by diffusion of K^+ out of the cells, down a gradient established by a K^+ pump. It is suggested that K^+ is accumulated by a primary K^+ pump and leaks out through a relatively specific channel that also passes Rb^+ and Ba^{2+} but rejects both larger and smaller cations as well as anions. Protons may be obligatory counterions for the ATP-driven pump, as proposed for yeasts (Malpartida and Serrano, 1981).

Potassium movements have been implicated in the operation of circadian rhythms, the molecular mechanisms of which have remained elusive (Lloyd, Poole and Edwards, 1982). The first reagent known to shift the phase of the rhythm was the K^+ ionophore valinomycin. One of many models proposes that a K^+ gradient (across an unspecified membrane) oscillates in a manner determined by the interaction of pumps and leaks modulated by external stimuli such as light (Harold, 1977).

3.12.3 Magnesium transport

(a) In ascomycete fungi

There are many parallels between Mg^{2+} transport systems in bacteria (Section 3.4) and fungi (Jasper and Silver, 1977). Thus, the concentration of free intracellular Mg^{2+} is high (3.3–20 mM) in fungi, and energy-dependent, high-affinity systems that transport Mg^{2+} as well as other divalent cations have been identified. Magnesium (Mg^{2+}) may also be transported by the system for K^+ and other monovalent cations, although Jasper and Silver (1977) suggest that the early reports of this may be attributable to less specific uptake by the yeast cell surface. Magnesium (Mg^{2+}) counteracts both the mutagenic and genetic-inhibitory actions of Mn^{2+} in *Saccharamyces cerevisiae*, perhaps by competing for the same carrier.

In contrast to the requirement in certain bacteria for nitrate as a charge-compensating co-transported anion, phosphate appears to be necessary for energy-dependent Mg^{2+} uptake in *S. cerevisiae* and *Penicillium chrysogenum*. Phosphate depletion increases the proportion of intracellular free magnesium by comparison with the amount bound as the orthophosphate.

The intracellular binding of magnesium, which is largely restricted in prokaryotes to ribosomes and certain polyanions, is complicated in eukaryotic cells by the existence of additional intracellular compartments, especially vacuoles, and by the deposition in many fungi of polymeric magnesium orthophosphate. Formation and breakdown of this complex may serve to regulate the level of free intracellular Mg^{2+} (Okorokov *et al.*, 1974).

In certain yeasts, magnesium is localized predominantly in intracellular vacuoles, where its concentration in the osmotically free and orthophosphate forms may exceed the bulk cytoplasmic levels by 30 or 100-fold, respectively. This implies the existence of a vacuole membrane-associated carrier that could (i) serve to regulate the free Mg^{2+} in the cytosol, and (ii) allow neutralization of intravacuolar polyphosphates.

(b) In algae

A system similar to the bacterial Mg^{2+} transport systems has been described (Jasper and Silver, 1977). The uptake is active and energy-dependent, operating against a concentration gradient, and may also transport Mn^{2+} and Co^{2+} albeit with lower affinity than for Mg^{2+}.

3.12.4 Manganese transport

Manganese is accumulated in *S. cerevisiae* by a divalent cation-transporting system, the affinity series being Mg^{2+}, $Ca^{2+} > Zn^{2+} > Mn^{2+} > Ni^{2+} > Co^{2+} > Sr^{2+}$ (Rothstein *et al.*, 1958). Experiments at much lower Mn^{2+} concentrations may reveal a more specific, high-affinity system such as that described in bacteria (Section 3.5).

3.12.5 Calcium transport

The calcium transport systems of eukaryotes share many features (e.g. classical saturation kinetics with K_m values in the µM range, and sensitivity to inhibitors) with those in bacteria. For example, in the fission yeast *Schizosaccharomyces pombe* the K_m is 45 µM Ca^{2+} and the uptake is competitively inhibited by high concentrations of Sr^{2+}, Mn^{2+} and other divalent cations, as well as by uncouplers and Ruthenium Red (Boutry, Foury and Goffeau, 1977). Efflux systems have also been identified, many of which are energy-dependent (Pitt and Ugalde, 1984).

The mitochondria of most higher organisms possess a chemiosmotically driven calcium transport system (Lehninger *et al.*, 1978), in which the cation enters the mitochondrion on an electrogenic uniport down a potential gradient. In contrast, as we have seen, bacteria actively expel calcium in the same direction as proton-pumping. Mitochondrial calcium accumulation is important in regulating the calcium content of the cytoplasm. A high-affinity calcium uptake system, however, is missing in the mitochondria of certain micro-organisms such as *Neurospora* and yeasts (Lloyd, 1974). The uptake of calcium by other membrane systems in eukaryotes is described by Silver (1977). An important feature of calcium uptake by mitochondria is its sensitivity to Ruthenium Red. Calcium accumulation by *Achlya* and the calcium uptake system of fission yeast (Boutry, Foury and Goffeau, 1977) are also Ruthenium Red-sensitive. The chemical nature of the inhibitor target is unknown.

3.12.6 Uptake and transport of zinc

Rather more is known about zinc accumulation by yeasts and fungi than by bacteria, where the exceedingly small requirements for the metal have frustrated study. A second reason is that algae have been studied for their potential to accumulate various radionuclides, including ^{65}Zn. Concentration factors (expressed as the ratio of the radionuclide concentration per g of biological material of the concentration per ml of solution) for zinc are tabulated by Failla (1977) and in some cases reach values of 6×10^4. Although some of this accumulation may be attributable to surface binding, high intracellular levels of zinc are attained, for example, by the marine alga *Eutrepia*, where the metal is localized in the chloroplasts (Failla, 1977) and in cytoplasmic granules in ciliates isolated from the rumens of sheep adapted to a high-zinc diet (Bonhomme, Quintana and Durand, 1980).

The zinc transport systems studied to date (Table 3.6) differ in their selectivity for zinc and other metals; Ca^{2+}, Cd^{2+} and Mn^{2+} are thus sometimes competitive inhibitors. More recent estimates (Mowll and Gadd, 1983) of the kinetic parameters of zinc transport in yeast reveal gross differences in the K_m in *S. cerevisiae* (5 mM Zn^{2+}) and the Zn^{2+}-sensitive *Sporobolomyces roseus* (0.09 mM Zn^{2+}). At sub-toxic zinc concentrations, potassium efflux does not accompany Zn^{2+} uptake in either species.

Failla and Weinberg (1977) have studied the varying capacity for Zn^{2+} uptake during batch growth of *Candida utilis*. The factors influencing the high rates of Zn^{2+} uptake early in the growth cycle and again in the late-exponential phase of growth have not been identified but could be due in part to changes in the proportions of cells at various cell cycle stages during these periods. Indeed, synchronous cultures of *Chlorella*

Table 3.6 Characteristics of energy-dependent transport systems for zinc in eukaryotic microbes

Organism	Inhibitors	Kinetic parameters
Candida utilis	CN^-, CCCP, EDTA, pCMB, NEM, Cd^{2+}, phosphate	$K_m = 1.3–1.8$; $V_{max} = 0.21$
Neocosmospora vasinfecta	N_3^-, UO_2^{2+}, Mn^{2+}, phosphate, N_2	$K_m = 200$
Saccharomyces cerevisiae	?	$K_m = 1000–1300$; $V_{max} = 0.043$
	Ca^{2+}	$K_m = 10$
Chlorella fusca	Ca^{2+}	$K_m = 5.7–8.7$; $V_{max} = 0.015$

Adapted from Failla (1977).
Units of K_m are μM, and for V_{max}, nmol/min/mg dry weight.

fusca provide evidence for just such a cell cycle-dependency (Wihlidal and Broda, 1977). During the activated period of uptake (by cells in the late exponential phase of growth) massive quantities of zinc are taken up within the cell, representing about 0.8% of the dry yeast weight. The mechanisms of protection against these high metal levels are unclear, but in *Neocosmospora*, electron-dense regions that probably contain zinc have been observed near the plasma membrane and within the nucleus (Paton and Budd, 1972). *Candida utilis*, grown in the presence of 100 μM zinc, produces a zinc-binding protein fraction resembling animal cell metallothionein. Synthesis of this fraction could explain in part the observed requirement for protein synthesis prior to initiation of Zn^{2+} uptake (Failla, Benedict and Weinberg, 1976).

3.12.7 Sodium transport

In higher organisms, there is abundant evidence that a circulation of Na^+ is generated by the Na^+, K^+-ATPase and supports metabolite transport across cell membranes and epithelial tissue (Crane, 1977).

Yeasts accumulate little Na^+, except in media of very high $[Na^+]$. Sodium enters the cell through the K^+ carrier (Section 3.12.2) and possibly also through Na^+-substrate symports. Accumulation of the cation is avoided, as in bacteria (Section 3.8) and algae (Raven, 1976) by an efflux pathway, probably a Na^+/H^+ antiport (Rothstein, 1974; Rodriguez-Navarro and Ortega, 1982).

3.12.8 Transport of other metals

There is scattered information on the uptake of other metals by eukaryotic microbes and much of this derives from studies with fungi (Ross, 1975). Uranium, cobalt, copper and nickel are all rapidly and reversibly bound to the yeast cell surface, from whence they may be transported by as yet poorly characterized mechanisms, sometimes of high affinity (de Rome and Gadd, 1987a). In *Sacch. cerevisiae*, Co^{2+} and Cd^{2+} are accumulated by a general cation uptake system, with Ni^{2+}, Zn^{2+} and Mn^{2+} as alternative substrates, and accompanying loss of K^+ (Norris and Kelly, 1977). Copper uptake following binding by yeast is energy-dependent (de Rome and Gadd, 1987a) and not mediated by the manganese uptake pathway (Parkin and Ross, 1985). The binding of mercury differs in being slower; desorption is also slow and very temperature-dependent, suggesting that it binds to groups different from those that bind other cations. Mercury uptake is accompanied by loss of other ions, notably K^+. Indirect confirmation that mercury is taken up is the ability of some fungi to methylate mercury by a mechanism that is thought to be intracellular. Silver is rapidly taken up and appears to cause loss of phosphate.

In *Penicillium* and *Aspergillus*, molybdate appears to share an energy-dependent transport system with selenate, SO_4^{2-} and $S_2O_3^{2-}$ (Tweedie and Segal, 1970). The supply of Cu and Zn to plants is enhanced by infection of the root system with mycorrhizal fungi (Tinker and Gildon, 1983). To date, it has only been established that these, and perhaps other, metals are absorbed by the fungus; the mechanism remains uncharacterized.

3.13 SUMMARY

Each metal required by microbial cells is transported across the membrane by one or more carriers with a high degree of specificity and avidity, providing the cell with adequate supplies of the metal even under conditions of low metal concentrations in the external environment. Most metal ions are actively transported against considerable concentration gradients. The metal transport system *par excellence* is that mediated by siderophores for transport of Fe(III), almost insoluble at neutral pH. The iron, bound most frequently by bidentate hydroxamate or catechol moieties of the ligand is transported across the membrane by virtue of receptors and released as Fe(II) intracellularly. Many bacteria possess multiple transport systems for iron and potassium, emphasizing the essentiality of the metals, and the complexity of outer and inner membrane components required for iron transport is now being unravelled by molecular genetic approaches. A degree of non-specificity is exemplified by transport systems for a number of metals. Thus iron uptake pathways may transport Al(III) or Ga(III), whilst Tl(I) is accumulated by K^+ transport systems in both *E. coli* and *Streptococcus faecalis;* Co(II) and Mn(II) are substrates for many bacterial Mg^{2+} transport systems. Little is known of the uptake systems for manganese, zinc, nickel, molybdenum and copper. This is attributable in some cases to the unavailability of useful radioisotopes but also to the low metal concentrations required for growth and the difficulty of distinguishing between surface binding and intracellular transport.

The transport systems for Ca^{2+} and Na^+, metals for which a requirement can rarely be demonstrated, involve active export as well as import. Studies with membrane vesicles reveal that such metal ions are excreted via H^+/metal antiports (and thus by the proton motive force) or by direct ATP involvement. In *Bacillus* special calcium transport systems allow calcium accumulation during sporulation. Sodium is second only to the proton as a coupling ion in bioenergetics and Na^+ gradients may be generated by Na^+ transport decarboxylases and respiration-driven pumps or by Na^+/H^+ antiports. A number of solute-Na^+ symporters, flagellar motors and ATPases utilize the Na^+ gradient thus generated.

Potassium transport across the plasma membranes of eukaryotic fungi, as in bacteria, is driven by an electrochemical H^+ gradient or by direct coupling to ATP hydrolysis. For magnesium, too, parallels exist between bacterial systems and those in fungi and algae. Thus high-affinity, energy-dependent systems have been identified to maintain high intracellular concentrations but the intracellular distribution of the metal is complicated by its accumulation in intracellular vacuoles. A notable feature of calcium transport in eukaryotes is active calcium accumulation by mitochondria. Zinc is also specifically localized in intracellular organelles and zinc-binding proteins.

Increasing appreciation of the biotechnological applications of metal uptake studies may be expected to improve our very incomplete picture of metal transport, as will improved methodologies for metal assays.

4

The IA and IIA cations

Sir Humphrey Davy
Abominated gravy.
He lived in the odium
Of having discovered Sodium

(Edmund Clerihew Bentley)

4.1 INTRODUCTION

In this chapter attention will be focused on the role in micro-organisms of Na^+, K^+, Mg^{2+} and Ca^{2+}, together with Mn^{2+} when it activates Mg^{2+}-dependent enzymes. It will be recalled from Chapter 1 that the IA and IIA cations are distributed selectively, with K^+ and Mg^{2+} concentrated inside the cell and Na^+ and Ca^{2+} concentrated outside the cell. Many examples will be given in this chapter to show the underlying importance of this distinction.

The biochemistry of these four cations, especially the IA cations, represents a difficult area of research but significant advances have been made in the last few years.

4.2 THE IA CATIONS, Na^+ AND K^+

The asymmetric distribution of these cations across the cell membrane is a general feature of living cells, although there is a remarkably wide range of cellular potassium and sodium cation concentrations in microbial cells. In the freshwater *Stentor coeruleus*, intracellular concentrations are $K^+ = 13$ mM and $Na^+ = 0.13$ mM at extracellular concentrations of $K^+ = 1$ mM and $Na^+ = 2$ mM (Song, 1981). For the extremely halophilic *Halobacterium halobium*, intracellular concentrations are $K^+ = 4.5$ M and $Na^+ = 1.4$ M at extracellular concentrations of $K^+ = 0.03$ M and $Na^+ = 4$ M

(Christian and Waltho, 1962). More typical values for intracellular K^+ are about 0.25 M.

The high levels of potassium within the cell are not necessary for activation of intracellular enzymes. Many K^+-dependent enzymes are known but they usually require concentrations of K^+ of around 10 mM for maximum activity (Suelter, 1970). Indeed, only a small fraction of the intracellular potassium is associated with sites on macromolecules: the remaining K^+ is associated with small anions, allowing potassium to play a major role in the control of osmotic pressure in the cell. In bacteria, cell osmolarity is maintained at a level above that of the medium. This difference (the turgor pressure) appears to be essential for growth of bacteria and of other cells with walls. The proposal that K^+ has a function in controlling osmotic pressure is supported strongly by the effect brought about by osmotic changes on the movement of potassium in and out of the cell. Increase in the osmolarity of the medium stimulates uptake of potassium while decrease in the osmolarity of the medium leads to efflux of K^+.

In summary, the functions of K^+ include osmotic control, stabilization of intracellular structures and the activation of intracellular enzymes. The role of Na^+ is less understood: a small number of enzymes are activated by Na^+, but the major functions of Na^+ in bacteria are linked to the presence of the concentration gradient across the cell membrane and include the co-transport of solutes and the maintenance of pH values in alkalophilic bacteria (Skulachev, 1987). These have been discussed in Section 3.8 and will only be summarized at this point.

4.2.1 Na^+-gradients and their exploitation

These gradients may be set up by several processes (Section 3.8.1), including Na^+/H^+ antiport for *Escherichia coli* (Bassilana, Damiano and Leblanc, 1984; Borbolla and Rosen, 1984), an ATP-driven sodium-extruding and potassium-accumulating pump for *Streptococcus faecalis* (Heefner and Harold, 1980, 1982; Kakinuma and Harold, 1985), and a halorhodopsin-catalysed coupling of Na^+ extrusion to light for *Halobacterium halobium* (MacDonald *et al.*, 1979). For several organisms, including *Klebsiella aerogenes* and *Veillonella alcalescens*, sodium efflux is linked to the decarboxylation of acids (Dimroth, 1982, 1987; Buckel and Semmler, 1982).

These sodium gradients are exploited in the co-transport of nutrients into microbial cells, for example, in the transport of isoleucine in the methanogenic, archaebacterium *Methanococcus voltae* (Jarrel, Bird and Sprott, 1984), and in the uptake of melibiose (Tsuchiya, Oho and Shiota-

Niiya, 1983) and glutamate in *Escherichia coli* (Hasan and Tsuchiya, 1977). Once inside the cell, the Na^+ dissociates from the complex and is expelled from the cell. The photosynthetic evolution of oxygen in the marine diatom *Phaeodactylum tricornutum* is dependent upon sodium, which cannot be replaced by other singly charged cations. Sodium is known to facilitate the utilization of bicarbonate by the diatom, and it seems probable that this is a further example of a nutrient being taken into the cell in association with Na^+.

The Na^+-gradient is also exploited in alkalophilic bacteria in the transport inwards of Na^+ in association with non-essential solutes. The resulting efflux of Na^+ via the Na^+/H^+ antiport is essential for maintaining a cytoplasmic pH that is up to 2 pH units lower than the pH of the medium, which is about 10.5 (Krulwich, Federbush and Guffanti, 1985). The carrier molecules associated with the Na^+/H^+ antiport are not known, although the Na^+/H^+ antiporter from the alkalophilic *Bacillus firmus* has been extracted from membranes and reconstituted into liposomes (Seto-Young, Garcia and Krulwich, 1985). Krulwich (1983) has reviewed Na^+/H^+ antiporters.

The presence of Na^+ in the medium is essential for the swimming action of certain alkalophilic *Bacillus*, suggesting that the flagellar motors are powered by the transmembrane electrochemical potential gradient of Na^+ (Hirota, Kotada and Imal, 1981).

4.2.2 Sodium cations as enzyme activators

A small number of enzymes, in addition to the Na^+, K^+-activated ATPase of *Streptococcus faecalis*, have a specific requirement for sodium cations. These include the oxaloacetate decarboxylase of *Aerobacter aerogenes* (Sachan and Stern, 1971), while the oxidation of NAD-linked substrates by the aerobic marine bacterium *Alteromonas haloplanktis* is dependent upon Na^+ (Khanna *et al.*, 1984). The Mg^{2+}-dependent ATPase in the isolated plasma membranes from amoebae of the slime mould *Dictyostelium discoideum* can be stimulated by Na^+ or K^+ (Blanco, 1982). Little is known of the function of Na^+ in these examples. It is probable that Na^+ binds to the enzyme and controls its conformation.

4.2.3 Potassium and sodium cations as structure stabilizers

A substantial fraction of intracellular K^+ (and Mg^{2+}) in bacteria is associated with the ribosomes. Both cations are known to affect the polymerization and stability of ribosomes *in vivo* and *in vitro*. It appears that the $[K^+]$: $[Mg^{2+}]$ ratio is important as well as the absolute concentration levels. A high ratio results in the breaking up of the ribosomes into subunits and the unfolding of the RNA chain. The role of potassium in maintaining the structural integrity of the ribosomes accounts for the requirement for K^+ in protein synthesis, but K^+ also appears to have a direct role in the incorporation of amino acids into the polypeptide chain by transfer from RNA.

Although divalent cations bind more strongly to nucleic acids, it is clear that nucleic acid structure is also affected by the alkali metal ions, which interact with the phosphate groups. Detailed reviews are available on the effect of metal ions on the structure and function of nucleic acids (Eichhorn and Marzilli, 1981). The effects of alkali metal cations are illustrated by the following studies relating to the double-helical form of DNA. Alkali metal cations interact with and progressively neutralize the array of negative charge associated with the phosphate groups, leading to stabilization of the double helix with respect to the single-stranded form. As the negative charge is neutralized so further binding of cations becomes weaker, while excess cation will destabilize the double helix. Binding of K^+ also affects the stabilities of the various conformations of the DNA double helix. The common form (the B form), in which the base pairs are arranged perpendicular to the helical axis, is stabilized by K^+ (and Li^+).

4.2.4 Potassium cations as enzyme activators

Many enzymes are specifically activated by K^+ (Suelter, 1970). The better-studied examples are those isolated from mammalian sources, and often only limited information is available on K^+-dependent enzymes of microbial origin. Some of these are listed in Table 4.1. It is probable that the true number of K^+-dependent enzymes is much higher than this, as often potassium is not excluded during the characterization of a newly-investigated enzyme. In some cases, over certain concentration ranges, K^+ may be replaced by Rb^+ or NH_4^+ (Bostian *et al.*, 1982), while the thallium cation has been used as a probe for the potassium-binding site.

An examination of Table 4.1 shows that K^+ activates the enzymes involved in protein synthesis, other biosynthetic enzymes and catabolic enzymes. Very little is known about the precise mechanisms by which K^+ exerts these catalytic effects. There is often good evidence from UV, CD and NMR spectroscopy that binding of K^+ to proteins may stabilize a conformation or bring about conformational changes to generate the active site. This could account for the presence of the high ratios of

Table 4.1 Some enzymes activated by K^+

Enzyme	Reference
Adenylate cyclase (*E. coli*)	Bostian *et al.*, 1982
Aldehyde dehydrogenase (yeast)	Wampler and Westhead, 1968
Aspartate kinase (*E. coli*)	Paulus and Gray, 1964
Aspartate kinase (*Bacillus polymyxa*)	Rosner and Paulus, 1971
Aspartate kinase (*Bacillus* spp.)	Villafranca and Raushel, 1982
Carbamoyl-phosphate synthetase (*E. coli*)	Mildvan and Loeb, 1981
DNA polymerases	
3-hydroxyisopropyl malate dehydrogenase (*Salmonella typhimurium*)	Burns, Umbarger and Gross, 1963
L-malate dehydrogenase (*Lactobacillus arabinosus*)	Nossal, 1951
Na^+, K^+-ATPase (*Streptococcus faecalis*)	Kakinuma and Harold, 1985
Peptidyl transferase (*E. coli*)	Pestka, 1972
Peptidyl transferase (*Bacillus stearothermophilus*)	Fahnestock, Erdman and Nomura, 1974
Phosphate acyltransferase (*Clostridium kluyver*)	Kyrtopoulos and Satchell, 1972
rec BC enzymes (*E. coli*)	Hermanns and Wackernagel, 1977
Serine deaminase (*E. coli*)	Dupourque, Newton and Snell, 1966
Tryptophan synthase (*Bacillus subtilis*)	Schwartz and Bonner, 1964
Tryptophan synthase (*E. coli*)	Crawford and Ito, 1964
Tryptophanase (*E. coli*)	Newton and Snell, 1964

$[K^+]$: [enzyme] often observed. If potassium has a role of this type then multiple binding of K^+ will take place on the enzyme rather than the substrate. It is also possible that K^+ could bridge enzyme and substrate to control finely the orientation of the substrate with respect to the protein. This may well be the role of K^+ in pyruvate kinase (Section 4.2.4(c)).

(a) The recBC *enzyme of* Escherichia coli

This enzyme has important DNA unwinding and nuclease metabolism. The *in vitro* degradation of DNA by purified *recBC* enzyme ceases 2 to 3 minutes after the start of the reaction, but activity may be restored by addition of potassium salts. Hermanns and Wackernagel (1977) have suggested that the enzyme undergoes a conformational change to give an inactive form, and that this is reversed by K^+.

(b) Aldehyde dehydrogenase from yeast

This enzyme catalyses the oxidation of aldehydes to acids. Activity is lost on removal of K^+ and restored on addition of K^+. Addition of K^+ to an inactive preparation of the enzyme which has stood for some minutes results in two conformational changes, a slow one and a fast one, which are promoted by binding K^+ at separate and independent sites on the enzyme. The use of equilibrium dialysis techniques indicates that there are four sites for Rb^+ (and presumably K^+) on each of the four subunits of the enzyme. In contrast there are six sites for Tl^+. This will account for the ability of Tl^+ to inhibit the enzyme at higher concentration, by binding to these additional sites (Bostian *et al.*, 1982).

(c) Pyruvate kinase

This is a particularly well-studied enzyme having a requirement for K^+. Muscle pyruvate kinase has been subjected to a wide range of instrumental techniques and illustrates well the use of NMR spectroscopy in obtaining distance measurements between certain substrate nuclei and metal centres. Pyruvate kinase has also been isolated from yeast, but this discussion will deal with the information obtained using the muscle enzyme to show the value of these techniques. Pyruvate kinase is a tetrameric enzyme (M_r 237 000) which catalyses the reversible phosphorylation of adenosine diphosphate from phosphoenol pyruvate to give ATP. Both K^+ and Mg^{2+} are required for activity (Equation 4.1).

$$\text{ADP} + \text{phosphoenol pyruvate} \underset{Mg^{2+},\ K^+}{\rightleftharpoons} \text{ATP} + CH_3COCO_2^- \qquad (4.1)$$

Many studies have been carried out using Tl^+ and Mn^{2+} to replace K^+ and Mg^{2+}, and with substitution-inert Cr^{III}ATP, in order to obtain distances between certain substrate nuclei and the probe metal ions by observing the effect of the magnetic centre on NMR relaxation rates. Nineteen distances have been determined (Dunaway-Mariano *et al.*, 1979). The model of the active site consistent with these distances suggests the presence of molecular contact between the γ-phosphoryl phosphorus of ATP and the carbonyl oxygen of pyruvate (Figure 2.5). This, in turn, leads to the conclusion that direct phosphoryl transfer from ATP to pyruvate occurs without the intermediacy of a phosphoenzyme species. The divalent cation probably activates the transferred phosphoryl group indirectly through a water ligand. The role of the monovalent cation (normally K^+) is not fully resolved, but may be to assist in the orientation of the carboxyl group of pyruvate during the phosphoryl transfer reaction (Van Divender and Grisham, 1985).

(d) Carbamoyl-phosphate synthetase from Escherichia coli

This enzyme requires K^+ and Mg^{2+} activators and catalyses the reaction shown in Equation 4.2 (Villafranca and Raushel, 1982).

$$\begin{aligned} &2\text{ATP} + \text{HCO}_3^- + \text{glutamine} \xrightarrow[\text{K}^+,\ \text{Mg}^{2+}]{} \\ &2\text{ADP} + \text{P}_i + \text{carbamoyl-P} + \text{glutamate} \end{aligned} \tag{4.2}$$

(e) DNA polymerases

Many DNA polymerases are stimulated by K^+ up to 50 mM concentration, but inhibition occurs at higher levels (Mildvan and Loeb, 1981).

4.3 MAGNESIUM AND MANGANESE

4.3.1 Introduction

Attention has been drawn to the high intracellular requirements of micro-organisms for Mg^{2+}, usually thought to be about 10–20 mM. A much higher figure of about 100 mM has been claimed for exponentially growing *Escherichia* B (Moncany and Kellenberger, 1981). Intracellular magnesium is bound to ATP, ADP and nucleic acids, while a substantial fraction is bound to ribosomes. The high requirement for Mg^{2+} is manifested in essential functions for this action inside the cell. Notable amongst these is the interaction of Mg^{2+} (and K^+) with nucleic acids, in stabilizing the structures of ribosomes and nucleic acids and in activating enzymes concerned with synthesizing DNA, RNA and proteins. The function of Mg^{2+} in many of these enzymatic processes involves the hydrolysis and transfer of phosphate or phosphocompounds.

Manganese as Mn^{2+} is included in this discussion as it can often substitute for Mg^{2+}, sometimes simply in response to the availability of the two cations (perhaps even in the assay medium). In other cases, the competition between Mg^{2+} and Mn^{2+} may have regulatory significance as in the $NADP^+$-specific malic enzyme of *Escherichia coli* (Brown and Cook, 1981). In a few cases there is a specific requirement for Mn^{2+} that cannot be met by Mg^{2+}. This general similarity between Mg^{2+} and Mn^{2+} is shown by the transport of Mn^{2+} into cells via Mg^{2+}-transport systems, although cells with a specific requirement for Mn^{2+} have transport systems for this cation which are unaffected by very large excesses of Mg^{2+} (Silver and Jasper, 1977). Section 3.5 should be consulted for more details.

Detailed understanding of the mechanisms by which Mg^{2+} exerts its catalytic effect in enzymes has been considerably advanced by the application of magnetic resonance techniques. These have usually involved the replacement of Mg^{2+} by paramagnetic Mn^{2+} as a spectroscopic probe. However, it should be noted that the ionic radii of these two cations are rather different from each other (Mn^{2+}, 0.80×10^{-10} m; Mg^{2+}, 0.65×10^{-10} m), and it is possible that Mn^{2+} acts in a slightly different way from Mg^{2+} in effecting catalysis. The enzyme pyruvate kinase, discussed in Section 4.2.4(c), illustrates this point. The Mg^{2+}-activated enzyme is inhibited by Ca^{2+} and Li^+, but the Mn^{2+}-activated enzyme is inhibited by Ca^{2+} and activated (although poorly) by Li^+. This is probably due to the poor fit for Li^+ at the active site, which is adventitiously accommodated to a limited extent by the larger size of the manganese cation. Again, Mn^{2+} can substitute for Mg^{2+} in many of its interactions with nucleic acids, but this results in inaccuracy. With Mn^{2+} and ribosomes, errors in translation occur, and DNA polymerase incorporates ribonucleotides as well as deoxyribonucleotides.

4.3.2 Structural effects from binding of Mg^{2+} to nucleic acids

The essential role of cations in controlling the structure and reactivity of nucleic acids has been emphasized. The folding of such polyanionic species can only occur if the negative charges on the phosphate groups are neutralized or largely neutralized. In addition, more specific effects can arise from binding cations at certain sites. The cations Mg^{2+} and the charged amine spermine ($NH_3^+(CH_2)_3NH_2^+(CH_2)_4NH_2^+(CH_2)_3NH_3^+$), are of greatest physiological importance, although the role of K^+ is also significant. In a few instances the effects of $[Mg^{2+}]$ may be mimicked by high $[Na^+]$. The competition between Na^+ and a doubly-charged polyamine for DNA has been studied by ^{23}Na NMR spectroscopy (Padmanabhan *et al.*, 1988).

There are many manifestations of the role of Mg^{2+} in controlling the structure of nucleic acids. Magnesium binds more strongly to double-helix DNA than to the single-stranded form, so denaturation of DNA is associated with release of Mg^{2+}. The structures of nucleosomes depend upon salt concentration. One structural change occurs as the ionic strength is reduced from 26 to 0.1 mM, at about 1.3 mM salt. At high salt concentration the nucleosome is in a compact, contracted form, but the low salt concentration probably does not allow neutralization of the phosphate charge, and so repulsion between these groups causes the structure to expand. This effect is fully reversed by 100 μM Mg^{2+} (Wu *et al.*, 1979).

Some specific illustrations of structural effects due to Mg^{2+} are given in the following sections.

(a) Transfer ribonucleic acid (tRNA)

Interactions between magnesium cations and nucleic acids have been best characterized for yeast phenylalanine tRNA (Teeter, Quigley and Rich, 1981). The function of tRNa is dependent upon Mg^{2+}. Addition of Mg^{2+} to denatured tRNA restores its ability to bind to amino acids. The binding of Mg^{2+} to tRNA is classified in terms of 'weak binding', responsible for the general folding of tRNA, and 'strong binding', which preserves specific features in the conformation. The numbers of binding sites are 23 ± 5 and 4–6 respectively. Structural studies on yeast tRNA have identified three classes of pockets formed by sharp bends or loops, which provide metal-binding sites. It is essential that the charge be neutralized at these positions as repulsion effects are enhanced by the folding. The three classes are distinguished by the ligands available, namely: Class 1–phosphate groups; Class 2–phosphate plus base; Class 3–mainly base. Mg^{2+} binds mainly to Class 1 sites, including the 5′ terminus, the P10 bend (as $Mg(H_2O)_6^{2+}$, linked indirectly to four phosphates), and to P20 + P21 (as $Mg(H_2O)_4^{2+}$). One Class 2 site has been identified (P19 + P21 + base), where the cation is bound as $Mg(H_2O)_5^{2+}$. Binding of Mg^{2+} to the 5′ terminal phosphate may be involved in the interaction of tRNA both to the ribosome and its synthetase. Binding of Mg^{2+} to the weak sites is thought to link tRNA to the ribosome.

Further binding of Mg^{2+} occurs at higher concentrations (Labuda, Nicoghosian and Cedergren, 1985), and has been termed 'low-affinity' binding. At appropriate $[Mg^{2+}]$, where Mg^{2+} binds at these cooperative low-affinity sites, the structure of tRNA appears to respond to environmental changes such as aminoacylation or codon binding.

(b) Ribosomes

It was noted in Section 4.2.3 that K^+ and Mg^{2+} stabilize ribosomal structure. Prolonged incubation of a suspension of *Escherichia coli* in a magnesium-free medium results in a progressive disintegration of the ribosomes in the cells. On addition of Mg^{2+}, the cells resume synthesis of ribosomal particles before resuming normal growth. Magnesium starvation also causes release of ribosomal proteins, but these are not usually reincorporated during recovery of the cells (Adachi and Sells, 1979).

The detailed effects of Mg^{2+} on the association and dissociation equilibria and on the kinetics of these processes have been well studied for ribosomes from *Escherichia coli*. Type A ribosomes associate at about 2.2 mM Mg^{2+}, and convert to type B 'loose' couples at 1 mM Mg^{2+} (Gorisch, Goss and Parkhurst, 1976; Goss, Parkhurst and Wahba, 1980). These two types of ribosomes bind 5–10 and 7–9 mol Mg^{2+} respectively. It is important that a mechanism exists for dissociation of the ribosomes into subunits in order to initiate protein synthesis. In *Escherichia coli* the subunit equilibrium is regulated by the initiation factor 3, IF3 (Goss, Parkhurst and Wahba, 1980).

The effect of Mg^{2+} on the structure of the 5S ribosomal ribonucleic acid from *Escherichia coli* has been monitored by the NMR spectrum of the imino proton. The transition between the physiological and low $[Mg^{2+}]$ forms of 5S RNA has a midpoint at 0.3 mM Mg^{2+}. Changes in the NMR spectrum at low $[Mg^{2+}]$ have been attributed to structural adjustments in specific regions of the RNA (Leontis, Ghosh and Moore, 1986).

4.3.3 Structural effects from binding of Mg^{2+} to cell walls and membranes

(a) Stabilization of cell envelopes by Mg^{2+}

Pleomorphic *Halobacterium* strains from the Dead Sea require about 75 mM Mg^{2+} for growth. When suspended in media containing lower concentrations of Mg^{2+}, cells lose their native shape within minutes, becoming spherical. At least 75 mM Mg^{2+} is needed to stabilize *Halobacterium marismortui* and *Halobacterium volcanii* cells for 20 h at 25 and 35°C, respectively. At 0°C the structure is stable for only 2 h. These changes are irreversible. The transition from pleomorphic to spherical shape in *Halobacterium volcanii* is accompanied by discharge of a glycoprotein-like material into the medium. The stabilizing effect of

Mg^{2+} results from surface binding (Cohen, Oren and Shilo, 1983). Similar stabilizing effects produced by surface binding of Mg^{2+} are known for other marine bacteria (Macleod and Matula, 1982).

(b) Effect of Mg^{2+} *on membranes and lipopolysaccharides of gram-negative bacteria*

The outer and cytoplasmic membranes of gram-negative bacteria differ radically in their associated counter-ions. The outer membranes of *Escherichia coli* have a 3-fold enrichment in divalent cations compared to the cytoplasmic membrane. The ratio $[Mg^{2+}]/[Ca^{2+}]$ in the outer membrane is about 5:1. The anionic lipopolysaccharide (LPS) on the outer monolayer and the underlying peptidoglycan provide sites for Mg^{2+}. The porin (channel-forming) proteins are thought to be linked to the outer membrane by Mg^{2+} bridges. This has been supported by analytical data which confirms the porin-LPS-peptidoglycan complex to be enriched in Mg^{2+} compared to LPS or the intact outer membrane. Magnesium is also the major cation involved in the stabilization of LPS domains in the outer membrane, and in the linking of LPS to the membrane proteins. This accounts for the ability of chelating agents to release the bulk of the lipopolysaccharide from whole cells (Coughlin, Tonsager and McGroarty, 1983).

4.3.4 Control of mitochondrial structure

The structural integrity of mitochondria depends upon magnesium ions (Lloyd, 1974; Dow, Walton and Fleischer, 1970). Growth of *Schizosaccharomyces pombe* in a magnesium-depleted medium results in structural changes in mitochondria, which become small and rounded. This appears to be due to a switch from a fermentative to a respiratory mode of metabolism, which has a lower requirement for magnesium (Walker *et al.*, 1982).

4.3.5 Mg^{2+} and Mn^{2+} as enzyme activators

Table 4.2 gives a representative selection from the many enzymes which are activated by Mg^{2+} and/or Mn^{2+}. The relative efficiency of these cations as activators varies from enzyme to enzyme. The concentration of Mn^{2+} inside cells is much lower than that of Mg^{2+}, so it is often questionable whether Mn^{2+} would have a physiological role. It may

Table 4.2 Some enzymes activated by Mg^{2+} and/or Mn^{2+}

Enzyme	Metal specificity	Reference
Pyruvate kinase (yeast)	Mg^{2+} (Mn^{2+}) + K^+	
Phosphofructokinase-2 (*E.coli*)	Mg^{2+}	Guixe and Babul, 1985
Hexokinase (yeast)	Mg^{2+}	Pederson and Gupta, 1979
Inorganic pyrophosphatase (yeast)	Mg^{2+}	Welsh *et al.*, 1983
F_1-ATPase (*E.coli*)	Mg^{2+},Mn^{2+}	Smith, Latchnay and Senior, 1985
F_1-ATPase (*R.rubrum*)	Mg^{2+},Mn^{2+}	Gromet-Elhanan and Khananshvili, 1984
Anthranilate synthase-phosphoribosyl-transferase (*Salmonella typhimurium*)	Mg^{2+},Mn^{2+}	Robinson, Nowak and Levy, 1979 Robinson and Levy, 1979
Carbamoylphosphate synthetase (*E.coli*)	Mg^{2+} + K^+	Villafranca and Raushel, 1982
Ribonucleotide reductase (*Brevibacterium ammoniagenes*)	Mn^{2+}	Lammers and Follmann, 1983
Phosphoglycerate phosphomutase (*Bacillus subtilis*)	Mn^{2+}	Watabe and Freese, 1979
L-threonine dehydrogenase (*E.coli*)	Mn^{2+}	Craig and Dekker, 1986
Enolase (yeast)	Mg^{2+} (Mn^{2+})	Rose, Dickinson and Westhead, 1984
* NAD^+-specific malic enzyme (*E.coli*)	Mn^{2+}/Mg^{2+}	Milne and Cook, 1979
* $NADP^+$-specific malic enzyme (*E.coli*)	Mn^{2+}/Mg^{2+}	Brown and Cook, 1981
* NAD^+-specific isocitrate dehydrogenase (*N.crassa*)	Mn^{2+}/Mg^{2+}	Barratt and Cook, 1978
Glucose isomerase (*S.griseofuscus*)	Mg^{2+}	Kasumi, Hayashi and Tsumura, 1982a
(D-xylose ketol isomerase) (*L.brevis*)	$Mn^{2+} > Mg^{2+}$	Yamanaka, 1961
(*B.coagulans*)	$Mn^{2+} > Mg^{2+}$	Danno, 1970
Citrate lyase (*K.aerogenes*)	Mg^{2+} (Mn^{2+})	Sivaraman and Sivaraman, 1979
Enzyme I (PEP-dependent phosphotransferase) (*E.coli*)	Mg^{2+} (Mn^{2+})	Hoving, Koning and Robillard, 1982

(continued)

* Glutamine synthetase (*E.coli*)	Mg^{2+}/Mn^{2+}	Villafranca and Balakrishnan, 1979
(*Salmonella typhimurium*)	Mg^{2+}/Mn^{2+}	Balakrishnan *et al.*, 1979
(*A.vinelandii*)	Mg^{2+}/Mn^{2+}	Siedel and Shelton, 1979
DNA polymerase	Mg^{2+}/Mn^{2+}	Mildvan and Loeb, 1981
RNA polymerase	Mg^{2+}/Mn^{2+}	Mildvan and Loeb, 1981

* Alternative activators, with regulatory implications.

well be that Mn^{2+}-stimulation of an enzyme is an artefact of the assay conditions. However, there are clear cases where enzymes have a requirement for Mn^{2+}, and, in one example, regulation of enzymatic activity involves the conversion of an Mg^{2+}-specific site into an Mn^{2+}-specific site.

Magnesium cations activate many enzymes that catalyse the transfer of the terminal phosphoryl group of ATP, either to water or to a second substrate. The $MgATP^{2-}$ complex frequently serves as a substrate in these cases. The binding of Mg^{2+} serves to weaken the bond to the terminal phosphoryl group. Nucleophilic attack upon the γ phosphorus atom will take place by an S_{N2} mechanism (Equation 4.3). Such an associative pathway allows greater control of the stereochemistry of the substitution, and the rates of such processes are more susceptible to catalysis by metal ions.

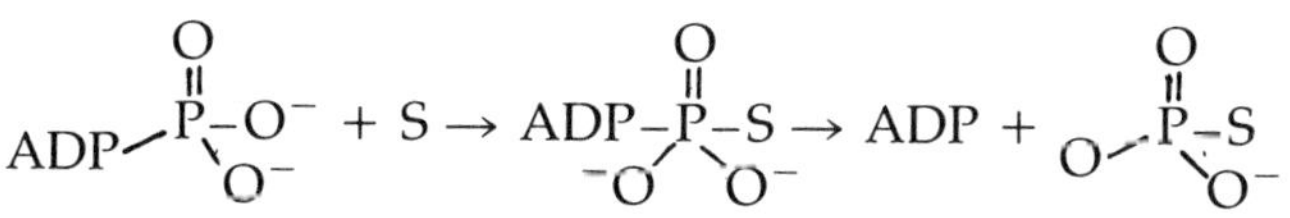

(4.3)

In the following sections, some examples will be discussed to illustrate the main types of enzymatic reactions catalysed by Mg^{2+} or Mn^{2+}. It should be noted that many processes discussed in other chapters have a requirement for $MgATP^{2-}$ (e.g. nitrogenase) or Mg^{2+} (e.g. alkaline phosphatase).

(a) Yeast inorganic pyrophosphatase

This enzyme, which catalyses the hydrolysis of pyrophosphate to phosphate, is dimeric with two identical subunits. Two magnesium ions are bound per subunit in the absence of pyrophosphate, and three in the presence of pyrophosphate. The substrate is $Mg(H_2O)_4PP^{2-}$, where PP = pyrophosphate. The enzyme also catalyses the hydrolysis of ATP.

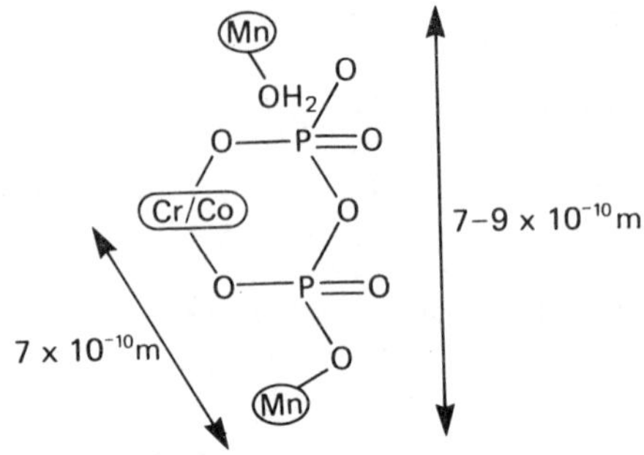

Figure 4.1 Distances between metal centres in yeast inorganic pyrophosphatase (Knight *et al.*, 1984).

Other divalent cations show catalytic activity but Mg^{2+} is much more effective, with $Mg^{2+} \gg Zn^{2+} > Co^{2+} > Mn^{2+} \gg Cd^{2+}$. The catalytic efficiency of Mg^{2+} has been attributed to the rapid release of phosphate from the enzyme in the presence of Mg^{2+} (Welsh *et al.*, 1983). Distance measurements obtained for the Mn^{2+}-activated enzyme show the two Mn^{2+} cations interact magnetically, with a separation of $10–14 \times 10^{-10}$ m. This separation decreases to $7–9 \times 10^{-10}$ m on binding diamagnetic Co(III)-containing substrate analogues. The use of paramagnetic Cr(III) substrate analogues gives an Mn^{2+}–Cr(III) distance of 7×10^{-10} m. The arrangement of cations is depicted in Figure 4.1 (Knight *et al.*, 1984).

(b) Proton-translocating ATPase of bacteria

A Mg^{2+}-dependent ATPase has been isolated from mitochondrial, bacterial and chloroplast membranes. It is made up of two parts, F_1 and F_o. The F_o portion is an intrinsic proton-translocating protein, while the F_1 component is the Mg^{2+}-activated, extrinsic membrane ATPase. The enzyme may function in reverse to synthesize ATP from ADP and P_i, by exploiting the electrochemical potential of the proton gradient.

Purified F_1 ATPase from *Escherichia coli* binds two moles of Mg^{2+} per mole of enzyme (Senior *et al.*, 1980). The Mg^{2+} may be replaced by Co^{2+} or Mn^{2+} to give fully active enzymes, which bind to F_1-depleted membranes and catalyse the ATP-driven proton pumping. The nature of the bound metal (Mg^{2+}, Mn^{2+} or Co^{2+}) does affect the rate of inactivation by the inhibitor *N,N′*-dicyclohexyl-carbodiimide, which is known to interact with the β subunits of the enzyme. It appears that metal binding affects the conformation of the β subunit. Isolated α subunits bind all three metal ions stoichiometrically, while the β subunit does not, suggesting that the tight binding sites for M^{2+} are on the α subunit (Smith, Latchnay and Senior, 1985). The α subunit also has a high-affinity site for ATP. Trypsin hydrolysis of the subunit gives different products in the presence and absence of ATP, indicative of ATP-induced conformational changes (Senda *et al.*, 1983).

The isolated β subunit from *Rhodospirillum rubrum* has two binding sites for nucleotides. One binds ATP or ADP independent of Mg^{2+}, and the other binds ATP in the presence of Mg^{2+} (Gromet-Elhanan and Khananshvili, 1984).

(c) Glutamine synthetase

Equation 4.4 gives the reaction catalysed by glutamine synthetase.

$$\text{ATP} + \text{L-glutamate} + NH_3 \xrightleftharpoons{2M^{2+},\ M^+} \text{ADP} + P_i + \text{L-glutamine} \qquad (4.4)$$

This is a key reaction in nitrogen metabolism. The enzymes from *Escherichia coli, Salmonella typhimurium* and *Azotobacter vinelandii* are dodecameric, and involve a symmetrical aggregate of the subunits in two hexagonal layers. The enzyme is regulated by a covalent adenylylation of a tyrosine residue in each subunit, which converts Mg^{2+}-specific site into Mn^{2+}-specific site with decrease in activity. Each subunit has two cation-binding sites, termed n_1 and n_2. Metals bind first at the n_1 site, which may involve the reorientation of a glutamate carboxyl group. The n_2 site is the catalytic site and may bind metal ion or metal–nucleotide complex. The distance between the n_1 and n_2 sites has been estimated as $7 \pm 2 \times 10^{-10}$ m (Balakrishnan and Villafranca, 1979). The active site of glutamine synthetase is represented in Figure 4.2 (Villafranca and Raushel, 1982).

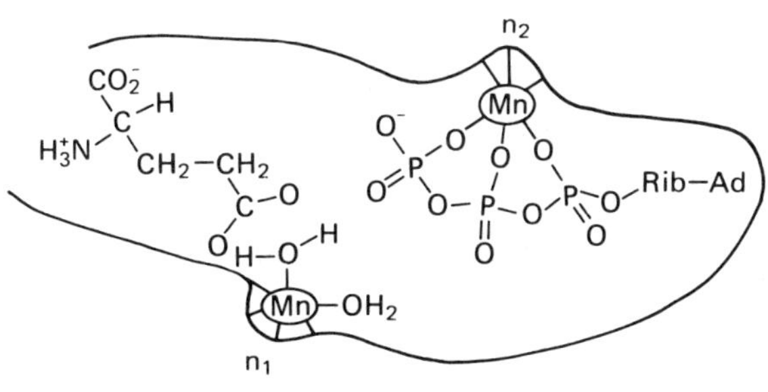

Figure 4.2 The active site in glutamine synthetase. (Bala Krishnan and Villafranca, 1979).

(d) DNA and RNA polymerases

These have a specific requirement for Mg^{2+} or Mn^{2+}. Both enzymes were once thought to have a requirement for Zn^{2+} as well, but it is now appreciated that this is the case only for the RNA polymerase.

The DNA polymerase catalyses the accurate copying of a polynucleotide template. It catalyses nucleophilic attack of the 3′-OH terminus on the α-phosphorus atom of a deoxynucleoside triphosphate, with displacement of pyrophosphate and incorporation of the nucleoside into the elongating DNA chain, as represented in Figure 4.3.

Distance measurements for the binary Mn–dTTP complex and the ternary DNA polymerase–Mn–dTTP complex (Loeb and Mildvan, 1981) show that all three phosphoryl groups of dTTP are coordinated to Mn^{2+} in the binary complex, but that only the γ group is coordinated in the ternary complex. The cation links the γ phosphoryl group of the substrate to the enzyme, so assisting the leaving of the pyrophosphate group.

The RNA polymerases show greater complexity, containing 2 mol Zn^{2+} which may have structural and catalytic functions. RNA polymerases from various sources have different requirements for Mg^{2+} and Mn^{2+}.

(e) Kinases

These enzymes catalyse the transfer of the terminal phosphoryl group of ATP directly to an acceptor. A distinction may be made between kinases where the metal ion is bound only to the substrate (Type 1, M–S–E) and those where it bridges enzyme and substrate (Type 2, S–M–E). The Type 2 situation has more exacting requirements in terms of the identity

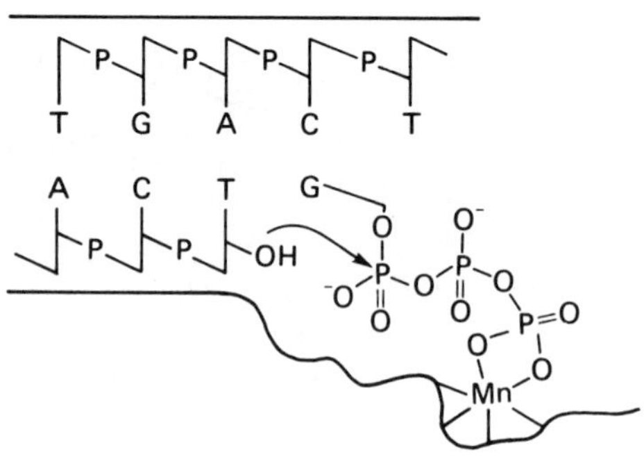

Figure 4.3 The DNA polymerase reaction.

of the metal ion; an example is pyruvate kinase, discussed briefly in Section 4.2.4(c). Kinases of low metal specificity usually belong to Type 1, where the metal usually has the role of charge neutralization.

Hexokinase catalyses the phosphorylation of glucose by ATP. Distance measurements on the complex of CrADP, glucose-6-phosphate and yeast hexokinase confirm the absence of direct coordination of the phosphoryl group of glucose-6-phosphate to the nucleotide-bound metal, but there is Van der Waal's contact between a phosphoryl oxygen and the hydration sphere of the metal ion (Pederson and Gupta, 1979).

Phosphoglycerate kinase is an enzyme of the glycolytic pathway, which catalyses the conversion of 1,3-diphosphoglycerate into 3-phosphoglycerate with production of ATP from ADP. Unlike most Type 1 kinases, there is substantial synergism in the binding of metal ion and nucleotide to the enzyme in the ternary complex. The metal nucleotide binds to the enzyme about two orders of magnitude more tightly than does the free nucleotide. High resolution X-ray diffraction data for the enzyme from horse muscle (Banks et al., 1979) and from yeast (Watson *et al.*, 1982) show it to consist of a single polypeptide chain organized into two widely separated globular domains. The binding sites for the Mg^{2+}-nucleotide complex and the phosphoglycerate are on separate domains, and it is suggested that the two domains come together through a hinge-bending mechanism. This expels the water from the site and allows the two substrates to interact. This mechanism has been investigated by biochemical (Adams and Pain, 1986) and structural techniques (Scheffler and Cohn, 1986) as well as by site-directed mutagenesis (Wilson *et al.*, 1987; Mas, Resplandor and Riggs, 1987).

Recently, ^{31}P NMR studies on enzyme-bound substrate complexes of 3-phosphoglycerate kinase have led to the suggestion that the high concentrations of ammonium sulphate in the crystals used for the structure determination may have altered the structure of the Mg^{2+}-nucleotide complexes of the enzyme, especially that of E. MgATP (Ray and Rao, 1988a). The structures of the various enzyme–substrate complexes have also been explored by observing the effect of paramagnetic Mn^{2+} and Co^{2+} on the relaxation rates of the ^{31}P nuclei (Ray and Rao, 1988b). This work shows that the Co(II) is directly coordinated to all the phosphate groups in the nucleotides in the complexes of CoADP and CoATP with the enzyme. The results for ADP are in accord with the X-ray data, but these data indicate that the Mg^{2+} is coordinated only to the γ phosphoryl group in ATP. This may be one manifestation of the effect of high levels of sulphate on the structure. Experiments on the quaternary complex E.MnADP. 3-P-glycerate give the distance between Mn(II) and the ^{31}P of 3-phosphoglycerate to be $11 \pm 0.3 \times 10^{-10}$ m, suggesting there is a gap between the oxygen atoms of the γ phosphoryl group of ATP and the carboxyl group of the 3-phosphoglycerate that is to receive the transferable phosphoryl group. This shows that the hinge is not closed under these conditions. It would be of great interest to measure this distance in the E.MnATP.3-P-glycerate complex, for which hinge bending should have taken place.

(f) Phosphoryl transfer other than from nucleotides

Many key steps in bacterial transport and biosynthetic processes involve phosphoryl group transfer and so depend upon Mg^{2+}. The *Escherichia coli* phosphoenolpyruvate-dependent phosphotransferase system is responsible for the concomitant translocation and phosphorylation of various sugars across the cytoplasmic membrane. Enzyme I catalyses the transfer of the phosphoryl group from phosphoenolpyruvate to a phospho-carrier protein HPr. The enzyme I-catalysed reaction proceeds by a phosphoenzyme I intermediate.

The active form of enzyme I is a dimer, whose stability and activity depends upon Mg^{2+} or Mn^{2+}. Dimer stability is also enhanced by univalent cations. One phosphoryl group is bound per dimer, but the mechanism by which the dimer is activated is not resolved. The reactions are described in Equations 4.5–4.7 (Hoving, Koning and Robillard, 1982; Misset and Robillard, 1982), where m and d represent monomeric and dimeric forms of the enzyne.

$$2E_I^m \overset{M^{2+}}{\rightleftharpoons} ME_I^d \underset{-PEP}{\overset{PEP}{\rightleftharpoons}} ME_I^d\ PEP \rightleftharpoons ME_I^d\ P + \text{pyruvate} \qquad (4.5)$$

$$ME_I^d\ P + HPr \rightleftharpoons ME_I^d + PHPr \qquad (4.6)$$

$$PHPr + \text{sugar}_{out} \underset{Mg^{2+}}{\overset{\text{Enzyme II}}{\rightleftharpoons}} HPr + P\text{–sugar}_{in} \qquad (4.7)$$

(g) Yeast enolase

Yeast enolase is a dimeric protein requiring divalent metal cations for activity. Two Mg^{2+} bind tightly to the apoenzyme in the absence of substrate, producing conformational changes, as shown by effects on the u.v. and CD spectra of the enzyme (Brewer, 1981). Two additional Mg^{2+} bind in the presence of substrate, 2-phosphoglycerate, resulting in catalytic activity. Other divalent cations will substitute for Mg^{2+} at both 'conformational' and 'catalytic' sites, with varying degrees of activity. Recent work has concentrated on the use of metal-for-magnesium probes.

Terbium(III) binds to yeast enolase in the absence of substrate, but does not give an active enzyme. Brewer *et al.* (1981) have shown that the fluorescence of the lanthanide is not readily observed on exciting the terbium enzyme at 280 nm, showing the absence of tyrosine or tryptophan ligands in the coordination sphere. However, the fluorescence is enhanced by excitation using 488 nm radiation, indicative of the involvement of carboxylate ligands. Other studies lead to the conclusion that three moles of water remain on the bound terbium, implying that the metal binds to six donor groups from the enzyme, provided that the terbium remains nine-coordinate.

The six-coordinate conformational site is confirmed by measurements on the Co(II) substituted enzyme (Rose, Dickinson and Westhead, 1984), which show that Co(II) ions at the conformational and catalytic sites are distorted octahedral and tetrahedral geometry respectively. Treatment with hydrogen peroxide leads to oxidation of the Co(II) at the catalytic site, but Co(II) in the conformational sites appears to be inaccessible to the oxidant.

(h) Glucose isomerase

Most glucose isomerases require a divalent cation for activity, but the requirement is rather non-specific, in that some are activated most effectively by Mg^{2+}, others by Mn^{2+}, and a third group by Co^{2+}. This does not appear to be of fundamental significance as the preference for metal changes with substrate in some cases. Kinetic studies on the enzyme from *Streptomyces griseofuscus* S–41 (Kasumi, Hayashi and Tsumura, 1982a) show that Mg^{2+} is the most efficient activator. Mg^{2+} and Co^{2+} are competitive towards each other, suggesting that they bind at the same site or very close together. However, Co^{2+} was most effective in stabilizing the enzyme against inactivation by heat or acid (Kasumi, Hayashi and Kumura, 1982b). The enzyme binds four Co^{2+} per molecule, of which three may be removed by treatment with EDTA without loss of activity. The additional binding of Co^{2+} could contribute to the enhanced stability. Such stabilization by Co^{2+} is also known for glucose isomerases from *Streptomyces phaeochromogenus* and *Streptomyces albus*, while Mn^{2+} stabilizes glucose isomerase from *Lactobacillus brevis* or *Bacillus coagulans*.

NMR studies (Young, Schray and Mildvan, 1975) on the activation of the *Lactobacillus brevis* enzyme by Mn^{2+} show that substrate interacts with the enzyme–Mn^{2+} complex.

(i) Enzymes with a specific requirement for Mn^{2+}

These are listed in Table 4.2. Notable among them is the phosphoglycerate phosphomutase from *Bacillus subtilis* (Watabe and Freese, 1979), which is associated with the germinating spore. Developing spores of *Bacillus megaterium* accumulate the substrate of this enzyme, 3-phosphoglyceric acid. The maintenance of dormancy may be due to the inactivation of this enzyme, possibly caused by low levels of Mn^{2+}. The transport of Mn^{2+} in *Bacillus* is discussed in Chapter 3.

The L-threonine dehydrogenase from *Escherichia coli* is involved in threonine utilization by the cell. It has four subunits, and is activated by Mn^{2+} and not by Mg^{2+}, with half-maximal stimulation at 9 μM Mn^{2+}. Each subunit binds one Mn^{2+}. The activation is dependent on a thiol (Craig and Dekker, 1986). In ESR studies on the Mn^{2+} content of *Staphylococcus aureus* cells, it has been shown (Ezra, Lucas and Russell, 1984) that manganese exists as Mn^{2+} under anaerobic conditions, and as Mn^{3+} in the presence of oxygen. Addition of 2-mercaptoethanol to an oxygenated suspension of *Staphylococcus aureus* cells results in the reduction of Mn^{3+} to Mn^{2+}. Similar results have been obtained by Archibald and Fridovich (1982a) for lactic acid bacteria. It appears, therefore, that the requirement for a thiol for activation by Mn^{2+} of L-threonine dehydrogenase may be due to the need to keep manganese as Mn^{2+}.

4.3.6 Magnesium and the control of the cell cycle

Calcium is usually regarded as the most important metal in the control of cell division but a case has been presented for assigning this role to Mg^{2+} in *Schizosaccharomyces pombe* and *Kluyveromyces fragilis* (Walker and Duffus, 1980, 1983). Synchronization of cell division can be induced by addition of magnesium to magnesium-exhausted cultures. Analysis of Mg^{2+} in cultures of both organisms synchronous shows a fall in magnesium concentration with cell growth, and a rapid influx of Mg^{2+} just before cell division. Walker and Zeuthen (1980) have reported that Mg^{2+} (and Ca^{2+}) are not responsible for triggering cell division in *Tetrahymena*.

4.3.7 Magnesium and manganese in photosynthesis

Magnesium has a unique involvement in chlorophyll, in which light energy is trapped to drive the chemical reduction of CO_2 to give carbohydrates. Chlorophyll is present in green algae (and higher plants), and in the cyanobacteria (or blue-green algae), which also have light-harvesting coloured proteins, the phycobiliproteins (Glazer, 1982). Bacteriochlorophyll functions in green bacteria and purple bacteria. The structures of chlorophyll and bacteriochlorophyll are shown in Figure 4.4, where it may be seen that Mg^{2+} lies in the centre of a tetrapyrrole, chlorin ring, slightly out of plane. The chlorin ring is unique among porphyrin-like ligands in having an alicyclic ring.

Carotenoids are usually found associated with bacteriochlorophyll. Their function is to protect from photooxidative damage by quenching triplet bacteriochlorophyll and singlet oxygen species. They also harvest light energy.

Photosynthetic centres are associated with membrane structures, the thylakoids, found in chloroplasts of plant cells and green algae, while in bacteria the photosynthetic membranes are derived by complex invaginations of the cytoplasmic membrane.

Bacteria such as *Halobacterium halobium* develop purple patches on their cytoplasmic membranes when grown in strong light, due to the

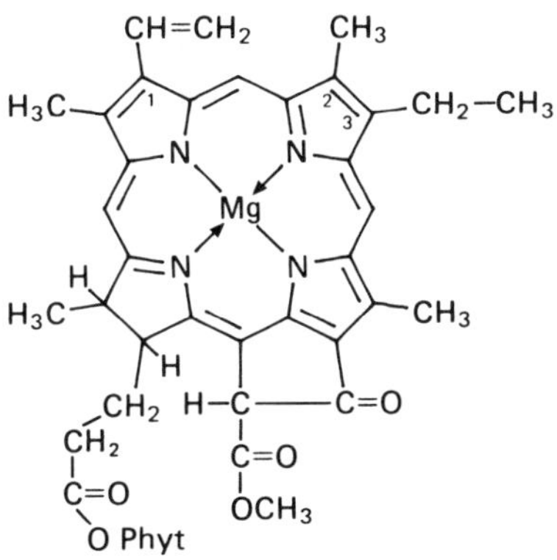

Figure 4.4 The structures of chlorophyll and bacteriochlorophyll.
Chlorophyll *b* : CHO at position 2
Bacteriochlorophyll *a* : $COCH_3$ at position 1
Bacteriochlorophyll *b* : $COCH_3$ at position 1, $CH{=}CH{-}CH_3$ at position 3.

synthesis of bacteriorhodopsin, which mediates the light-dependent formation of ATP. They are unique in having no chlorophyll pigments.

(a) Anoxygenic photosynthesis

The green and the purple bacteria use H_2S or organic compounds as a source of hydrogen atoms and so do not produce oxygen. The green bacteria have up to 2000 molecules of bacteriochlorophyll plus carotenoids at each reaction centre to act as antennae to trap photons. Their large antennae allow them to grow well in weak light. The purple bacteria have as few as 60 bacteriochlorophyll molecules in each antenna. The absorption of photons raises the chlorophyll to an excited state, and this excitation energy is then transferred *via* a sequence of photoactive molecules until it arrives at the photochemical reaction centre in an integral membrane protein (Glazer, 1983).

When cells of the chlorobiacean green bacteria *Chlorobium* and *Prosthecochloris* are sonicated at pH 10, a water-soluble bacteriochlorin *a*-protein is released. The structure has been determined at 2.8×10^{-10} m resolution (Matthews *et al.*, 1979). The protein from *Prosthecochloris aestuarii* is made up of three identical subunits. Each consists of seven bacteriochlorophyll *a* molecules inside a 'bag' of protein. The phytyl side chains of the bacteriochlorophylls form an inner hydrophobic core. The position and orientation of each bacteriochlorophyll molecule is determined by interaction with the protein through the phytyl side chains and the magnesium ions. The Mg^{2+} centres are thus five-coordinate – five of them are bound to histidyl residues, one to a backbone carbonyl oxygen and one to a water molecule. It is not certain if this type of structure is of general application to the organization of chlorophyll in photosynthetic membranes.

The reaction centre complexes of purple bacteria are typified by that of *Rhodopseudomonas sphaeroides*. It is made up of three protein subunits, which span the membrane, four bacteriochlorophyll *a*, two bacteriopheophytin molecules (a derivative of bacteriochlorophyll having no Mg^{2+}), one carotenoid, one ubiquinone, plus one Fe(II). The reaction centre from *Rhodopseudomonas viridis*, containing bacteriochlorophyll *b*, has been crystallized. This contains four protein subunits (Michel, 1982).

(b) Oxygenic photosynthesis

Cyanobacteria, green algae and higher plants have oxygen-evolving photosynthetic processes. They all contain Photosystems I and II at the reaction centre, which appear to be similar for all of these organisms.

Photosystems I and II respond to light of different wavelengths, and operate in concert, through the 'Z' scheme. Work on PSII has been concerned with the nature of the oxygen-evolving site, which is thought to be a manganese protein (Dismukes, 1986). ESR studies on spinach chloroplasts have led to the proposal that a pair (or possibly two pairs) of manganese ions, separated by a bridging ligand, are involved in oxygen evolution and that these undergo redox reactions. It seems reasonable that two water molecules should be bound to the dimeric manganese centre, for ready formation of an O–O bond, *via* bridged peroxide and superoxide. The arrangement of intrinsic and extrinsic proteins in PSII is still uncertain (Murata and Miyao, 1985).

4.4 CALCIUM

4.4.1 Introduction

Calcium is an extracellular cation; its exclusion from the cell is necessary to avoid precipitation of various calcium salts. Recently, the intracellular concentration of free calcium has been measured directly for *Escherichia* coli (Gangola and Rosen, 1987) and found to be 90 ± 9 nM. This value is very similar to that usually quoted for eukaryotic cells. The intracellular concentration of free calcium is maintained at this value for metabolically active cells irrespective of the external concentration of calcium, which was varied from 10 μM to 10 mM. This implies a regulatory role for intracellular Ca^{2+} in prokaryotes.

The role of calcium in micro-organisms is largely considered to be an extracellular one, and to involve such functions as the stabilization of cell walls and the activation of extracellular enzymes. However, it is now appreciated more fully that Ca^{2+} also has an essential intracellular role in the regulation or triggering of a range of cell functions (Harold, 1986). The appearance of relatively low levels of calcium ions in the cytoplasm, either by translocation from outside the cell or by release from internal stores, results in the activation of various cellular processes. In higher eukaryotes, the trigger action of calcium is usually mediated by a calcium-binding protein, which undergoes a conformational change on binding Ca^{2+}. Examples of such binding proteins are calmodulin, troponin C and S–100. It is of major significance, therefore, that such calcium-dependent regulatory proteins have now been identified from a range of yeasts and fungi. Their calmodulin-like properties have been confirmed by their ability to activate myosin light chain kinase and pea NAD^+ kinase (Nakamura *et al.*, 1984).

The ability of Ca^{2+} to bind well to biological molecules is exploited in its interaction with regulatory proteins, as noted above, and also in its role in stabilizing a range of structures. Calcium-binding proteins of mammalian origin sometimes contain the γ-carboxyglutamate residue. These are the 'gla proteins', such as those involved in blood coagulation. It is noteworthy that cytosolic and ribosomal fractions from *Escherichia coli* contain proteins with γ-carboxyglutamate residues, even though their function is as yet unknown. It appears probable that they too represent binding sites for calcium (Van Buskirk, Low and Kirsh, 1979).

There is a considerable literature on the biochemistry of calcium (Cheung, 1983; Kretsinger, 1980; Siegel *et al.*, 1980; Spiro, 1984), while the functions of Ca^{2+} in bacteria (Devés and Brodie, 1981) and fungi (Pitt and Ugalde, 1984) have been reviewed. Calcium ions may interfere with processes involving sodium ions in view of their similar ionic radii and the greater polarizing power of Ca^{2+}.

4.4.2 Structural effects produced by binding of calcium

(a) Effects on cell walls and membranes

Although several divalent cations are found in bacterial cell walls, Ca^{2+} seems to be of particular importance in stabilization, probably because of its special ability to bridge carboxylate groups. Thus, Ca^{2+} may be removed from the cell walls of *Pseudomonas aeruginosa* by treatment with EDTA to give osmotically fragile species, which may be restored by the addition of calcium. Chelation of calcium ions with EDTA or EGTA also allows previously excluded hydrophobic molecules to permeate into gram-negative bacterial cells. The Ca^{2+} ions are probably involved in cross-linking the lipopolysaccharide within the outer membrane. The binding of external Ca^{2+} to membranes can cause major changes in structural properties.

Calcium ions inhibit swarming in *Proteus mirabilis*. Swarming is associated with the development of elongated (60–80 μm), highly flagellate swarm cells. At calcium concentrations causing swarm reduction, the cells are not as elongated but are motile. Cells grown in Ca^{2+} are non-motile, being devoid of flagella. Flagella development during swarmer formation occurs before cell elongation. Presumably Ca^{2+} ions have cross-linked lipopolysaccharide, thus preventing flagella development (Cole and Smith, 1981).

Lactobacillus bulgaricus, the micro-organism used in the manufacture of dairy products, grows as long filamentous chains in the presence of

Mg^{2+} and the absence of any other cations. Dechaining of these cells to short rods occurs on the addition of Ca^{2+} (or Mn^{2+}, but not other divalent cations), although only at pH values less than 7.0. It is not clear if Ca^{2+} promotes the activity or binding of dechaining enzymes to teichoic acid sites (Wright and Klaenhammer, 1983). The morphology of the *Lactobacillus acidophilus* cell appears to be an important factor in the long-term stability of cultures at low temperatures. Cell types susceptible to freeze-damage have blebs protruding from the cell wall at regular intervals, which are membrane evaginations through the cell wall and likely targets for structural damage during freezing. Cells of *Lactobacillus bulgaricus* grown under low [Ca^{2+}] give cells showing such blebs, but addition of calcium to the medium eliminates bleb formation. Cells grown in the presence of calcium are more resistant to adverse effects on freezing. The most probable explanation is that calcium minimizes or repairs cell wall damage in *Lactobacillus* spp. (Wright and Klaenhammer, 1981).

(b) Stabilization of enzymes by Ca^{2+}

The best-known example of the stabilization of an enzyme by Ca^{2+} is thermolysin, isolated from the thermophilic organism *Bacillus thermoproteolyticus*. This enzyme is a zinc-requiring endopeptidase, but the thermal stability of the enzyme arises from the presence of four calcium ions per mole of enzyme. The Ca^{2+}-free protein shows a thermal transition at about 48°C, but the native enzyme is stable up to about 80°C. The Ca^{2+} ions prevent the unfolding of the protein by providing cross-links although it is not fully understood why the stabilizing effect is so marked in thermolysin. The structure of thermolysin has been determined by Matthews and Weaver (1974a) and Matthews, Weaver and Kester (1974b). The ligands are mainly glutamate and aspartate residues, with several residues bridging the two Ca^{2+} ions forming the 'double site' (in which the calcium ions are 3.8×10^{-10}m apart) (Table 4.3). This site is probably particularly important in stabilizing the protein structure.

Calcium also has an as yet undefined role in the stabilization of other enzymes. The enzyme nitrogenase is inactivated by oxygen, even though some nitrogen fixers, such as the cyanobacteria, actually evolve O_2. A number of strategies have been adopted to protect nitrogenase, but the unicellular cyanobacterium *Gloeocapsa* seems to have a mechanism involving calcium, in that treatment with EDTA (to deplete the cyanobacterial cells of Ca^{2+}) leads to loss of ability to reduce acetylene (which is a marker of nitrogen-fixing properties) (Hamadi and Gallon, 1981).

Calcium (Ca^{2+}) is essential for the retention of maximal O_2-evolving

Table 4.3 Some calcium sites

Enzyme/protein	Function of calcium	Site	Ligands
Calmodulin (internal)	Trigger	I	Asp-20, Asp-22, Asx-24, Thr-26, Thr-28, Glu-31
		II	Asp-56, Asp-58, Asx-60, Thr-62, Asp-64, Glu-67
		III	Asp-93, Asp-95, Asx-97, Tyr-99, Ser-101, Glu-104
		IV	Asn-129, Asp-131, Asp-133, Glu-135, Asn-137, Glu-140
Staphylococcus nuclease	Active site		Asp-19, Asp-21, Thr-41, $3H_2O$
Thermolysin	Structural	1	Asp-138, Glu-177, Asp-185, Glu-187, Glu-190, H_2O
		2	Glu-177, Asp-183, Asp-185, Glu-187, Glu-190, H_2O
		3	Asp-57, Asp-59, Glu-61, $3H_2O$
		4	Tyr-193, Thr-194, Ile-197, Asp-200, H_2O

activity in Photosystem II particles extracted from the cyanobacterium *Anacystis nidulans*. Photosystem II preparations extracted in Ca^{2+}-containing buffer have three polypeptides present, which are absent or much-decreased in preparations extracted in Mg^{2+}-containing buffer. These are probably Ca^{2+}-binding proteins. Thus the effect of Ca^{2+} in the retention of maximal O_2-evolving activity in the Photosystem II particles appears to be associated with the retention of these polypeptides within the particles, and in turn, possibly, to some structural role for Ca^{2+} (England and Evans, 1983).

4.4.3 Calcium as a regulator of cellular processes

Several requirements have to be fulfilled before Ca^{2+} can serve as a regulator of cellular processes: the intracellular calcium concentration must be kept at a low and constant value, calcium ions must be released into the cytoplasm in response to some stimulus, a calcium-binding protein is necessary to moderate the effect of the Ca^{2+} and finally there must be a pump to remove the Ca^{2+} back into the stores. All these requirements have been well studied in mammalian systems, and are now being clarified for microbiological examples.

(a) Control of intracellular calcium concentration

As noted in Chapter 3, most bacterial cells maintain low (<0.1 μM) intracellular levels of Ca^{2+} compared to the medium by the operation of specific Ca^{2+}-extruding transport systems. These involve Ca^{2+}/H^+ antiport, Ca^{2+}/Na^+ antiport, and, for *Streptococcus faecalis*, an ATP, Mg^{2+}, Ca^{2+}-dependent ion pump. In higher eukaryotic cells, a number of organelles act as calcium stores to help buffer cytoplasmic calcium levels (Carafoli and Crompton, 1978). It is essential that cytoplasmic Ca^{2+} homeostasis is maintained in micro-organisms, despite the possible wide variation in calcium concentration outside the cell. It appears that mitochondria in *Saccharomyces cerevisiae* are not able to take up calcium, and that the question is not yet resolved for fungi. Eilam, Lavi and Grossowicz (1985) have shown that most of the cellular Ca^{2+} in the yeast is precipitated or bound within the vacuole, being taken up by Ca^{2+}/H^+ antiport across the vacuolar membrane.

Attempts have been made to identify Ca^{2+}-rich subcellular compartments by fractionation and direct analysis for Ca^{2+}, but this approach has inherent uncertainty as Ca^{2+} may be released during the fractionation process. Pools of Ca^{2+} have been found in mitochondrial and microsomal fractions of the slime mould *Physarum polycephalum* by Holmes and Stewart (1979), and in vacuoles of the same species by the use of X-ray microprobe analysis (Kuroda and Kuroda, 1980). The main storage site for calcium in *Tetrahymena* appears to be membrane-bound volutin granules, which are made up of calcium-magnesium pyrophosphate (Nilsson and Coleman, 1977).

(b) General comments on calmodulin-like trigger proteins

The identification of calcium-binding trigger proteins in microbial cells is not surprising in view of their widespread distribution in other cells. A discussion of mammalian calmodulin will provide useful background material.

Calmodulin is a protein of molecular weight about 17 000, which regulates the activity of a number of calcium-dependent enzymes. Like other calcium-binding proteins, calmodulin has high binding constants for calcium and much lower constants for magnesium, with a discrimination for Ca^{2+} of about 10^3 to 10^4. This is essential, as Mg^{2+} is present in cells at high concentration. Calmodulin binds four calcium ions at rates close to the diffusion-controlled limit, apparently with the induction of sequential conformational changes, as shown by the application of

NMR, fluorescence, CD and Raman spectroscopy, leading to the exposure of a hydrophobic region on the surface of calmodulin (Burger *et al.*, 1984).

The likely ligands for the four Ca^{2+} ions have been assessed by NMR data and the use of sequence analogies (Levine and Williams, 1982). The calcium sites (Table 4.3) involve carboxylate and other oxygen functions from amino acid residues very close together in the protein. This situation holds for many trigger proteins, but is not the case for the calcium sites in phospholipase A2 and *Staphylococcus* nuclease (Section 4.4.4) where the Ca^{2+} is bound at the active site, This distinction must be significant. A useful comparison may be made with zinc enzymes, where the Zn^{2+} (which nearly always has a catalytic function) is also bound to donor groups from residues at widely differing parts of the protein, which have come together as a result of protein folding around the zinc. This explains why the Zn^{2+} does not dissociate from the protein. In contrast, the site for Ca^{2+} in the trigger proteins is already present in a loop of the protein. This means that Ca^{2+} will bind readily and dissociate easily from the protein – an essential feature for a trigger system, where response must be fast.

The binding of Ca^{2+} to calmodulin triggers major conformational changes, which are relatively slow, and the Ca^{2+}-calmodulin complex can then bind to and activate enzymes. The use of calmodulin allows a much more precise control mechanism, as calmodulin–enzyme interaction necessitates many points of contact.

The detailed nature of the conformational changes has been much studied (Dalgarno *et al.*, 1984), including the function of the inhibitor trifluoperazine, which functions by binding to helical surfaces, hindering the conformational change.

Calmodulin from *Tetrahymena* has eleven amino acid substitutions and one deletion compared to bovine brain calmodulin. These changes are found near or at the Ca^{2+}-binding sites, III and IV but ^{43}Ca NMR studies show that they have very little effect on the mobility of Ca^{2+} (Shimizu *et al.*, 1982).

Calcium-dependent regulatory proteins have been isolated from the yeasts *Saccharomyces cerevisiae* and *Candida albicans* (Hubbard *et al.*, 1982), the fungi *Agaricus campestris, Coprinus lagopus* (Nakamura *et al.*, 1984) and *Neurospora crassa* (Cox *et al.*, 1982), the slime mould *Dictyostelium discoideum* (Clarke, Bazari and Kayman, 1980) and *Escherichia coli* (Iwasa *et al.*, 1981). Amino acid sequences have been determined for calmodulin from *Saccharomyces cerevisiae* (Davis *et al.*, 1986) and *Schizosaccharomyces pombe* (Takeda and Yamamoto, 1987).

(c) *Calcium and calmodulin-like proteins in the regulation of reproduction and cell division*

The involvement of calmodulin with Ca^{2+} ions in the regulation of cellular processes is sometimes deduced indirectly from the observation that these processes are inhibited by trifluoperazine (TFP), the inhibitor of calmodulin.

Regulatory roles of calcium in yeast cells are well known. The ionophore A23187 blocks nuclear division in budding and fission yeasts (Penman and Duffus, 1975), while calcium controls the activity of a chitin synthetase in *Saccharomyces cerevisiae* (Fernandez, Correa and Cabib, 1982). Arising from the reports of a calmodulin-like protein in this yeast (Hubbard *et al.*, 1982), and the results of work on a Ca^{2+}-dependent mutant and TFP-dependent suppressor mutants (Ohya, Ohsumi and Anraku, 1984, 1986), it appears that Ca^{2+} ions and calmodulin play essential roles in the yeast cell division cycle. The haploid cell type MAT_{a} of *Saccharomyces cerevisiae* accumulates Ca^{2+} ions on incubation with a mating pheromone, via the specific induction of Ca^{2+}-transport activity. This Ca^{2+}-uptake may trigger localized cell-surface growth to give a mating tube (Ohsumi and Anraku, 1985). Calcium has also been implicated in the mating ability of *Kluyveromyces lactis* (Mas, Glender and Pina, 1980).

A Ca^{2+}/calmodulin control system has been implicated in the emergence and elongation of mating tubes in cells of *Rhodosporidium toruloides*, following the release of the mating pheromone Rhodotorucine A (Miyakawa *et al.*, 1986) and also in the development of reproductive structures in *Phytophthora* (Elliott, 1986). Divalent cations have been suggested to play a critical role in sexual signalling in *Chlamydomonas* (Snell, Buchanan and Clausell, 1982).

The identification of calmodulin in *Neurospora crassa* (Cox *et al.*, 1982), and the importance of calcium as a branching signal in the presence of A23187 (Reissig and Kinney, 1983) suggests that Ca^{2+}-calmodulin interactions are important. The branching response is reduced markedly by 10–30 mM cyclic AMP, suggesting that Ca^{2+} and calmodulin are involved with the Ca^{2+}-dependent cyclic nucleotide phosphodiesterase. Cyclic nucleotide levels are associated with cell division.

(d) *Calcium and the regulation of encystment of zoospores in fungi*

Zoospores are motile, flagellate cells, which act as the dispersive elements during the vegetative life cycle of fungi belonging to both the chytrids and the oomycetes. The dispersive phase of this cycle is terminated when the zoospore undergoes the processes of cyst formation (with loss of flagella and conversion to a spherical form) and germination to form a new mycelium. Zoospores are very sensitive to the cationic environment. Swimming biflagellate zoospores of *Phytophthora cinnamomi* encyst and germinate on exposure to Ca^{2+} (at mM concentration), while other cations induce encystment only (Byrt, Irving and Grant, 1982). Calcium (and other cations) could function by binding to the zoospore plasma membrane, leading to stabilization or destabilization, or cause an increase in the entry of calcium into the cell and trigger various intracellular events. The addition of the calcium ionophore A23187 results in Ca^{2+} stimulation of zoospore differentiation at much lower concentrations than that observed in its absence, suggesting that Ca^{2+} functions by altering the internal Ca^{2+} balance rather than by acting directly on the plasma membrane (Irving and Grant, 1984).

Monoflagellate zoospores of *Blastocladiella emersonii* respond differently to cations, with Ca^{2+} inhibiting and K^+ producing encystment. The use of ESR spectroscopy and spin labels has shown (Leonards and Haug, 1981) that Ca^{2+} interacts strongly with the glycolipids of the plasma membrane from zoospores of *Blastocladiella emersonii*. These effects are reversed by the addition of ATP and K^+ to isolated plasma membranes or by K^+ to zoospores *in vivo*, possibly as a result of the K^+-induced depolarization of the membrane potential. Thus, Ca^{2+} and K^+, in this example, may regulate zoospore differentiation by effects on the structural and electrical properties of the plasma membrane.

Encystment of zoospores is associated with loss of flagella and fusion of the plasma membrane with peripheral vesicles derived from the zoospore's gamma particles. The gamma particles contain the last enzyme of the biochemical pathway for chitin synthesis. The glycolipids comprise the major fraction of the plasma membrane and the gamma particles, and are also thought to be the intermediate for chitin synthesis. Thus, changes in the properties of the glycolipids at the cell surface could be important in regulating the encystment process.

(e) Bacterial sporulation and germination

Large amounts of calcium (up to 4% dry weight) are accumulated during sporulation in *Bacillus*, together with some Mn^{2+} and Zn^{2+} (Hogarth and Ellar, 1978, 1979). The calcium is thought to be present as a calcium dipicolinate precipitate in the spore, so providing a concentration gradient to assist the entry of calcium. The location of nearly all the total divalent cations in the core region has been demonstrated by high-resolution electron probe X-ray microanalysis (Johnstone, Ellar and Appleton, 1980; Stewart *et al.*, 1980a). The Mn^{2+} may also be bound to dipicolinate (Johnstone *et al.*, 1982).

The bacterial spores have a remarkable resistance to heat and various deleterious agents, and are able to germinate within minutes in response to agents such as L-alanine, heat and manganese (Levinson and Hyatt, 1955; Chapter 3). Specific changes in membrane structure and function occur during the transition from an actively-growing vegetative cell into a spore, and also during germination of the spore. Membrane fluidity changes during sporulation and germination of *Bacillus megaterium* have been examined by ESR and NMR measurements (Stewart *et al.*, 1980a). This shows that lipids in dormant spore membranes are in a highly ordered, rigid state, possibly due to interaction with calcium. Germination is associated with change to the more fluid environment characteristic of vegetative cells, which is accompanied by the release of calcium. A role for Ca^{2+} in controlling membrane fluidity requires an excess of calcium over the picolinic acid.

(f) Calcium and the regulation of locomotion

Paramecium usually swims forwards, driven by the beating of the cilia that cover its surface. On meeting an obstacle, the protozoan swims backwards briefly and then resumes forward motion in a different direction. This is brought about by an influx of Ca^{2+} ions through the excitable ciliary membrane. Permeability to Ca^{2+} is increased, and Ca^{2+} runs down the concentration gradient into the cell. The higher concentration of Ca^{2+} triggers an energy-dependent mechanism for reversing the direction of the ciliary beat. Normal forward swimming resumes when the intracellular $[Ca^{2+}]$ levels return to their usual values. Organisms whose outer membranes have been disrupted, swim forward at $[Ca^{2+}]$ $< 10^{-7}$M and reverse at higher values (Harold, 1986).

The target of the Ca^{2+} entering the cilia is a membrane-bound guanylate cyclase, which also involves calmodulin. The calmodulin can be dissociated from the guanylate cyclase by low levels of La^{3+} (Schultz and Klumpp, 1982). Deciliation of *Paramecium* releases proteins, including a Ca^{2+}-dependent ATPase (Riddle, Rauh and Nelson, 1982), which appears to be similar to a membrane-bound ATPase isolated by other workers, and which could be associated with the ciliary beating or with the Ca^{2+}-extrusion mechanism. A further role for a Ca^{2+}-ATPase could be the extrusion of trichocysts. These are exocytotic vesicles, lying just underneath the *Paramecium* surface membrane, which fuse with the plasma membrane in response to certain stimuli. This results in their contents being expelled. Calmodulin is a major component of extruded trichocysts (Rauh and Nelson, 1981).

Chemotactic behaviour in bacteria is controlled by calcium. Certain bacteria swim towards some chemicals and away from others. When heading in an undesired direction, bacteria show a tendency to tumble. This results in random changes in direction and the possibility that the organism will then swim in the desired direction. Tumbling is caused by clockwise rotation of the flagella, and swimming by anti-clockwise rotation. Tumbling is produced by influx of calcium (Ordal, 1977, 1980) as demonstrated by studies with A23187 for *Bacillus subtilis*. Added Ca^{2+} also regulates reversal of movement in chemotaxis (Baryshev, 1982) and phototaxis (Murvanidze and Glagolev, 1981) in *Halobacterium halobium*. Onset of the Ca^{2+}-induced tumbling or reversal modes requires methionine and a methylation of membrane proteins (Baryshev, Glagolev and Skulachev, 1982; Schimz, 1982).

Calcium ions are also involved in the control of direction of movement in other organisms.

4.4.4 Calcium and extracellular enzymes

Some extracellular enzymes are included in Table 4.3. They are usually digestive enzymes and have an absolute requirement for Ca^{2+}, to ensure that they are not activated prematurely inside the cell by Mg^{2+} after synthesis. Other proteins have to withstand the action of hydrolytic enzymes outside the cell. This is accomplished by protecting the peptide bond through protein folding brought about by binding of Ca^{2+}. Examples include carboxypeptidase from *Streptococcus griseus* and α-amylases from *Bacillus subtilis*, which catalyse the hydrolysis of glucosides. The phospholipids of intact bacteria are resistant to the action of both endogenous and exogenous phospholipases. However if *Escherichia coli* is exposed to bactericidal concentrations of the membrane-active antibiotic polymixin B, the envelope phospholipids are promptly hydrolysed by the cell's own phospholipases (A_1) in the outer membrane, which are activated by the Ca^{2+} released from the lipopolysaccharides on disruption of the membrane by polymixin B (Eisbach, Weiss and Kao, 1985).

Staphylococcus nuclease is an unusual Ca^{2+}-dependent enzyme in that the Ca^{2+} appears to have a catalytic role rather than a structural one.

(a) Staphylococcal nuclease

This extracellular 5′-phosphodiesterase of *Staphylococcus aureus* catalyses the hydrolysis of DNA or RNA to give 3′-mono and dinucleotides. The structure has been determined at 1.5×10^{-10}m resolution by Cotton Hazen and Legg (1979) for the ternary enzyme–Ca^{2+}-inhibitor complex with thymidine 3′,5′bisphosphate (pdTp). The Ca^{2+} has an octahedral environment, with three positions filled by donor atoms from Asp-21, Asp-40 and Thr-41, two by water molecules and the sixth by the 5′-phosphate of the inhibitor. One water molecule serves as a bridge between the carboxylate group of Glu-43 and one of the 5′ phosphoryl oxygen atoms. The Glu-43 is proposed to function as a general base, promoting reaction by the abstraction of a proton from the attacking water molecule. One of the coordinated water molecules lies between the Ca^{2+} and the Glu-43 carboxylate, so preventing it from acting as a ligand (Figure 4.5). The X-ray structure also shows that a number of hydrogen bonding interactions occur which suggest that pdTp and Ca^{2+} are precisely and rigidly locked into position at the hydrolytic site.

Site-specific mutagenesis techniques have been used to replace the Asp-40 residue in staphylococcal nuclease by glycine (Serpersu, Shortle and Mildvan, 1986). The binding of Ca^{2+} and Mn^{2+} to the wild-type and mutant enzyme D40G has been investigated by ESR, NMR and kinetic methods. The D40G mutant forms a (5 ± 2)-fold weaker binary complex with Ca^{2+}, and binds Mn^{2+} three times weaker than does the wild-type enzyme, as expected from the structural change at residue 40. The binding of substrate DNA amplifies these differences in metal affinity.

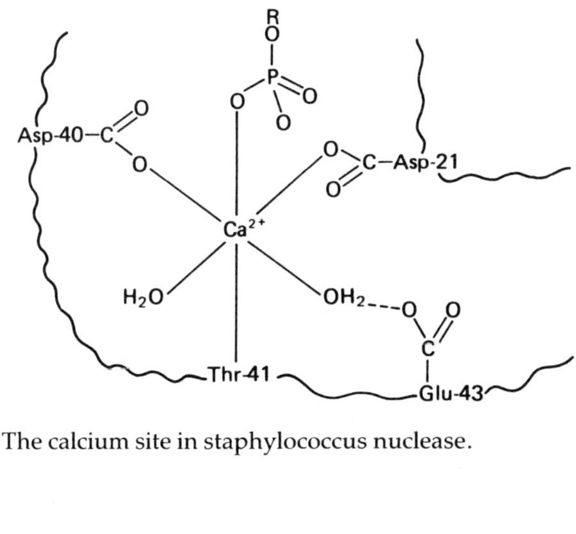

Figure 4.5 The calcium site in staphylococcus nuclease.

The maximal velocity at saturating levels of both Ca^{2+} and DNA was 31-fold lower for the mutant enzyme, probably due to a relaxation of the structure of the active ternary enzyme-Ca^{2+}-DNA complex. Decrease of activity could result, for example, from an increase in the distance from Ca^{2+} either to the phosphodiester group or to the Glu-43 residue.

(b) Bacterial calmodulin-sensitive adenylate cyclase

The adenylate cyclases from mammalian sources and certain bacteria present an interesting contrast. The calmodulin-sensitive adenylate cyclase from bovine cerebral cortex requires Ca^{2+} for calmodulin stimulation (Andreasen *et al.*, 1983), whereas the adenylate cyclase released by *Bordetella pertussis*, the whooping cough pathogen, does not require Ca^{2+} as an obligatory participant in activation by calmodulin. This suggests that calmodulin can activate enzymes through interaction with different binding domains. Calcium enhances the sensitivity of the cyclase to calmodulin, probably through occupancy by Ca^{2+} of the two C-terminal Ca^{2+}-binding sites, as shown by studies using tryptic fragments of calmodulin (Wolff, Newton and Klee, 1986).

The adenylate cyclase from *Bordetella pertussis* has been characterized, and shown to be the most active cyclase known so far (Shattuck, Oldenburg and Storm, 1985). The release of this enzyme contributes to the pathogenesis of the disease, as mutants deficient in it are avirulent. The adenylate cyclase may enter animal cells and increase intracellular cyclic AMP (adenosine cyclic 3′,5′-phosphate) levels. The *Bacillus anthracis* oedema factor is also a calmodulin-stimulated adenylate cyclase, that appears to enter animal cells and raise intracellular cAMP levels (Leppla, 1982).

4.4.5 Deposition of calcium compounds: biomineralization

The importance of the interactions between calcium ions and anionic polymers such as polysaccharides has been stressed. Such interactions also appear to be important in the deposition of calcium minerals in certain micro-organisms, which is also a consequence of the low solubility of many calcium compounds.

Biomineralization may involve precipitation of calcium compounds with cell metabolites or the deposition of calcium compounds, usually carbonate, on an organic structural framework such as a polysaccharide matrix. This matrix-dependent biomineralization process sometimes leads to the formation of minerals having unique crystal habits and some remarkably complex, but symmetrical, structures. Matrix-mediated deposition processes are also referred to in Chapter 5 (for magnetite in magnetotactic bacteria) and Chapter 7 (the laying down of minerals).

The coccolithophorids, a group of photosynthetic marine algae, are characterized by the formation of calcified body scales of coccolith plates composed of calcium carbonate. These organisms have contributed largely to the formation of deep sea calcium carbonate ores and, in the geological past, to the formation of vast rock deposits.

The morphology of the coccolith plate is species-specific, and a wide range of complex structures have been described. The difference in crystal structure from normal calcite produced by inorganic precipitation results from the ability of the cell to regulate the crystallization of calcium carbonate. *Coccolithus (Emiliania) huxleyi* has an intracellular apparatus for coccolith formation located in a large, long vesicle close to the nucleus. Small rhombohedral crystals are deposited inside the vesicle in the early stages of coccolith formation, while the coccolith plate obtains its final shape within one hour and is then excreted. Coccoliths treated with EDTA to solubilize $CaCO_3$ leave a residue made up of a soluble polysaccharide and an insoluble glycoprotein. The polysaccharide is complex, containing at least fourteen different monosaccharides, and binds Ca^{2+} strongly. This polysaccharide appears to inhibit crystal growth, and must differ from the intracellular polysaccharide matrix (Bowman, Allen and Slayman, 1983). Acidic polysaccharides have been isolated from coccoliths from other organisms. Biological and geological perspectives of biomineralization provide a topic of increasing interest (Westbroek and De Jong, 1983).

4.5 SUMMARY

The IA and IIA cations Na^+, K^+, Mg^{2+} and Ca^{2+} are distributed selectively, with K^+ and Mg^{2+} concentrated within the cell and Na^+ and Ca^{2+} concentrated outside the cell. Concentrations of K^+ inside the cell vary considerably from organism to organism but are usually about 0.25 M. Potassium cations act as activators of a range of intracellular enzymes, where their role often involves the control of enzyme conformation, and they are also involved in the stabilization of intracellular structures such as ribosomes, sometimes in association with Mg^{2+}. However, only a small part of the total potassium concentration is bound to sites on these structures: the major part is effectively free in solution and is necessary to maintain the osmotic pressure inside the cell at values slightly greater than that of the medium.

The sodium gradients (low inside) across the cell membrane are set up by several processes, including Na^+/H^+ antiport, an ATP-driven sodium-extruding pump and other pathways. These gradients are used by the cell in the co-transport of nutrients into the cell. Alkalophilic bacteria import Na^+ in association with non-essential solutes so that the resulting efflux of Na^+ via the Na^+/H^+ antiport maintains a lower intracellular pH than that of the medium.

The intracellular concentration of Mg^{2+} lies in the range 10–20 mM. It is bound largely to ADP, ATP, nucleic acids and ribosomes, appropriate for its major roles in the activation of enzymes involving hydrolysis and transfer of the phosphate group and in the stabilization of ribosomes, nucleic acids and other intracellular structures. Sometimes Mn^{2+} is present as an activator rather than Mg^{2+}, often reflecting their relative concentrations in the assay medium, but some enzymes have a specific requirement for Mn^{2+} and in a few cases competition between these two cations is of regulatory significance for enzymes. Both magnesium and manganese have well-known roles in photosynthesis, with Mg^{2+} present in chlorophyll and Mn^{2+} involved in the oxygen production site.

The concentration of Ca^{2+} within *Escherichia coli* has recently been determined as 10^{-7} M, a value very similar to that found for eukaryotic cells. This confirms the role of calcium as a regulatory and trigger element in bacteria, as is well known in other systems. Calcium is involved, for example, in the regulation of cell division, sporulation locomotion and chemotaxis, and may well exert this control through the mediation of calcium-binding proteins. Calcium also has important extracellular roles, in the stabilization of cell walls and extracellular enzymes. *Staphylococcus* nuclease is an example of an extracellular enzyme where calcium has a rare catalytic role rather than a structural one. Calcium ions are readily precipitated as carbonate or phosphate. In the case of certain marine algae this process takes place on cellular matrices such as polysaccharides, giving regular, beautifully complex structures known as coccolith plates. This process is of geological significance, and has been responsible for the laying down of vast calcium carbonate deposits.

5

Special roles for the transition elements and zinc

Gold is for the mistress – silver for the maid – Copper for the craftsman cunning at his trade! 'Good!' said the Baron, sitting in his hall; 'But iron – cold iron – is master of them all.'

Rudyard Kipling

Nuntio vobis gandium magnum. Habemus ferrum. *I announce to you a great joy. We have iron.*

Primo Levi
The Periodic Table (1984)

5.1 INTRODUCTION

In this chapter, we examine the diverse roles played by iron, copper, manganese, cobalt, molybdenum, nickel and zinc in microbial biochemistry and physiology. The most striking biological functions of the transition elements are in the binding and activation of small molecules and generally in redox catalysis. The concept of trace element catalysis is not new; at the beginning of this century, Bertrand and Javallier ascribed biocatalysis to trace elements, shortly before the convincing demonstration of the necessity of trace metals for microbial growth (Chapter 2) and long before our current concepts of enzymes and coenzymes and the burgeoning interest in them by inorganic biochemists.

With the exception of iron, the metals considered here are, and always have been, of low abundance on Earth, by comparison with those metals such as sodium, potassium, magnesium and calcium, which form more soluble compounds (Chapter 4). Consideration of the special properties of transition metals (Chapter 1), their abundance and availability in nature and the microdomains in which they are segregated in biological systems leads to the view that their properties are exploited in full in biology (Williams, 1981). For example, iron is the metal usually selected in electron transfer, not only because of its abundance, but also for its ability to interact with O_2, H_2 and H^-/H^+. Molybdenum is associated with oxo group transfer (but also with nitrogen fixation, where its appropriateness is less clear and may have been over-stressed), manganese with O_2 release in photosynthesis and zinc has become the major Lewis acid catalyst. Examples of two or more different metals having overlapping functions are rare: however iron and copper both function in oxygen binding and electron transport; copper, iron and manganese act in superoxide elimination; and molybdenum, vanadium and iron perform similar functions in nitrogen fixation.

In enzymes, the transition metal is generally trapped, deeply buried in the pocket of a protein, such as in redox catalysts and centres of energy transduction, but with access for small molecules such as H_2O or gases. The properties of the metal may be profoundly influenced by its binding by biological ligands. Thus, metals bound in unusual or distorted geometries, or in more open sites, will have enhanced reactivity. The properties of a metal may be finely 'tuned' by the protein, as is well illustrated by the case of iron, which displays such diversity in enzyme catalysis (Williams, 1981). Although many biological molecules contain a single metal atom (e.g. iron in myoglobin and the subunit of haemoglobin, in some cytochromes, and zinc in carbonic anhydrase), others of great importance and interest possess functional centres containing two or more metal atoms placed close (0.5–2 nm) together, often sharing a bridging ligand. The centres involved are generally haem iron, FeS clusters, Mn and Cu sites and their disposition allows sequential or concerted electron transfer to a bound ligand, particularly in the activation of small molecules such as gases (Hughes, 1985).

In addition to these roles in enzyme catalysis, some transition metals play direct nutritional roles. Thus iron (arguably again due to its abundance in nature) has been adopted as both terminal oxidant and primary reductant in the respiration of some bacteria. Manganese(II) is also oxidized by certain bacteria but the nutritional significance of this is unclear. The most recently described role for iron in micro-organisms, that of biological compass, further illustrates the diversity of functions adopted by this metal.

5.2 PROTEINS CONTAINING IRON AS HAEM

5.2.1 Diversity and synthesis

With the possible exception of iron-sulphur proteins, haem proteins are the most ubiquitous and well-understood of electron transfer components. Haem is composed of four pyrrole rings joined by –CH=bridges to form a complex planar molecule, porphyrin, at the centre of which is a single iron atom (Figure 5.1). The metal cycles between the reduced (Fe(II)) and oxidized (Fe(III)) states and transfers one electron. The iron centre is usually six-coordinate, with the four equatorial coordination sites occupied by the nitrogens of the pyrrole ring, and the fifth and

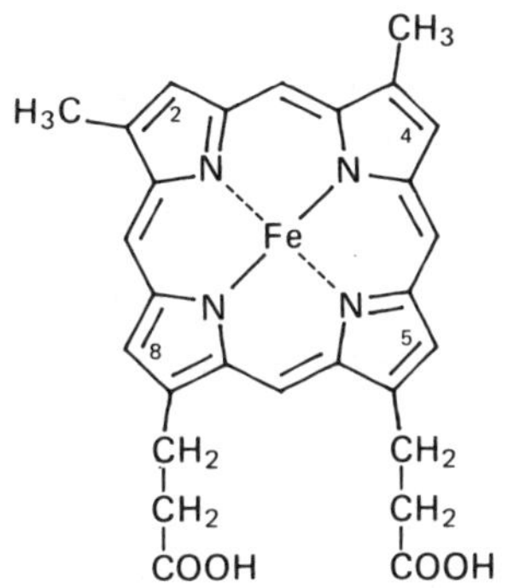

HAEM	C-2	C-4	C-5	C-8
a	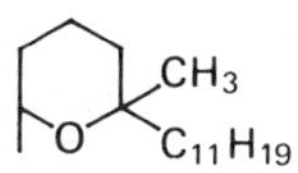	$-CH{=}CH_2$	–	$-CHO$
b	$-CH{=}CH_2$	$-CH{=}CH_2$	$-CH_3$	$-CH_3$
c	$-CH(S-)-CH_3$	$-CH(S-)-CH_3$	$-CH_3$	$-CH_3$
d	$-CH(OH)-R$	$-CH{=}CH_2$	$-CH_3$	$-CH_3$

Figure 5.1 The haem structures of the cytochromes. Note that haem *d* is a dihydroporphyrin in which the 8, 9 double bond is saturated. Reproduced with permission from Jones and Poole (1984).

Table 5.1 Classification of haemoproteins by function

Class/function	Examples
Cytochromes	6-coordinate electron transfer proteins Cytochrome oxidases
Oxygen carriers	Myoglobin and haemoglobin
Oxygenases	Monooxygenases (e.g. cytochrome P-450) Dioxygenases
Hydroperoxidases	Peroxidases Catalases

sixth sites often filled with donor atoms from side chains in the protein. The important donors in these axial positions are the imidazole (N) and –SH groups of histidyl or cysteinyl residues, respectively. The iron is usually, but not always, in the low-spin configuration, reflecting the presence of these strong ligands. The low-spin configuration is an appropriate one for the facilitation of electron transfer (Chapter 1). Other positions around the periphery of the haem are available for substitution and it is these substituents that are primarily responsible for the spectroscopic and functional properties of the haemoprotein.

Haem proteins perform a variety of functions in electron transfer and ligand binding. It is notable that closely similar iron porphyrins, when combined with different proteins, catalyse diverse reactions indicating the role played by the protein in determining haem function. This functional diversity provides a convenient means of classification (Table 5.1) and a framework for the first part of the present chapter. An alternative, but less frequently accepted classification, is a structural one, with regard to the type of haem present and the spin state of the haem. A detailed description of bacterial haem proteins using the latter criteria has recently been presented by Wood (1984).

Iron(II) is inserted into protoporphyrin IX in the last step of haem synthesis, a reaction catalysed by ferrochelatase (protohaem ferrolyase; EC 4.99.1.1). A recent review of the enzyme is available (Jones, 1983). This enzyme is tightly bound to the cytoplasmic membrane of prokaryotes or the inner membrane of mitochondria and chloroplasts in eukaryotes. It has been studied in a number of micro-organisms; in some cases, e.g. the enzyme from *Spirillum itersonii* and yeast (Camadro and Labbe, 1982), it can also incorporate other metal ions, such as Co(II), Zn(II), Ni(II) or Mn(II) into the porphyrin. Zinc protoporphyrins are common in yeast (Pretlow and Sherman, 1967).

5.2.2 Cytochromes

(a) Discovery and nomenclature

Cytochromes were discovered almost a century ago by MacMunn but realization of their importance had to await their rediscovery some forty years later by Keilin. He demonstrated the presence and heterogeneous nature of 'cytochrome' ('cell pigment') in various animal, yeast and bacterial cells and pointed out the apparent lack of cytochromes in obligately anaerobic fermentative bacteria such as certain *clostridia*. Spectroscopic examination of the characteristic absorption bands of cytochromes showed that they were able to undergo reversible changes on oxidation and reduction and pointed to a role in aerobic respiration. Warburg, despite disputing this proposal at first, subsequently measured the relative ability of different wavelengths of light to relieve the inhibition of respiration by CO (Section 5.2.2(b)) and thus obtained the photochemical action spectrum, directly implicating a cytochrome as the oxygen-reactive species. The examination of eukaryotic and prokaryotic micro-organisms, especially yeasts and bacteria, played a seminal role in these early studies. Later, cytochromes were also shown to function in anaerobic respiration (i.e. electron transfer to terminal acceptors other than O_2, such as sulphate) and in photosynthesis.

Broadly following the original nomenclature of Keilin, four classes of cytochromes are generally recognized (Table 5.2). Detailed description of the spectral properties of the various types and the methods of analysis (Jones and Poole, 1984) are outside the scope of this book. In general the reduced minus oxidized difference spectrum of a cytochrome exhibits three major absorption bands in the visible region, namely the α band (in the approximate range 545–650 nm), the β band (520–560 nm) and the γ or Soret band (410–450 nm).

The E_m (pH 7.0) values of the cytochromes lie mostly in the range −100 to +500 mV but a few specialized cytochromes in bacteria have much lower or higher potentials (Section 5.2.2(c)i).

(b) Cytochromes of eukaryotic micro-organisms

Broadly speaking, the composition and sequence of cytochromes (and other components of the electron transport chains) involved in photosynthesis and respiration in eukaryotic micro-organisms are similar to those of higher organisms (Hipkins, 1983; Lloyd, 1974).

Only mitochondrial cytochrome *c* oxidase is covered in some detail

Table 5.2 Classification of cytochromes by haem type

Cytochrome	Haem	Examples
a	*a* (formyl side chain)	a_1, aa_3
b	*b* or protohaem	various non-autoxidizable cytochromes *b*, cytochrome *o*, cytochrome P-450
*c**	*c* or mesohaem, covalently bound to apoprotein	
Type I	Single haem attached to protein close to N-terminus end	*c*, c_2, c_4
Type II	Single haem attached to protein close to C-terminus; high spin	*c*′†
Type III	Multiple haems	c_3
Flavocytochromes		
d (a_2)	chlorin	cytochrome oxidase *d*
d_1	modified chlorin	cd_1

* The cytochrome *f* involved in non-cyclic electron transport is a *c*-type cytochrome.
† The primed letter indicates that, in both the oxidized and reduced forms, the haem prosthetic group is not in a haemochrome linkage (Enzyme Nomenclature, 1978). See Pettigrew and Moore (1987) for a review of *c*-type cytochromes.

here because of its complex complement of metal centres and the multidisciplinary approach that has characterized its study. The terminal step of the aerobic respiratory chain in most eukaryotes is the reduction of O_2 to water, with minimal release of the toxic products of the partial reduction of oxygen, O_2^- (superoxide) and O_2^{2-} (peroxide). This process is catalysed by cytochrome oxidase. This membrane-bound enzyme accepts electrons from cytochrome *c* in the respiratory chain and transfers them to oxygen, the oxidoreduction being accompanied by vectorial translocation of protons across the inner mitochondrial membrane (Figure 5.2). The great interest in cytochrome oxidase shown by inorganic biochemists stems from the 4-electron reduction of oxygen that it catalyses and the enzyme's complement of (at least) four redox-active metal centres, viz. two *a*-type haems and two copper ions. Cytochrome *a* and Cu_A are magnetically isolated and detectable by electron spin resonance (ESR) in the iron(III) and copper(II) states, respectively.

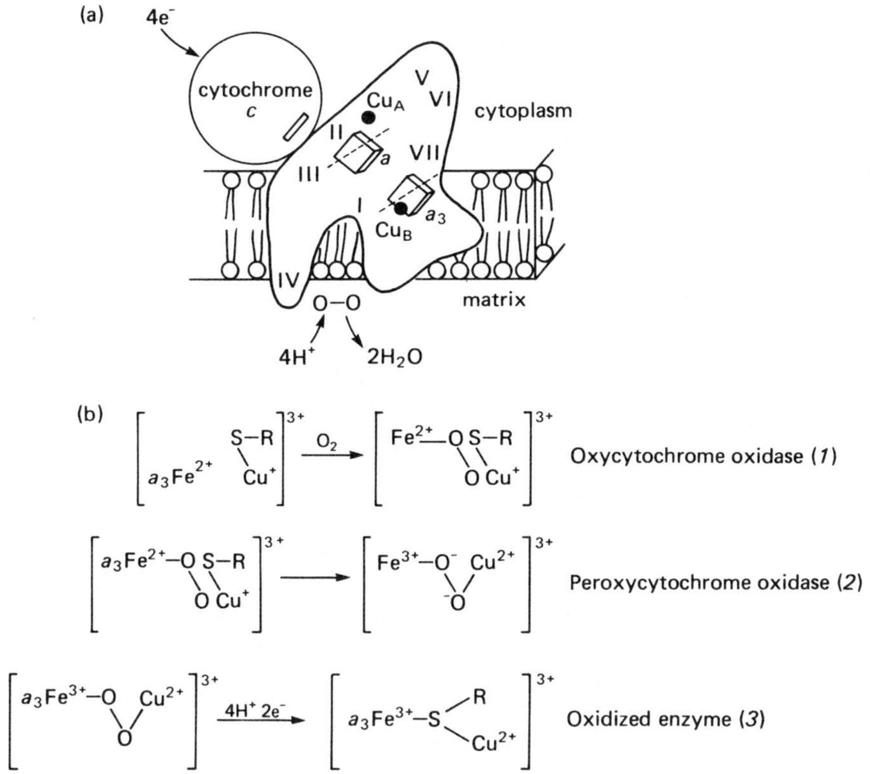

Figure 5.2 Mitochondrial cytochrome *c* oxidase. (a) shows the structure and topography of the enzyme embedded in the lipid bilayer of the mitochondrial inner membrane. The topography of the subunits (I to VII) is based on chemical and immunological labelling studies. The square planar structures are the *a* haems and the circles (●) the copper atoms. Cu_B and haem a_3 are in very close proximity, probably coupled by a bridging ligand. ESR studies show that haem *a* is close to the cytoplasmic side. Optical and ESR studies of vesicles and multilayers show that both haems are oriented with their planes perpendicular to the membrane plane. Estimates of haem *c*–haem *a* (25×10^{-10}m) and haem *a*–haem a_3 (10×10^{-10} m) have been made. (Slightly modified from Poole *et al.*, 1985; reproduced by permission of the Society for General Microbiology, publishers of *Microbial Gas Metabolism*, Special Publications Series, vol. 14.) (b) shows a possible sequence of events in the reduction of oxygen, ignoring haem *a* and Cu_A. The presence of Cu_B close to the iron of haem a_3 allows rapid 2-electron donation to the peroxide state (species 2). A further 2-electron donation gives the oxidized enzyme and water (species 3; see also (a)). Reproduced with permission from Greenwood and Hill (1982).

Cytochrome *a* is a six-coordinate low-spin complex and is the immediate recipient of electrons from cytochrome *c*. The second *a*-type haem, called a_3, and the second copper (Cu_B) are ESR-'silent' in the oxidized enzyme probably because they are very close and their spins are anti-ferromagnetically coupled. Cytochrome a_3 is the component that binds O_2 and the classical respiratory inhibitors such as CO and cyanide. It is five coordinate, and high spin in both redox states of the iron. The presence of a third copper, of unknown function, has recently been demonstrated in the cytochrome oxidases of *Paracoccus denitrificans* and ox heart. The mitochondrial complex also contains 1 Zn and 1 Mg (Bombelka *et al.*, 1986; Steffens, Biewald and Buse, 1987).

The reduced enzyme has a high affinity for O_2 ($K_m \leqslant 1\,\mu M$) and the reaction with O_2 is extremely rapid ($\approx 10^8\ M^{-1}\ s^{-1}$ at 20°C). Ingenious techniques involving stopped-flow rapid kinetic spectrophotometry and initiation of the reaction at sub-zero temperatures have led to the identification of a number of intermediates and steps in the reaction (for examples of these techniques applied to a yeast, see Poole, Lloyd and Chance, 1974, 1979). Thus, O_2 is first bound to the ferrous haem a_3 to give a first 'oxygenated' intermediate similar to oxyhaemoglobin. This may be followed by rapid electron transfer from a_3 and Cu_B to the ligand giving a μ-peroxo-bridged species. Such a two-electron transfer overcomes the energetic constraints of a one-electron transfer, avoids superoxide formation and also serves to trap the oxygen irreversibly. Unlike other metal-containing proteins that react with oxygen to form a peroxo intermediate (such as haemocyanin, the copper-containing O_2-transporting protein of molluscs and crustacea), cytochrome oxidase has 'electron ballast' in the form of haem *a* and Cu_A, which transfer two further electrons giving water (Figure 5.2).

The subunit composition of the enzyme is complex; at least seven and as many as thirteen different subunits have been reported. Controversy continues over the number and role of the subunits, but there is persuasive evidence that subunits I and II bind the redox-active haems and coppers. Interestingly, these are the subunits most resembling those found in the analogous bacterial oxidases (Poole *et al.*, 1985; Poole, 1988) and are encoded by genes on mitochondrial DNA.

Many eukaryotic micro-organisms and plants (Lloyd, 1974; Degn, Lloyd and Hill, 1977) display cyanide-resistant respiration, alternative to the main cyanide-sensitive, cytochrome oxidase-mediated respiratory pathways. Cyanide-insensitive respiration is well known to be inhibited by hydroxamic acids (e.g. salicylhydroxamic acid, SHAM) which are established metal chelators, especially of Fe(III). More direct evidence for an involvement of iron in such alternative oxidase pathways comes from

(i) the reversal by added Fe^{3+} (and other metals) of hydroxamate-inhibited respiration (Henry and Nyns, 1977) and (ii) the requirement for Fe(III) or, in some cases, Fe(II), in the biogenesis of the cyanide-insensitive pathway in yeast (Henry, Bonner and Nyns, 1977) and protozoa (Hryniewiecka, Jener and Michejda, 1980).

(c) Cytochromes in bacteria

(i) Overview Cytochromes play indispensable redox roles in most bacteria. The redox potential of the iron-containing haem can be so dramatically, yet finely, modified by the protein that some cytochromes have redox potentials appropriate, for example, to anaerobic electron transport to sulphate (the multihaem cytochromes c_3 having mid-point potentials close to -300 mV) or to O_2 reduction in certain autotrophs where the mid-point potential of the oxidase can be as high as $+700$ mV at pH 3.2. For reasons of space (and prejudice on the part of the authors), we deal here briefly only with the respiratory roles; Jones (1982) provides an excellent overview of cytochromes in both bacterial respiration and photosynthesis. Cytochrome P-450 (an oxygenase) is covered in Section 5.4.2(b).

The respiratory cytochrome systems of aerobic chemoheterotrophs and facultative phototrophs usually consist of several non-autoxidizable *b* and *c*-type cytochromes plus one or (more frequently) several CO-binding cytochromes that may be regarded as potential, autoxidizable cytochrome oxidases. There is considerable variation both amongst species and as a result of growth conditions in the complexity and organization of the aerobic pathways. All exhibit branching at the dehydrogenase level and most have branched cytochrome pathways as well. Figure 5.3 shows the potential for complexity in the respiratory chain of *Escherichia coli*, and the involvement of other metal-containing proteins such as the dehydrogenases (Section 5.3.5(d)), fumarate reductase and nitrate reductase (Section 5.3.5(c)).

(ii) Cytochrome oxidases At least five classes of cytochrome oxidase have been described in bacteria, cytochromes aa_3, a_1, d, cd_1 and o (Poole, 1983; Poole *et al.*, 1985; Poole, 1988). The oxidases constitute a distinguishing feature in bacterial classification and form an interesting comparison with the mitochondrial cytochrome *c* oxidase (Section 5.2.2(b)). They are a structurally and functionally diverse group of electron transport proteins (Table 5.3) with, by definition, the ability to reduce oxygen.

The aa_3-type oxidases are analogous to the mitochondrial cytochrome

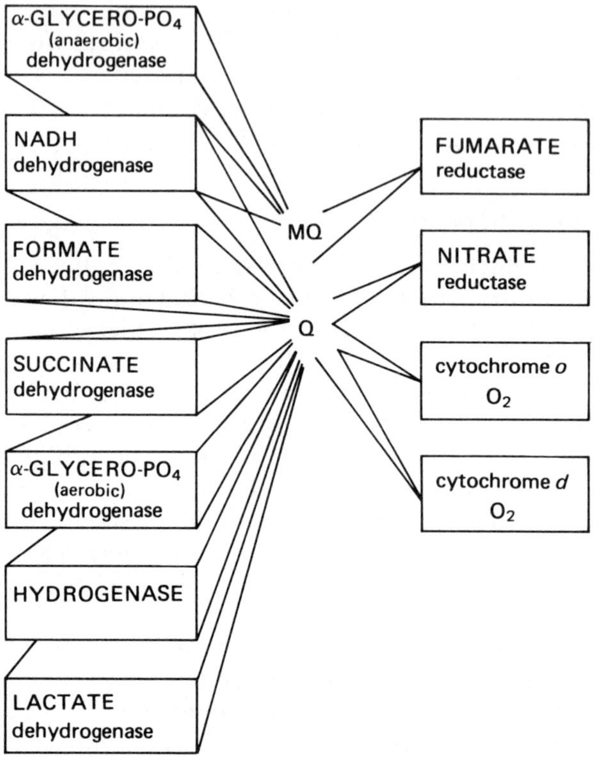

Figure 5.3 Branching in the respiratory chain of *Escherichia coli* and the involvment of metals in the electron transfer complexes. The main primary dehydrogenases are shown on the left. Of these, several contain non-haem iron, namely the anaerobic α-glycerophosphate dehydrogenase, possibly NADH dehydrogenase, formate dehydrogenase (which also contains molybdenum and selenium), succinate dehydrogenase, and hydrogenase. The major oxidase complexes on the right are those utilizing oxygen (terminated by cytochromes *o* and *d*, the former containing also copper), nitrate (the reductase containing cytochrome *b*, molybdenum and iron-sulphur centres) or fumarate (the reductase containing iron-sulphur centres). The dehydrogenases and oxidase complexes are shown linked by the quinones ubiquinone (Q) and menaquinone (MQ). Reproduced with permission from Ingledew and Poole (1984).

c oxidase in having the same redox-active centres, being membrane-bound and probably having similar functional properties with respect to proton translocation and oxygen reduction. Although the oxidase purified from *Paracoccus denitrificans* contains, in addition to two *a*-type haems and two or three coppers, substoichiometric amounts of Mn,

Table 5.3 Overview of the major classes of bacterial cytochrome oxidases

Oxidase	Redox centres	Visible absorption maxima (reduced form; nm)		No. of subunits	Oxygen reaction	
		α	γ		K_m (μM)* for O_2	Kinetic competence to act as terminal oxidase
aa_3, †caa_3	a; a_3; 2Cu + c	600-605	400–445	2–3	4–7	+
a_1	a-type haem (s); copper, probably absent	585–596	≈ 440	?	(0.33?) in doubt	variable
o	protohaem(s); copper	555–565	≈430	1–4	0.1–6.5	+
d	chlorin haem(s) with cytochrome b but not copper	≈ 630	w	2	0.018–0.35	+
cd_1	2c:$2d_1$ copper absent‡	550, 664 (br)	420, 460 (w)	4		+ (dominant role is nitrite reductase)

* Affinity probably underestimated in most cases
† In certain thermophiles
‡ But present in immediate electron donor, azurin
br, broad; w, weak
Modified from Poole (1983)

there is no evidence for a functional role for this metal. The most striking feature of the bacterial aa_3-types, however, is the relative simplicity of subunit structure and the resemblance of the subunits in the bacterial examples with the mitochondrially synthesized subunits in eukaryotes. Significant advances in our understanding of the complex mitochondrial enzymes may be expected from studies of the bacterial aa_3-type oxidases, in particular from the potential of applying genetic techniques, cloning and sequencing the genes involved, and completely characterizing the subunits and metal-binding sites (Raitio *et al.*, 1989). In addition, the ability to manipulate oxidase composition and function phenotypically, e.g. by limitation of either copper or iron in the growth medium, has been barely exploited. The application of Mössbauer spectroscopy to the enzyme from *Thermus thermophilus* is a good illustration of the advantages of microbial systems. The low natural abundance of the Mössbauer nucleide, ^{57}Fe, limits the usefulness of this technique with the mammalian oxidase, but growth of *Thermus* in a chemically defined medium, in which the naturally occurring Fe was replaced by ^{57}Fe has allowed detailed characterization of the c_1-aa_3 complex, particularly of the EPR-silent haem a_3 (Poole, 1983).

The a_1-type cytochromes are the most enigmatic group of putative oxidases. They are characterized by having α absorption bands (in the reduced state) at slightly shorter wavelengths than the aa_3-types (Table 5.3). In a few cases they are functional terminal oxidases (as in *Acetobacter* and *Thiobacillus* spp.) or part of a nitrite dehydrogenase complex (as in *Nitrobacter agilis*; Tanaka, Fukumori and Yamanaka, 1983). In *E. coli* and probably many bacteria, the a_1-like pigment is actually a high-spin *b*-type cytochrome (Poole *et al.*, 1984; Poole, 1988) called haemoprotein *b*-590, only superficially similar to cytochrome a_1, and may function as a hydroperoxidase (Section 5.2.3). The involvement of copper in the 'true' a_1-type oxidases is unknown.

Cytochrome *o* is a *b*-type haem protein, common as an oxidase ('o' for oxidase) in many bacteria. Several cytochromes *o* have now been purified (Poole, 1988), and, with one recent exception (Kita, Konishi and Anraku, 1984) have been shown to be devoid of copper. Two haems (1*b* and 1*c*, or 2*b*) are generally present and perhaps act in concert to donate a pair of electrons to the bound O_2. Low temperature photolysis of the CO complex of the *E. coli* enzyme has demonstrated the formation of a transient oxygenated intermediate ($Fe^{2+}.O_2$) in the reaction (Poole, 1988).

A striking feature of carboxydobacteria is their ability to grow with CO as sole carbon and energy source yet utilize oxygen as electron acceptor in respiration. The remarkable lack of toxicity has been attributed to the presence of a novel, CO-insensitive, '*o*'-type oxidase (Cypionka and

Meyer, 1983). It is difficult to envisage how CO cannot compete with O_2 for the ferrous haem of a cytochrome, although steric effects involving the 'forcing' of CO to adopt the bent $Fe-O_2$ structure might offer sufficient protection for the haem (Hughes, 1987). An alternative explanation is that the carboxydobacterial terminal oxidase is not a haemoprotein.

Cytochrome *d* has unusual optical properties that arise from the presence of the chlorin haem (Timkovich *et al.*, 1985). The α band of the reduced form is at about 630 nm and the Soret band indistinct. The cytochrome is clearly an oxidase and in some organisms has an unusually high K_m for O_2 (Table 5.3), possibly rationalizing the appearance of the enzyme under conditions of low O_2 availability. Nevertheless, the O_2-binding characteristics are unusual, forming a very stable oxygen-bound species with distinctive optical properties. EPR and optical studies have led to the hypothesis that the form absorbing at 650 nm is not the unliganded ferric haem *d*, but is an oxygenated intermediate, analogous to oxyhaemoglobin or oxymyoglobin. Indeed, resonance Raman spectroscopy supports this similarity and suggests the structure $Fe^{3+}O_2^-$ (Poole, 1988). Copper is absent from the purified enzyme, in accordance with the previous failure to detect optical changes in the near-infrared region of the spectrum during the reaction sequence. Recently, photometric and low temperature studies (Poole, 1988) have shown that the 'a_1'-like absorbance is attributable not to haem *a*, but to a high spin haem *b* which is believed to transfer electrons from low-spin cytochromes *b* to the chlorin.

The cytochrome cd_1-type of oxidase found in *Pseudomonas aeruginosa* shows some spectral similarities to *E. coli* cytochrome *d* but the conjugation of the chlorin tetrapyrrole ring is altered by the presence of further hydroxyl groups. The enzyme complex contains two *c*-type haems, covalently linked to the protein, as well as two haems d_1. The water-solubility of the complex has allowed intensive physical characterization of the composition and shape of the molecule and the chemical properties of the four metal centres. Both haems are low-spin in the oxidized protein as revealed by ESR and MCD spectroscopy and are probably six-coordinate even in the absence of ligands. The enzyme reduces O_2 to water and nitrite to NO; the latter activity is probably the most important function *in vivo*. Oxygen and nitrite each bind to haem d_1.

The diversity of types of cytochrome oxidase found in bacteria is puzzling in the absence of more complete information on the functional capabilities of the various types found. A partial explanation may lie in the ability of the various oxidases to utilize O_2 over a wide range of O_2 tensions and in the presence of potential inhibitors, particularly ligands to the haems.

(iii) Hydroxylamine oxidoreductase This enzyme is an example of an extremely complex haemoprotein. It catalyses the rapid, aerobic 4-electron oxidation of hydroxylamine to nitrite, an intermediate in the nitrification mediated by *Nitrosomonas europaea*. The enzyme (M_r 222 000) contains approximately 21 *c*-type haems and an unusual ligand-binding prosthetic group called P-460, in which the iron resides in a haem-like macrocycle (Andersson *et al.*, 1984). Three *c*-type haems are each associated with the three α subunits (M_r 11 000), whereas the three β subunits (M_r 63 000) each contain six haems *c* and one P-460. The *c*-type haems fall into at least four potentiometrically distinguishable groups. In a recent model for the mechanism, NH_2OH is oxidized at the P-460 centre and electrons transferred to external acceptors by the *c*-type haems (Lipscomb and Hooper, 1982). Mössbauer spectroscopy has been restricted to utilizing the low natural abundance of ^{57}Fe (2.2%) since the low yield of cells makes enrichment of the culture medium impractical. By quantifying the reduced P-460-CO species it has been shown that this haem is present at a concentration of approximately one unit per αβ dimer (Lipscomb *et al.*, 1982*a*, *b*).

5.2.3 Hydroperoxidases

Catalases and peroxidases, collectively referred to as hydroperoxidases, are responsible for removing the biologically deleterious alkyl and hydrogen peroxides (Saunders, Holmes-Siedle and Stark, 1964; Yonetani, 1974). Their activities are fundamentally similar: peroxidase reduces peroxide by electron transfer from a reduced donor, such as cytochrome *c*, whereas peroxides are disproportioned by catalase, the oxidative power of one peroxide molecule being used to oxidize another in an electron pair transfer.

$$H_2O_2 + SH_2 \xrightarrow{\text{peroxidase}} 2H_2O + S$$

$$H_2O_2 + H_2O_2 \xrightarrow{\text{catalase}} 2H_2O + O_2$$

Study of the intermediate compounds, named I and II, in the reaction of horseradish peroxidase with H_2O_2 was instrumental in providing direct experimental evidence for the Michaelis–Menten enzyme–substrate complex. Further structural information is now being provided by, for example EXAFS studies (Chance *et al.*, 1984).

Although some catalases are similar to the mammalian enzyme (Table 5.4) in comprising four identical subunits, each carrying one *b*-type

Table 5.4 Properties of selected microbial hydroperoxidases

Enzyme	Source	M_r (and number of subunits)	Haem(s)	Reference
Eukaryotic				
Catalase T ('typical')	*Saccharomyces cerevisiae*	248 000 (4)	haem *b*	Seah and Kaplan (1973)
Catalase A ('atypical')	*S. cerevisiae*	180 000 (4)	4 haem *b*	Seah, Bhatti and Kaplan (1973)
Catalatic nucleoprotein	*S. cerevisiae*	90 000 (4)	(4 non-haem iron)	Seah, Bhatti and Kaplan (1978)
Nitrate-induced catalase	*Neurospora crassa*	320 000 (4)	3.4 haem 'a_2'	Jacob and Orme-Johnson (1979)
Cytochrome *c* peroxidase	Yeast	34 100 (1)	1 haem *b*	Yonetani (1974)
Prokaryotic				
Catalase	*Micrococcus luteus*	232 000 (4)	4 haem *b* (?)	Herbert and Pinsett (1948)
Catalase (with peroxidase activity)	*Comamonas compransoris*	150 000 (2)	2 haem *b*	Nies and Schlegel (1982)
Hydroperoxidase I (*o*-dianisidine peroxidase and catalase activities)	*E.coli*	337 000 (4)	2 haem *b*	Claiborne and Fridovich (1979) Poole, Baines and Appleby (1986)
Hydroperoxidase II (predominantly catalase)	*E.coli*	312 000 (4)	2 haem *b* (plus 2 unidentified?)	Claiborne, Malinowski and Fridovich (1979)
Cytochrome *c* peroxidase	Ps. aeruginosa	53 500 (1)	2 haem *c*	Ellfolk and Soininen (1971)

haem, there is also considerable diversity in the microbial enzymes. The two 'green' catalases (from nitrate-induced *Neurospora crassa* and *E. coli* hydroperoxidase II) resemble each other in containing another type of haem prosthetic group. The catalase from the carbon monoxide-oxidizing bacterium *Comamonas compransoris* and catalase A from *S. cerevisiae* are of low M_r. Some of these also differ from the mammalian enzyme, characterized by a straight substrate (H_2O_2)–saturation curve, in exhibiting hyperbolic saturation kinetics; these are the *Comamonas* enzyme and both hydroperoxidases from *E. coli*. Some bacterial hydroperoxidases are dithionite reducible, unlike mammalian catalase.

Cytochrome *c* peroxidase is induced in yeast by O_2 and has been widely studied (Poulos *et al.*, 1980). It is readily crystallized, with or without its haem *b* group, which can be replaced by a Mn-porphyrin to give an enzyme of moderate activity.

The cytochrome *c* peroxidase of *Ps. aeruginosa* catalyses the peroxidation of *c*-type cytochromes and azurin, a copper protein. It is unusual amongst hydroperoxidases in having two *c*-type haems in a single polypeptide chain. One haem is in the high-spin state, the other low-spin. The former (of low potential, ≈ -330 mV) is the target of H_2O_2 in the peroxidatic reaction cycle. As with other hydroperoxidases, this cycle has been studied in some detail and several functional intermediates identified. The active form of the enzyme ($Fe^{3+}.Fe^{2+}$) reacts rapidly with H_2O_2 (Figure 5.4), the ferric high-spin haem forming a primary compound I. One oxidizing equivalent of O_2^{2-} is accepted by the

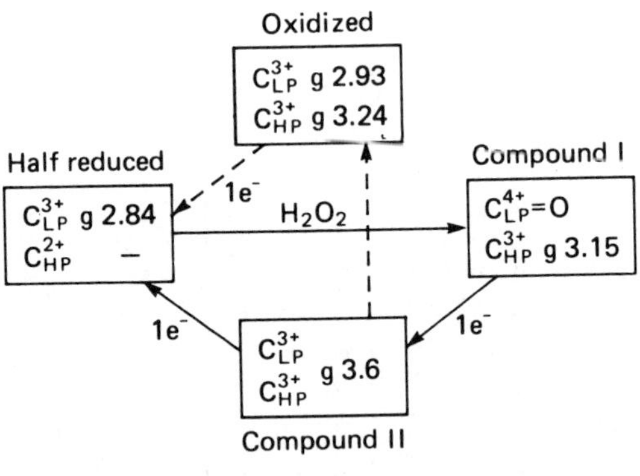

Figure 5.4 Reaction scheme for cytochrome *c* peroxidase. The two haem moieties are labelled LP (low potential) and HP (high potential). The g_z values for the haems in their ferric states are given. The intermediate position of g 3.6 in Compound II indicates the uncertainty in assigning this signal to a specific haem. The dashed arrows show the reactions not involved in the catalytic cycle. Reproduced with permission from Ellfolk *et al.* (1983).

high-spin haem, the other is accepted by the second haem. (Interestingly, in the yeast enzyme, the second equivalent is transferred to the protein side chain.) This intermediate, compound I, is converted to compound II by a one-electron reduction (Ellfolk *et al.*, 1983). In compound I, the iron is assumed to be in the oxyferryl ($Fe^{IV}=O$) state as in horseradish peroxidase.

It may be noted that a vanadium-containing bromoperoxidase has been isolated from the seaweed *Ascophyllum nodosum* (de Boer *et al.*, 1986). ESR studies suggest that the vanadium is bound by water and nitrogen and/or oxygen donor ligands (de Boer, Boon and Wever, 1988). Wever *et al.* (1987) have reviewed the involvement of vanadium in enzyme-catalysed redox reactions.

5.2.4 Microbial haemoglobins

Haemoglobin-like pigments have occasionally been reported in eukaryotic micro-organisms, including the protozoa *Tetrahymena pyriformis, Paramecium tetraurelia* and *P. caudatum* (Steers, Barnett and Lee, 1981), the fungi *N. crassa* and *Penicillium* and in the yeasts *S. cerevisiae, S. carlsbergensis* and certain *Candida* sp. (Lloyd, 1974). For the most part, these haemoproteins have been poorly characterized and the name 'haemoglobin' has sometimes been loosely applied to otherwise unidentified pigments observable in difference spectra.

In *Candida mycoderma*, however, spectral similarities with human haemoglobin have been confirmed as arising from yeast haemoglobin by measuring the half-oxygenation values of the pigment (Oshino *et al.*, 1973). The O_2 affinity is about 0.02 μM both in intact cells and for the isolated haemoglobin.

Leghaemoglobin, the O_2-carrying haemoprotein found in legume root nodules, is a product of the host plant and will not be described in detail. The reader is referred to the review by Appleby (1984). Its function is the facilitation of O_2 flux to the vigorously respiring phosphorylating, N_2-fixing *Rhizobium* bacteroids at a low (typically ≈10 nM) and stabilized O_2 tension. Significantly, the bacteroid cytochrome oxidase (Section 5.2.2(c)ii) operative under such conditions has not been identified.

A small (M_r 11 000) monomeric protein has been isolated from *Chromatium vinosum*. This high-spin haemoprotein is spectrally similar to haemoglobin and myoglobin and binds O_2 and CO, suggesting that it may allow such purple sulphur bacteria to exist under microaerobic conditions by regulating intracellular oxygen concentrations (Ondrias *et al.*, 1984). Much more convincing evidence for a bacterial haemoglobin comes from studies of a soluble haemoprotein in *Vitreoscilla*. This protein, previously thought to be an *o*-type oxidase but with unusual O_2-binding characteristics (Poole, 1988), shows extensive homologies with the haemoglobins of higher organisms and leghaemoglobin (Wakabayashi, Matsubara and Webster, 1986).

5.3 IRON-SULPHUR PROTEINS

5.3.1 Introduction

With respect to their ubiquitous distribution, their roles in diverse reactions in microbial biochemistry (H_2 metabolism, N_2 fixation, respiratory and photosynthetic electron transport) and the sheer number of centres in certain complexes, the iron-sulphur proteins are the most important group of proteins that mediate electron transfer. They range from simple molecules containing one type of Fe-S cluster to complexes containing several, sometimes different, Fe-S clusters, as well as other redox-active centres such as haem or molybdenum. Each Fe-S centre has at least two redox states, the difference between them, as in haem, corresponding to a one-electron change. With the exception of hydrogenases, the cluster is involved in electron transfer rather than substrate conversion. Unlike the haems, however, the iron-sulphur proteins generally function at negative oxidation-reduction potentials (-340 to -480 mV), although the so-called HiPIP proteins (high-potential iron proteins) have redox potentials of approximately $+350$ mV.

The prosthetic group of all iron-sulphur proteins is non-haem iron, bound to the peptide chain through cysteinyl-sulphur ligands. In the simplest group of non-haem iron proteins, the rubredoxins, a single such iron is present and inorganic sulphur (i.e. sulphide) is lacking. The lack of sulphur distinguishes the rubredoxins from the vast majority of non-haem iron proteins, which contain inorganic sulphur. The sulphur (sometimes designated S*) is present in the -2 valence state and is acid-labile, acid denaturation of the protein releasing the sulphur as H_2S. The sulphur atoms of certain synthetic and protein-bound iron-sulphur clusters are replaceable by the metalloid selenium (Moulis and Meyer, 1982).

The structures of the major types of clusters (one, two, three and four-iron types) found in the rubredoxins and ferredoxins were shown in Figure 1.7. The most familiar are the two-iron and four-iron-sulphur centres, in which each iron is surrounded by a tetrahedral array of sulphur donors, either cysteine residues or bridging sulphide. Note that each group, irrespective of the number of iron atoms, transfers only one electron. For example, the oxidized form of the two-iron protein involves two Fe(III) d^5 centres, and is diamagnetic (ESR-silent) by virtue of anti-ferromagnetic coupling between the d^5 centres. The reduced form contains Fe(II) and Fe(III) and is paramagnetic. The three-iron cluster (Figure 1.7) found in certain bacterial ferredoxins is still the subject of controversy.

5.3.2 [2Fe-2S] Ferredoxins and rubredoxins

These are found in certain anaerobic bacteria (being involved, for example, in pyruvate oxidation in *Clostridium pasteurianum*) and photosynthetic bacteria, where their function is not known. They are abundant, however, in O_2-evolving cyanobacteria and plants (Table 5.5); indeed, because they were once thought to be restricted to these organisms, they are still often referred to as 'plant-type' ferredoxins. They have also been implicated in camphor hydroxylation and α-keto acid oxidation in certain bacteria (Yoch and Carithers, 1979) and as electron transfer components in the hydrogenosomes of anaerobic protozoa (Gorrell, Yarlett and Müller, 1984).

The two-iron ferredoxins from cyanobacteria and plants have M_r values of about 12 000, whereas the proteins from *Clostridium* and *Azotobacter* are much larger (M_r in the range 21 000 to 24 000). Thus, all are larger than the simple 8- and 4-iron ferredoxins and generally contain only one cluster per molecule, perhaps reflecting the need for a more rigid arrangement of thiol ligands to prevent isomerization into [4Fe-4S] clusters. An X-ray structure of the ferredoxin from the cyanobacterium *Spirulina platensis* has been reported (Tsukihara *et al.*, 1981), confirming the general structure shown in Figure 1.7, which involves four cysteine ligands.

Oxidized [2Fe-S] clusters are diamagnetic and ESR-inactive but reduced [2Fe-2S] clusters give prominent ESR signals at $g = 1.94$. Evidence for the involvement of both the iron and the sulphide in the electron-accepting centre has come from reconstitution of *Ps. putida* ferredoxin with ^{57}Fe or ^{35}S. The strict conservation of the cysteine positions in the amino acid sequence of certain [2Fe-2S] proteins also suggests a crucial role for cysteinyl sulphurs.

The ESR spectra of FeS proteins are their most characteristic property. Generally, very low temperatures ($<77\,K$) are required for observation. Resolution of complex mixtures of clusters, e.g. in photosynthetic and respiratory electron transfer chains, may be achieved by appropriate choice of microwave power, temperature and by poising the sample at a

Table 5.5 Examples of iron-sulphur proteins in bacteria*

Cluster (name)	Source	M_r	E_m (mV)
No sulphide			
1Fe (rubredoxin)	*Clostridium pasteurianum*	6 000	−570
1Fe (rubredoxin)	*Peptococcus aerogenes*		
1Fe (rubredoxin)	*Desulfovibrio desulfuricans*	6 000	
2Fe (desulforedoxin)	*D. gigas*	7 900	−35
2Fe	*Pseudomonas oleovorans*		
[2Fe-2S]			
Rieske protein	*Thermus thermophilus*		
Putidaredoxin	*Ps. putida*	12 500	−240
	Azotobacter	21 000	−270
	Cl. pasteurianum	25 000	−300
	Escherichia coli	12 600	−360
	Spirulina platensis	10 648	−381
	Halobacterium	14 000	−345
[3Fe-4S]			
FdII	*D. gigas*	24 000	−130
	Hydrogenase (*D. desulfuricans*)		
	B. stearothermophilus		0
[4Fe-4S]			
Fd: $[C]^{2+/1+}$	*Bacillus stearothermophilus*	8 000	−280
	B. polymyxa	9 000	−422
	D. desulfuricans	6 000	−330
	Cl. thermoaceticum	7 300	
	Cl. formicoaceticum	6 000	
	Rhodospirillum rubrum	14 500	−430
HiPIP: $[C]^{3+/2+}$	*Chromatium vinosum*	10 000	+350
	Paracoccus		+360
2 [4Fe-4S]			
	Cl. acidiurici	6 000	−434
	Cl. pasteurianum	6 000	−395
	Chromatium vinosum	13 000	−490
	Peptococcus aerogenes	6 000	−400
[7Fe-nS]			
[4Fe-4S], [3Fe-3S]	*Azotobacter vinelandii*	14 500	+340, −420
[4Fe-4S], [3Fe-nS]	*T. thermophilus*		−530, −250
[4Fe-4S], [3Fe-nS]	*Ps. ovalis*		−600

* for references, see Hughes (1987).

known redox potential. In general, the paramagnetic (ESR-detectable) species are oxidized 3Fe proteins, oxidized HiPIP and reduced 2Fe and 4Fe ferredoxins. In the case of membrane-bound FeS clusters, ESR spectroscopy also allows an investigation of the spatial orientation of the clusters relative to the membrane plane (for a bacterial example, see Blum, Poole and Ohnishi, 1980). More recently, all four classes of FeS proteins have been distinguished by combining ESR with studies of the dependence on magnetic field of the shift in g-value induced by an externally applied electric field (Peisach *et al.*, 1983).

A [2Fe-2S] ferredoxin with unusual spectroscopic properties and high redox potential (150–300 mV) is associated with the cytochrome bc_1 complex of mitochondria, chloroplasts and certain bacteria. This Rieske iron-sulphur protein has been purified from *Thermus thermophilus* and shown to contain two identical [2Fe-2S] clusters, each coordinated by two cysteinyl residues and two other ligands. The favoured non-cysteine ligands are carboxylates which seem to be necessary for the high redox potential (Hughes, 1987).

The rubredoxins are the simplest type of iron-sulphur protein (Table 5.5). The iron is bound by four cysteinyl ligands in an approximately tetrahedral geometry. Their functions are poorly defined (Lovenberg, 1974).

The rubredoxins from *Desulfovibrio gigas* and *Ps. oleovorans* each have two iron centres. In the former, the electronic spectrum is not the sum of the spectra of two one-iron rubredoxins, suggesting interaction between the irons. In the latter, two non-interacting irons are present but one of the irons is labile and lost on isolation (Hughes, 1987).

5.3.3 3Fe-containing clusters

There is still much uncertainty about the structure and function of three-iron clusters. A 3-iron cluster was first discovered in *Az. vinelandii*. Ferredoxin I (FdI) was thought to contain two 4Fe-4S clusters but refined X–ray diffraction studies showed one such cluster plus a new structural type, a three-iron protein. Many such centres are now known (Table 5.5) (Moura *et al.*, 1982; Beinert and Thomson, 1983). In some cases these clusters are thought to have arisen from damage to a [4Fe-4S] cluster. Thus, aconitase has been isolated in an inactive 3-iron-containing form, which can be activated by the addition of Fe(II) or on standing. During such activation, the resonance Raman spectrum of the cluster changes to that typical of a [4Fe-S] centre (Johnson *et al.*, 1983). Similarly, the conversion of [3Fe-3S] clusters into [4Fe-4S] clusters in *Desulfovibrio gigas* ferredoxin, and the reverse process in *Bacillus stearothermophilus* have been demonstrated (Hughes, 1985).

The structures of the 3Fe clusters are still in doubt. The *Az. vinelandii* ferredoxin I is shown in Figure 1.7, but the Fe-Fe distance here (4.2×10^{-10} m) differs from that determined for the *D. gigas* ferredoxin II by EXAFS. It is unclear whether one of these structures is wrong, or whether the two 3Fe centres do differ. It has been suggested that 3Fe clusters could exist in either a [3Fe-3S] or [3Fe-4S] form. The former structure is relatively flat, but the latter would retain the essential cubic features of the 4Fe-4S clusters. The interrelationships of the 4S-containing clusters are shown in Figure 1.7.

The amino acid sequences of four 3Fe-3S ferredoxins have now been reported. The sequence of the protein from *Methanosarcina barkeri* (Hausinger *et al.*, 1982) is quite similar to that of the clostridial 4Fe-4S ferredoxin, indicating that similar sequences can accommodate either type of centre.

5.3.4 [4Fe-4S] ferredoxins

Small ferredoxins containing a single 4Fe-4S cluster are found in several bacteria. The eight-iron ferredoxins (Table 5.5) contain two such clusters in a small polypeptide. Iron and sulphide are arranged in a distorted cubic structure with each iron linked to the protein via cysteine residues. These ferredoxins generally have low midpoint potentials, close to that of the H^+/H_2 couple (−420 mV). The reduction process is a change from oxidation level +2 ($2Fe^{3+} + 2Fe^{2+}$) to +1 ($Fe^{3+} + 3Fe^{2+}$). The latter state gives a characteristic ESR spectrum around $g = 1.96$.

Ferredoxins with single [4Fe-4S] clusters were first isolated from *D. gigas* and *B. polymyxa*. They have M_r values between 6 000 and 10 000. This class also contains the high-potential iron proteins (HiPIPs) of photosynthetic bacteria, like *Chromatium*. Intriguingly these proteins have very much higher potentials (+350 mV) and operate between oxidation levels +2 and +3 ($3Fe^{3+} + 1Fe^{2+}$). The latter state gives an ESR spectrum with g-values greater than 2. However, despite detailed physical characterization of the *Chromatium* HiPIP, its cellular function remains unclear.

A number of ferredoxins have two [4Fe-4S] clusters that operate at about the same (negative) redox potential. In view of their distribution, they are called the clostridial-type ferredoxins but are also found in photosynthetic bacteria and *Peptococcus aerogenes*. They have M_r values of 6 000–9 000 and catalyse numerous two-electron transfer reactions

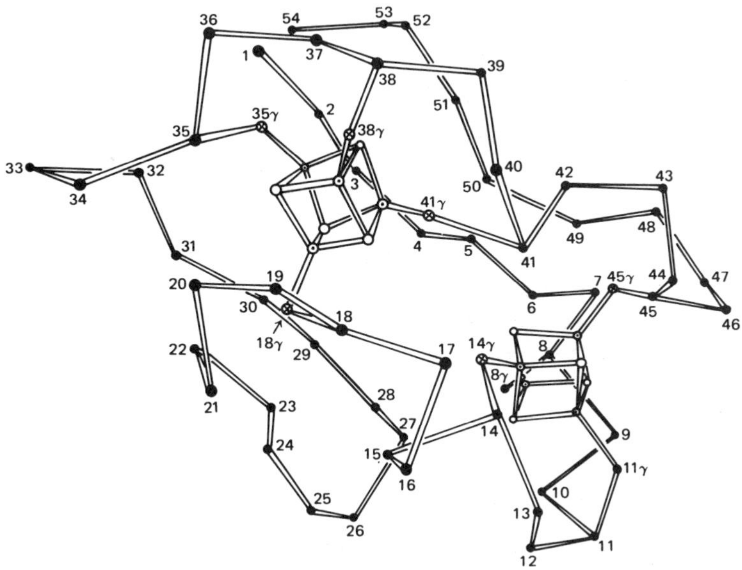

Figure 5.5 Structure of *Peptococcus aerogenes* ferredoxin obtained from X-ray crystallography. The iron and sulphur atoms are in two complexes, with 4Fe, 4 inorganic sulphur and 4 cysteine sulphurs each. The protein backbone is folded such that each of the 4Fe-4S clusters, which are about 1.2 nm apart, are bound by cysteine residues from both ends of the polypeptide; that is, cysteines 8, 11, 14 and 45 are coordinated to iron atoms in one complex, and cysteines 18, 35, 38 and 41 are coordinated to the irons in the second complex. Iron atoms are shown as ⊙, inorganic sulphur as ○, cysteine sulphurs as ⊗ and α-carbons as ●. Reproduced with permission from Adman, Sieker and Jensen (1973).

such as H_2 evolution and pyridine nucleotide reduction. The structure of the *Peptococcus aerogenes* ferredoxin, as determined by X-ray crystallography, is shown in Figure 5.5. The two [4Fe-4S] clusters are about 1.2 nm apart and spin-coupling between them may result in the complex ESR spectra characteristic of this class of protein.

Recently a class of ferredoxins, in which two iron-sulphur clusters operate at very disparate potentials, has been described. They are sometimes referred to as *Azotobacter*-type ferredoxins, being typified by ferredoxin I from *Az. vinelandii*. The function of the potentiometrically dissimilar centres (Table 5.5) is unclear.

Finally, the [4Fe-4S] ferredoxins of *D. gigas* consist of a 'building block' of a single polypeptide ($M_r \approx 6\,000$) containing one [4Fe-4S] cluster. Ferredoxins I and I′ are trimers of this unit, whilst ferredoxin II is a tetramer. The interactions of the polypeptides give proteins with a range of potentials from −30 (in I′) to 455 mV (I).

5.3.5 Examples of iron-sulphur enzymes

(a) Introduction

In addition to the simple ferredoxins, many enzymes employ iron-sulphur clusters in conjunction with haem, flavin or other cofactors (Table 5.6). In most cases, the FeS cluster serves to carry electrons from one group to another within a complex multi-centre system. Only in the case of hydrogenase is there evidence for a direct catalytic activity of the

Table 5.6 Examples of microbial iron-sulphur enzymes*

Metals and other prosthetic groups	Enzymes
Iron-sulphur	Hydrogenase, ω-hydroxylase, 4-methoxybenzoate-*O*-demethylase, iron-sulphur proteins bound to photosynthetic and respiratory membranes, glutamine phosphoribosyl pyrophosphate amidotransferase
Iron-sulphur-thiamine pyrophosphate	Pyruvate-ferredoxin oxidoreductase
Iron-sulphur-flavin	Succinate dehydrogenase, NADH dehydrogenase, *Pseudomonas* formate dehydrogenase, dihydroorotate dehydrogenase, glutamate synthase, adenylyl sulphate reductase, trimethylamine dehydrogenase, 5-10-methylenetetrahydrofolate reductase
Iron-sulphur-haem	Sulphite reductase (dissimilatory)
Iron-sulphur-molybdenum	Nitrogenase, nitrate reductase (dissimilatory), CO_2 reductase (formate dehydrogenase)
Iron-sulphur-nickel	Some hydrogenases
Iron-sulphur enzymes with two or more additional cofactors	Sulphite reductase (assimilatory), coliform formate dehydrogenase, xanthine dehydrogenase

* Reproduced with slight modification and with permission from Yoch and Carithers (1979).

iron-sulphur clusters. In a few enzymes, e.g. aconitase, the iron-sulphur cluster may have a regulatory rather than a redox role.

Table 5.6 shows the range of cofactors frequently associated with FeS clusters in enzymes and the diversity of reactions in which such enzymes participate. Only three illustrative cases will be described further. One of these involves iron-sulphur clusters in conjunction with molybdenum (respiratory or dissimilatory nitrate reductase), another with haem (assimilatory sulphite reductase) and the third with flavins (respiratory chain dehydrogenases). Nitrogenase, another iron-sulphur and molybdenum-containing enzyme is treated separately in Section 5.14.2, because of the considerable detail available on the molybdenum site.

(b) Assimilatory sulphite reductase

This enzyme catalyses the six-electron reduction of sulphite to sulphide, which is then incorporated during the biosynthesis of iron-sulphur cofactors and sulphur-containing amino acids. The enzyme from *E. coli* has been extensively studied by Siegel and others (for references, see Yoch and Carithers, 1979). It is a very large (M_r 670 000) and complex, soluble protein composed of eight larger (M_r 60 000) α subunits and four smaller (M_r 55 000) β subunits. Four FAD and four FMN are distributed over the α subunits, while the β subunits, sites of SO_3^{2-} reduction, bear 16 non-haem irons, the acid-labile sulphide and 4 unusual haems (sirohaem). These irons and sulphurs appear to exist as [4Fe-4S] clusters, possibly one per smaller subunit. A simplified scheme involves electron flow as follows:

$$\text{NADPH} \rightarrow \underbrace{\boxed{\text{FAD} \rightarrow \text{FMN} \rightarrow \text{Fe-S} \rightarrow \text{sirohaem}}}_{\text{sulphite reductase}} \rightarrow SO_3^{2-}$$

The Fe-S clusters thus serve to transfer electrons from flavin to haem.

(c) Respiratory nitrate reductase

Again, the best characterized enzyme is from *E. coli* (Ingledew and Poole, 1984). This enzyme is induced during anaerobic growth on nitrate and when traces of molybdenum are also available. The reaction catalysed,

$$NO_3^- + 2e^- + 2H^+ \rightarrow NO_2^- + H_2O$$

has a mid-point potential of +420 mV, allowing its use as a terminal oxidoreductase in a phosphorylating, anaerobic respiratory chain.

The enzyme is an integral membrane-bound protein and the necessity of detergent solubilization for study of the enzyme has generated some controversy over its exact composition. Nonetheless, there are three subunit types; these are the α (M_r 150 000), β (M_r 60 000) and γ subunits (M_r 20 000) generally agreed to be present in the ratio 2α:2β:4γ. The α-subunit is the molybdenum-containing catalytic portion and the γ-subunit is a cytochrome *b*. The purified enzyme contains 12 to 28 acid-labile sulphurs and non-haem irons per minimum stoichiometry (1α:1β:2γ). ESR studies reveal a HiPIP and at least two ferredoxins. In addition to these and the two *b*-type haems, there is a molybdenum, probably on the α subunit and associated with a pterin as it is in a wide range of molybdoproteins (Section 5.14). The molybdenum is ESR-detectable; although mid-point potentials for the Mo(V) to Mo(VI) and Mo(V) to Mo(IV) transitions have been measured, it is not possible at this stage to propose a detailed mechanistic scheme.

(d) NADH dehydrogenase

The respiratory electron transport chains in eukaryotic micro-organisms and bacteria contain high levels of Fe-S proteins. The mitochondrial inner membrane, for example, contains non-haem iron and acid-labile sulphur each at about twice the concentration of haem iron. The discovery in 1960 by Beinert and others of unusual 'g = 1.94' signals in mitochondria has led to extremely detailed studies of the numerous clusters found.

Avian and mammalian mitochondria contain three Fe-S centres in the succinate → ubiquinone segment of the respiratory chain, between four and six in the NADH → ubiquinone region and one in the cytochrome bc_1 region (Figure 5.6). Considerable impetus was given to the study of iron-sulphur proteins by (i) the use, by Ohnishi and others, of ESR at very low temperatures (<77 K) appropriate to the rapid relaxation time of these centres and (ii) controversy over the relationship between ESR-detectable FeS clusters, the variable capacity for oxidative phosphorylation and inhibitor sensitivity in the Site I region of the yeast respiratory chain. Thus, whereas ox heart mitochondrial particles exhibited both an NADH-elicited g = 1.94 ESR signal and Site I phosphorylation, both features were absent from *S. cerevisiae*. There is a confusing literature on the interrelationships between Site I, ESR-detectable Fe-S clusters, rotenone and piericidin-sensitivity and the growth conditions and species of yeast (Lloyd, 1974). However, studies of *Candida (Torulopsis) utilis* in a chemostat by Garland and others have demonstrated that iron-limitation decreases Fe-S signals to levels undetectable by ESR. These

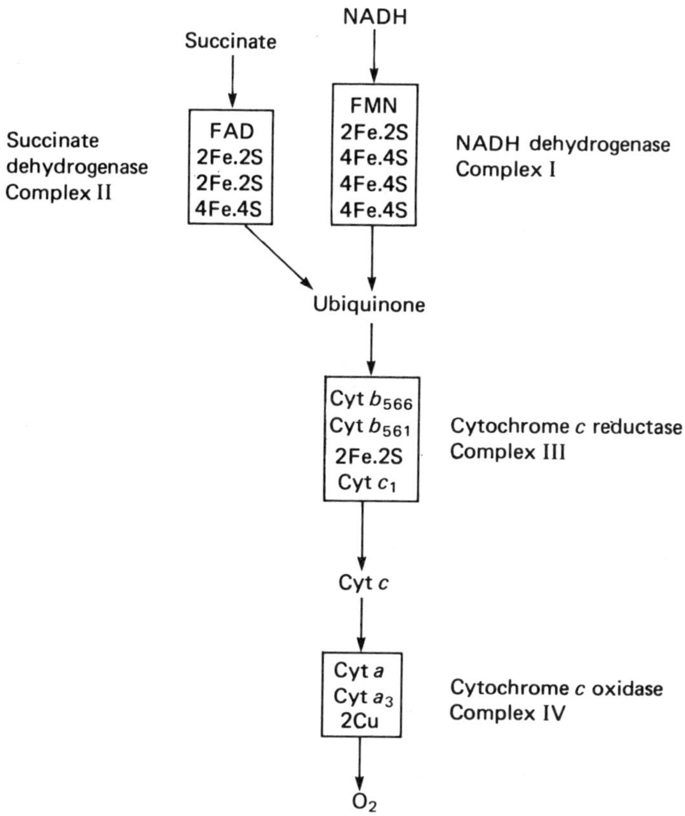

Figure 5.6 Organization of the mitochondrial respiratory electron transport chain.

centres are necessary, but not sufficient, for the coupling of ATP synthesis to the NADH dehydrogenase (Gray *et al.*, 1975).

Relatively little is known about bacterial NADH dehydrogenases. As is the case for other respiratory chain components of *P. denitrificans*, the ESR signals attributable to NADH dehydrogenase are similar to those observed in mitochondria. Sulphate-limitation causes loss of some of these signals (specifically those assigned to Fe-S centre N-2) and a decrease in NADH dehydrogenase activity. Similar correlations are seen in cells grown in the presence of rotenone (Meijer *et al.*, 1978). In *E. coli*, it is not clear whether the enzyme contains Fe-S clusters (Ingledew and Poole, 1984). Nevertheless, sulphate limitation causes loss of g = 1.94

ESR signals, attributable to ferredoxins, and loss of energy coupling in the NADH dehydrogenase region of the chain (Poole and Haddock, 1975).

The *Az. vinelandii* NADH dehydrogenase is interesting in that the form of the enzyme present depends on the iron concentration in the growth medium. Both forms have similar M_r values (56 500) but the enzyme isolated from cells grown with limiting iron contains 1 Mo:2 Fe-S:1 FMN, whereas the enzyme from cells grown with high iron concentrations contains 4 FeS:1 FMN (Yoch and Carithers, 1979).

5.4 OXYGENASES: IRON- OR COPPER-CONTAINING ENZYMES

5.4.1 Nomenclature

Hayaishi's demonstration that oxygen atoms from dioxygen could be incorporated directly into products by oxygenases represents a milestone in biochemistry. Previously, the accepted hypothesis was that advanced by Wieland at the beginning of the century, in which biological oxidations were thought to involve only the transfer of hydrogen atoms and that any oxygen incorporated into a product was derived from water.

There are two general classes of oxygenases. Monooxygenases (hydroxylases or mixed function oxidases) insert one O atom into a substrate and release the other as water. An obligatory second substrate supplies two electrons to reduce a metal to a form in which it can activate O_2. Dioxygenases, on the other hand, insert both O atoms into a substrate, often with the cleavage of a C–C bond. The substrates for both classes are often electron-rich phenolic compounds, which are themselves capable of reducing the catalytic metal. Although most dioxygenases appear to contain a metal as their cofactor, some monooxygenases are simple flavoproteins. In both monooxygenases and dioxygenases, the metal may be iron (as haem or non-haem iron); certain monooxygenases contain copper.

5.4.2 Monooxygenases

The monooxygenases are a diverse group (Table 5.7) including enzymes with copper as cofactor (e.g. tyrosinase and dopamine β-hydroxylase), iron and iron-sulphur centres, as well as the haem protein cytochrome P-450.

Table 5.7 Examples of microbial monooxygenases

Enzyme	Organism	Reaction catalysed	Metals or other redox groups
Putidamonooxin (O_2-activating component of 4-methoxybenzoate monooxygenase)	*Ps. putida*	4-methoxybenzoate → 4-hydroxy-benzoate + HCHO	1 [2Fe-2S]/subunit; 1 non-haem Fe/subunit
Cytochrome P-450$_{cam}$	*Ps. putida*	Hydroxylation of D-camphor	Protohaem
Methane monooxygenase*	*Methylococcus capsulatus*	Oxidation of methane to methanol	A: 2 to 3 non-haem Fe/mol B: none C: 1 FAD, 1 FeS/mol
Tyrosinase	*Neurospora crassa*	Mono and diphenol hydroxylation	2 Cu/mol

* Three compounds; A is the true oxygenase.
For references, see Nozaki *et al*. (1982) and Dalton and Leak (1985).

(a) Monooxygenases containing non-haem iron

Examples of these enzymes are shown in Table 5.7. Putidamonooxin, the oxygen-activating component of 4-methoxybenzoate monooxygenase, is unusual in having [2Fe-2S] clusters as well as mononuclear, non-haem iron. The former seems to serve as a one-electron storage centre and the latter as the oxygen-binding site.

Compound A of the methanemonooxygenase from *Methylococcus capsulatus* contains non-haem iron but no extrudable FeS centre. ESR studies (Woodland and Cammack, 1985) suggest that the iron may be present as an antiferromagnetically coupled pair of high-spin iron atoms, such as in haemerythrin and protein B_2 from ribonucleotide reductase.

(b) Cytochrome P-450

Cytochromes P-450 are so-called because of the characteristic absorbance at about 450 nm of the CO-liganded, reduced form. An unusual ESR spectrum arises from the fifth iron ligand being a thiolate anion of a cysteine residue. They are often involved in hydroxylating rather inert hydrocarbon groupings such as in drugs and fatty acids (White and Coon, 1980). In the well-characterized *Pseudomonas* haemoprotein, electrons are supplied from NADH via a flavoprotein and putidaredoxin, an iron-sulphur protein (Section 5.3). The reaction sequence typically involves binding of the substrate to the oxidized (Fe(III)) low-spin cytochrome, followed by uptake of one electron to give a high-spin Fe(II) complex. Reaction with O_2 gives a compound resembling peroxidase compound III (Section 5.2.3). Acceptance of a second electron from putidaredoxin is required to form the oxygenated product and reform the Fe(III) cytochrome P-450 (see Figure 5.7 and Malmström, 1982, for detailed information on mechanistic aspects). In many other microbes, cytochrome P-450 has been identified by its CO-binding characteristics but its function is unknown.

(c) Copper-containing monooxygenases

Tyrosinase is an interesting copper-containing monooxygenase that catalyses the oxidation of monophenols and has been extensively studied in *Neurospora*. The active site contains a bimetallic copper centre with antiferromagnetic coupling between the two Cu ions. The oxygenated species, formed by reaction of the Cu pair with O_2,

$$2\text{Cu(I)} \xrightarrow{O_2} 2\text{Cu(II)}(O_2^{2-})$$

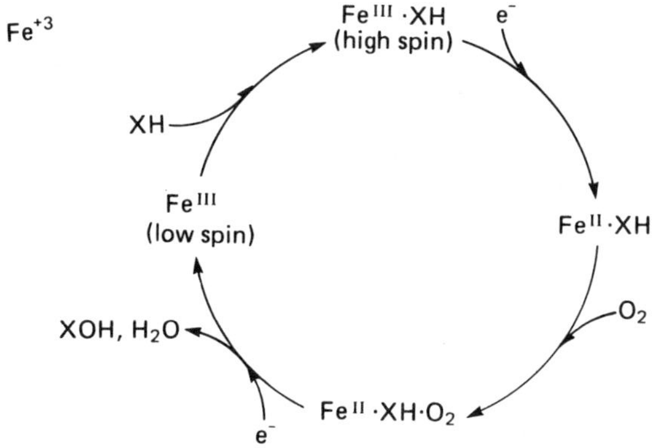

Figure 5.7 Simplified catalytic cycle for cytochrome P-450 (Malmström, 1982). The Fe(III) form binds substrate (XH) followed by reduction of the Fe(III) enzyme in a one-electron process. Binding of oxygen to the Fe(II) form ($Fe^{2+}.XH.O_2$) is probably followed by conversion to an intermediate, formally represented by $Fe^{3+}XH.O_2^-$, before receiving the second electron. Reproduced with permission from the *Annual Review of Biochemistry*, vol. 51, copyright 1982 by Annual Reviews Inc.

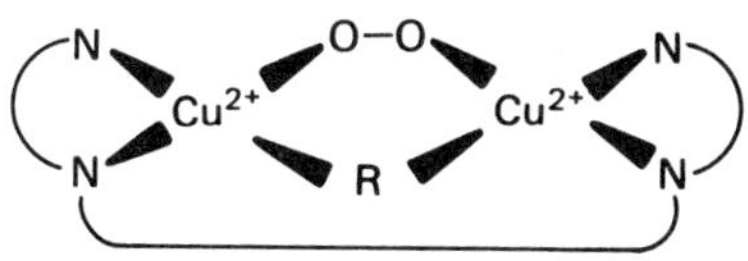

Figure 5.8 The active site of oxytyrosinase. N=nitrogen donor of histidyl residues.

has an optical spectrum similar to oxyhaemocyanin and has the structure shown in Figure 5.8. This oxytyrosinase is believed to be the hydroxylating species. Both coppers are replaceable by cobalt but with loss of enzyme activity (Rüegg and Lerch, 1981).

His-62 and His-189 appear to be copper ligands in tyrosinase from *Streptomyces glaucescens*, as their replacement by asparagine by site-directed mutagenesis gave an enzyme which binds only one, ESR-detectable, copper ion (Huber and Lerch, 1988).

5.4.3 Dioxygenases

Most dioxygenases contain a metal, not flavin, and the metal is generally iron. This may be part of a haem group, as in tryptophan 2,

Table 5.8 Examples of microbial dioxygenases

Enzyme	Organism	Reaction catalysed	Metals or other redox groups
Protocatechuate 3,4 dioxygenase	Various *Pseudomonas* spp.; similar in *Brevibacterium fuscum*	Protocatechuic acid → β-carboxymuconic acid	$1Fe^{3+}/90K\ M_r$ protomer
Catechol 2,3-dioxygenase (metapyrocatechase)	*Ps. putida*	Catechol → α-hydroxymuconic acid semialdehyde	1 Fe(II) per tetramer
Catechol 1,2-dioxygenase (poyrocatechase)	*Ps. arvilla*	Catechol → *cis, cis,* muconic acid	1 Fe(III) per dimer
Toluene dioxygenase	*Ps. putida*	Toluene → *cis*-toluene dihydrodiol	FeS
L-tryptophan 2,3 dioxygenase	*Ps. acidovorans*	L-tryptophan → L-formyl kynurenine	Protohaem IX
Lipoxygenases FL-1, FL-2	*Fusarium oxysporum*	Linoleate → 2 isomeric 9- and 13-hydroperoxides	Protohaem IX

See Nozaki *et al.* (1982).

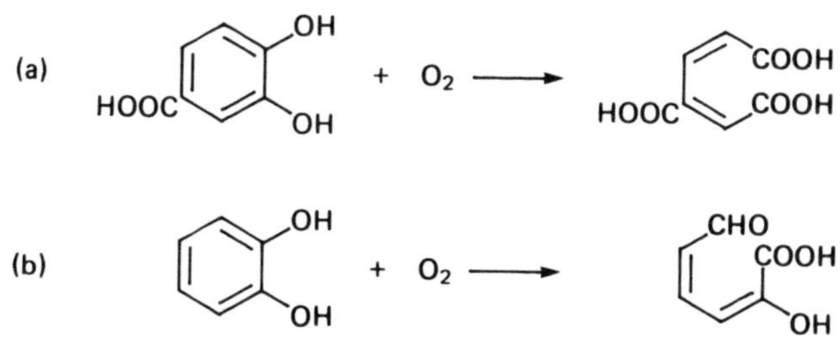

Figure 5.9 Reactions catalysed by intradiol (a) and extradiol (b) catechuate dioxygenases.

3–dioxygenase, or present as non-haem iron (for examples, see Table 5.8). Note that lipoxygenase is a protohaem-containing enzyme in the fungus *Fusarium*, whereas the soybean enzyme contains non-haem iron. The terminal oxygenase component ($1SP_{Tol}$) of toluene dioxygenase is unusual in being an iron-sulphur protein (Table 5.8).

The catechuate dioxygenases, common in fluorescent pseudomonads (Table 5.8) where the enzymes are plasmid-encoded, cleave the catechol ring. They vary in protein structure, but all contain labile non-haem, non-sulphur iron. Two kinds of cleavage are possible with respect to the positions of the hydroxyls on the ring. Protocatechuate 3, 4-dioxygenase (EC 1.13.11.3) is an example of an intradiol-type enzyme, cleaving the catechol ring between the hydroxyls (Figure 5.9(a); Que, 1983). Metapyrocatechase (EC 1.13.11.2) is a typical extradiol-type enzyme, catalysing the oxygenative cleavage of catechol immediately adjacent to only one of the hydroxyl groups (Figure 5.9(b)). Enzymes catalysing intradiol cleavage are red with high-spin Fe(III), while those catalysing extradiol cleavage are colourless. Extradiol-cleaving enzymes were once thought to be Fe(II)-containing enzymes but more recent evidence shows that the two irons in, for example, protocatechuate 4,5-dioxygenase are quite different (Lipscomb *et al.*, 1982b). Mössbauer spectra of the enzyme purified from bacteria grown on ^{57}Fe-enriched media show that one iron is high-spin Fe(II) and remains in this oxidation state in the anaerobic substrate complex. The second iron is high-spin Fe(III). However, most of the ferric iron in the native enzyme is ESR-silent (becoming ESR-active in the substrate complex), perhaps due to being in close proximity with a paramagnetic ferrous iron.

The mechanism of action of some dioxygenases has been studied in detail. In some, such as indoleamine 2,3-dioxygenase, the metal has to be reduced to the Fe(II) state before oxygen binding. Oxygen activation then involves electron transfer to the bound ligand. In this enzyme, an oxygenated intermediate (Fe(III)O_2^-) has been observed, similar to the so-called compound III of horseradish peroxidases and to oxymyoglobin (Uchida *et al.*, 1983). In other cases, however, the iron does not seem to undergo a redox change and is not directly involved in O_2 activation. Thus, in protocatechuate 3,4-dioxygenase, the single iron remains as high-spin Fe(III) in all forms of the enzyme, i.e. the resting state, the enzyme–substrate complex and the presumed tertiary complex. Since O_2 is not known to form a complex with high-spin Fe(III), it is believed that the iron serves to activate the organic substrate rather than O_2, and O_2 appears to be attached only to the substrate, not to the iron (Bull, Ballun and Otsuka, 1981).

The mechanism of action of a *meta*-cleaving enzyme (chloridazon-catechol dioxygenase) from *Phenylobacterium immobilis* is striking. The enzyme (M_r 200 000) consists of six identical subunits and contains one non-haem iron atom per subunit. However, in this state it is inactive and requires the addition of Fe(II) ions or ascorbate for activity, a novel feature (Müller, Schmitt and Lingens, 1982).

5.5 IRON-STORAGE PROTEINS

In eukaryotes, from fungi (Spartalian, Oosterhuis and Smarra, 1975) to man, ferritin acts as a temporary store for iron, thus preventing a potentially toxic accumulation of free iron and acting as a long-term mobilizable reserve (Harrison and Hoare, 1980; Theil, 1987). The iron (as a hydrous ferric oxide–phosphate complex) is surrounded by a protein shell (apoferritin), rendering the Fe(III) soluble in the cytoplasm. Up to 4500 Fe atoms per molecule may be present, representing at least 30% by weight of the ferritin molecule.

Although the mechanisms for iron acquisition by bacteria are well-known (Chapter 3), intracellular iron-storage molecules have only recently been described. Stiefel and Watt (1979) have shown that a protein from *Az. vinelandii*, previously described as cytochrome $b_{557.5}$, closely resembles mammalian ferritin in some aspects. It has an iron content of 13 to 20% by weight, a subunit weight of 17000, a molecular diameter comparable to animal ferritin and an electron-dense core of 55×10^{-10} m. It contains weakly-coupled high-spin Fe(III) reminiscent of ferritin iron, but the presence of a second type of iron (as protoporphyrin IX; one per two protein subunits) is unique. An iron-storage molecule has also been described in *E. coli* (Yariv, 1983). It is indistinguishable from 'cytochrome b_1' in size, redox potential and optical spectra. A novel iron-containing protein from *Desulfovibrio gigas* is characterized by extreme intractability of the metal. These features and its ability to be reconstituted after dissociation to iron-free monomers are all suggestive of a ferritin-type protein (Smith and Emery, 1982).

5.6 OTHER ROLES FOR IRON IN ENZYMES

Certain enzymes require iron for activity, even though the metal is not a prosthetic group. Familiar examples are aconitase and phosphoenolpyruvate carboxykinase (Harrison and Hoare, 1980; Sagers, 1974). More recent microbial examples are the Fe(II)–(or Co(II)–dependent D-(-)-tartrate dehydratase of certain pseudomonads (Rode and Giffhorn, 1982), the Fe(III)-activated acid phosphatase of *Saccharomyces rouxii* (Arnold and Garrison, 1979) and possibly an iron-stimulated cellulase in *Clostridium thermocellum* (Johnson and Demain, 1984). Iron is also found in the bacterial enzyme 3-deoxy-D-arabino-heptulosonate-7-phosphate synthase which is involved in the biosynthesis of aromatic compounds (McCandliss and Herrmann, 1978).

5.7 MICROBIAL OXIDATION OF IRON

Microbes play important roles in the biogeochemical cycling of iron, including the anaerobic corrosion of iron in buried pipes, the reduction of Fe(III) to Fe(II) (Section 5.8) and the aerobic oxidation of Fe(II) to Fe(III).

5.7.1 Distribution

Bacteria that oxidize iron fall into two groups, (i) the acidophilic iron-oxidizers such as the well-characterized *Thiobacillus ferrooxidans* and *Sulfolobus* spp. and (ii) the 'neutral pH iron bacteria'. It is still debated whether members of the latter group actually oxidize iron and grow autotrophically or merely deposit the iron in an oxidized form. *Gallionella ferruginea*, for example, deposits $Fe(OH)_3$ (Nealson, 1983b). The sheathed bacteria (the *Spherotilus-Leptothrix* group) and some of the budding bacteria (*Hyphomicrobium* and *Pedomicrobium*) accumulate manganese as well as iron.

5.7.2 *Thiobacillus ferrooxidans*

The oxidation of ferrous iron has been investigated in detail only in *Th. ferrooxidans*, a gram-negative acidophilic rod, growing optimally at pH

2.0–3.5, where Fe(II) is soluble. The E'_0 value of the Fe(III)/Fe(II) couple (+ 780 mV, pH-independent) is so close to that of $\frac{1}{2}O_2/H_2O$ that insignificant energy is available using ferrous iron as reductant at neutral pH (Ingledew, 1982). However, even at an extracellular pH of approximately 2.0, the internal pH remains close to neutrality, providing a transmembrane Δ pH of 4.5 units (270 mV), sufficient to drive ATP synthesis via a proton-translocating ATPase. The function of the respiratory chain seems to be to remove these translocated protons and maintain the internal pH and ΔpH.

The respiratory chain contains several *c*-type cytochromes and a cytochrome a_1 with two haem *a* centres of unusually high potential (E_m +420, +500 mV, respectively, at pH 7.0), and which is probably the terminal oxidase. In addition, there is a novel copper protein called rusticyanin (E_m + 680 mV) which is closely related to azurin (Section 5.10.4).

5.7.3 Functions

The molar growth yields on ferrous iron alone are very low. This results from the requirement for at least 6 Fe(II) to be oxidized for each NAD^+ reduced by reversed electron flow. In environments where iron oxidation occurs, however, sulphur is almost always available as sulphide and this is also oxidized. In iron bacteria other than the acidophilic species, autotrophy using ferrous iron as energy source has not been conclusively demonstrated and it is interesting to speculate on the possible functions of iron oxidation and deposition (Nealson, 1983b). Iron oxidation may have developed as a mechanism for removal of potentially toxic oxygen. Such an extracellular O_2 sink may have been of value during the evolution of other iron-based oxygen-production mechanisms such as superoxide dismutases (Section 5.11). Alternatively, deposition of insoluble oxides of iron or other metals may confer mechanical advantages, such as stabilization of the holdfast or the provision of a protective cell covering.

5.8 MICROBIAL REDUCTION OF IRON

5.8.1 Distribution

With a standard reduction potential (E′ at pH 7) of +780mV, ferric iron is a powerful oxidant and has been exploited as a useful electron acceptor in a few biological electron transport systems.

The ability of bacteria to reduce iron oxides was first described in 1927 but the accumulated information is scattered and incomplete and the mechanisms of iron reduction, of which there may be many, until recently could not be described in detail for any one bacterium. The genera reported to reduce Fe(III) include *Bacillus, Pseudomonas* and *Proteus* (Nealson, 1983b; Jones, Davison and Gardener, 1984). Fungi (Ottow and Von Klopotek, 1969), the roots of higher plants and isolated mitochondria also reduce iron. Iron reduction is frequently assayed by the use of ferrozine (the disodium salt of 3-(2-pyridyl)-5,6-bis(4-phenyl sulphonic acid)-1,2,4-triazine), which specifically traps ferrous iron to give a coloured complex (λ_{max}, 562 nm).

5.8.2 Pathways

A common feature of some iron-reducing bacteria is the presence of nitrate reductase; field studies of iron reduction also support the idea that facultative nitrate reducers are involved (Sørensen, 1982; Jones, Gardener and Simon, 1983). Both nitrate and chlorate (which is reduced to chlorite by nitrate reductase) diminish iron reduction by *Aerobacter aerogenes*, but only when nitrate reductase is present, and nitrate reductase-deficient mutants exhibit a decreased capacity for iron reduction (Ottow, 1970).

Recent findings, however, have weakened the argument that nitrate reductase is essential for iron reduction. First, bacteria have been described (Jones, Davison and Gardener, 1984) that reduce NO_3^- but not Fe(III) (e.g. a strain of *E. coli* and *Thiobacillus denitrificans*), whereas others reduce Fe(III) but not NO_3^- (e.g. *Desulfotomaculum nigrificans*). These results suggest the presence of an alternative enzyme or ferri-reductase. Secondly, an alternative explanation for the inhibitory effect of nitrate appears to apply to a pseudomonad and to *E. coli* (Williams and Poole, 1987; Figure 5.10), namely that nitrite, the product of nitrate reduction, inhibits Fe(III) reduction or chemically reoxidizes the Fe(II) formed (Obuekwe, Westlake and Cook, 1981).

5.8.3 Functions

The role of iron reduction in these bacteria is unclear, but there are several possibilities.

1. Iron reduction may be linked to energy conservation by acting as terminal electron acceptor in a phosphorylating respiratory chain (Brock and Gustafson, 1976; Jones, Gardener and Simon, 1983);

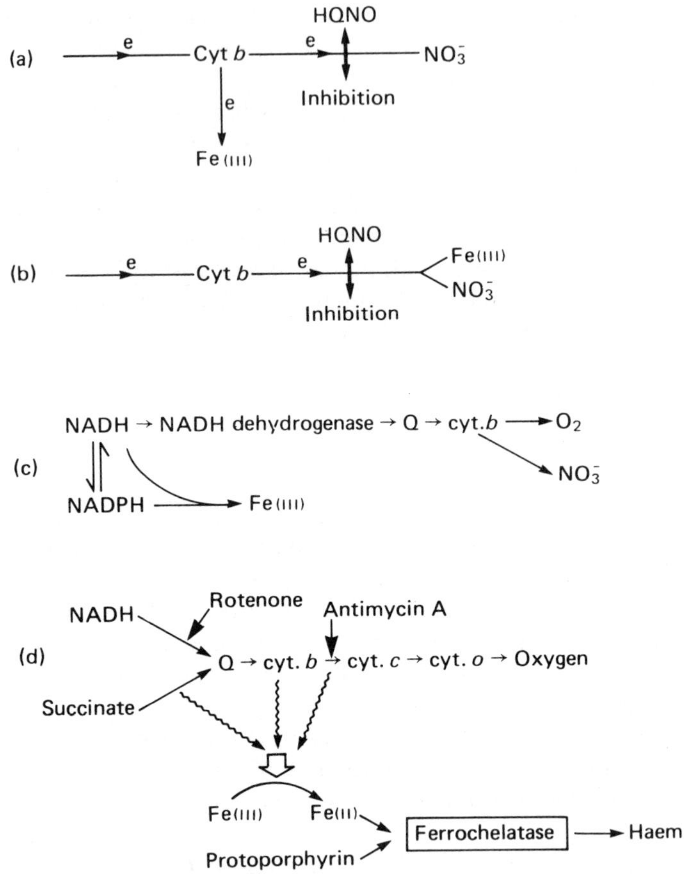

Figure 5.10 Pathways of iron(III) reduction by bacteria. (a) shows the scheme proposed by Lascelles and Burke (1978) for *S. aureus*, (b) for a pseudomonad (Obuekwe, Westlake and Cook, 1981), (c) for *E. coli* (scheme based on data in Williams and Poole, 1987) and (d) for *Spirillum itersonii* (Dailey and Lascelles, 1977). In the pseudomonad and *E. coli*, the apparent inhibitory effect of nitrate on Fe(III) reduction may be attributed to Fe(II) reoxidation by nitrite. Cyt = cytochrome; Q = quinone.

2. Very recently, an anaerobic bacterium has been described (Lovley *et al.*, 1987) that oxidizes organic acids while reducing ferric oxide (Fe_2O_3) to extracellular magnetite (Fe_3O_4). Although the product may have no specific function, the density of magnetite might serve as an anchor to keep cells in intimate contact with ferric oxide precipitates. Ancestors of such bacteria might also have been instrumental in forming Fe_3O_4 salts in ancient rock formations;

3. The reduction might provide the Fe(II) required as a substrate for the ferrochelatase-catalysed synthesis of haem from iron and porphyrin (Section 5.2.1). This may be the case in *Spirillum itersonii* (Dailey and Lascelles, 1977; Figure 5.10(d)) and in mitochondria, and could be necessitated by the instability of Fe(II) under physiological conditions. Alternatively, the redox state and/or compartmentalization of iron may influence the specificity of ferrochelatase activity, since depletion of the Fe(II) pool in yeast seems to allow incorporation of 'foreign' metals such as Zn(II) (Camadro and Labbe, 1982);
4. Iron reduction activity could reflect ferrisiderophore metabolism. As described in Chapter 3, siderophores chelate Fe(III) with extraordinary avidity, but bind Fe(II) only weakly. Thus, iron reduction is an attractive mechanism for acquisition by the cell of iron from the siderophore–metal complex;
5. Iron(III) could serve solely as a 'sink' for excess reducing equivalents, allowing reoxidation of, say, NADH;
6. Iron reduction may be necessary for mobilization of Fe(III) stored in the form of bacterioferritin (Section 5.5).

Some of the organisms studied as iron-reducers also reduce other metals. For example, iron can be reduced by a *Bacillus* previously shown to reduce MnO_2 and other managanese oxides in ferromanganese nodules (De Castro and Ehrlich, 1970) and a wide range of metals can be reduced by iron-reducing freshwater isolates (Jones, Davison and Gardener, 1984). Any commonality between the mechanisms of reduction of iron and other metals is obscure at present.

5.9 IRON IN MAGNETOTACTIC BACTERIA

5.9.1 Discovery

In recent years, a novel tactic response in bacteria has been discovered and has attracted great interest. Certain aquatic bacteria are now known to be magnetotactic, tending to swim along magnetic field lines and influenced by the earth's magnetic field. They synthesize and harbour small crystals of magnetite (Fe_3O_4), the oxide of iron known to early navigators as lodestone. Many higher organisms are known to sense the earth's magnetic field and the ferromagnetic detection system in bacteria has interesting parallels in geomagnetically responsive animals.

The phenomenon was discovered in 1975 by Blakemore and has been extensively reviewed since (Blakemore and Frankel, 1981; Blakemore, 1982; Frankel, 1984). Microscopic examination of enrichment cultures of bacteria revealed highly motile bacteria that migrated almost unidirectionally, irrespective of light or the location of the microscope. That the response was magnetic in origin was vividly demonstrated by bringing a magnet near the slide. Cells were always attracted by the end of the magnet that also attracted the north-seeking end of a compass needle and were repelled by its opposite end. The organisms themselves were thus north-seeking, a fact that facilitates the detection and isolation of such bacteria from freshwater sediments. However, organisms from the Southern Hemisphere swim southward. It is thought that magnetotaxis may serve to direct bacteria downwards, away from surface waters and into sediments.

5.9.2 Magnetosomes

The magnetotactic bacteria described to date possess intracellular iron grains observable by electron microscopy. In *Aquaspirillum magnetotacticum*, there are approximately twenty such grains, opaque to the electron beam and usually cubic or octahedral in form, arranged in a chain parallel to the cell's long axis (Figure 5.11). Each particle, or magnetosome, is sheathed by a membrane that is adjacent to, and may be contiguous with, the cytoplasmic membrane. The particles are within the single magnetic domain size range for magnetite, being about 50 nm in width; it is possible that this size is regulated by the envelope surrounding each particle. Such single magnetic domains will be constrained to lie in a chain with their magnetic axis parallel, so that the total magnetic energy of the magnetosome chain is the sum of the individual magnetic moments of the particles. The magnetic energy of a cell has been calculated to be 1.3×10^{-19} J/G, quite sufficient to account for the organism's ability to align in the 0.5 G geomagnetic field.

5.9.3 Synthesis and properties of magnetite

Magnetite appears to be synthesized by *A. magnetotacticum* only when O_2 is growth-limiting and in the presence of a suitable iron source. Ferric quinate is satisfactory in a chemically-defined medium and in the natural habitat may be present as a plant-derived product. These magnetotactic cells lack catalase (Section 5.2.3), explaining their sensitivity to high oxygen levels. Because iron salts are well known to catalyse decomposition of H_2O_2, it is possible that magnetite synthesis might protect the cell from H_2O_2 toxicity by preventing its accumulation.

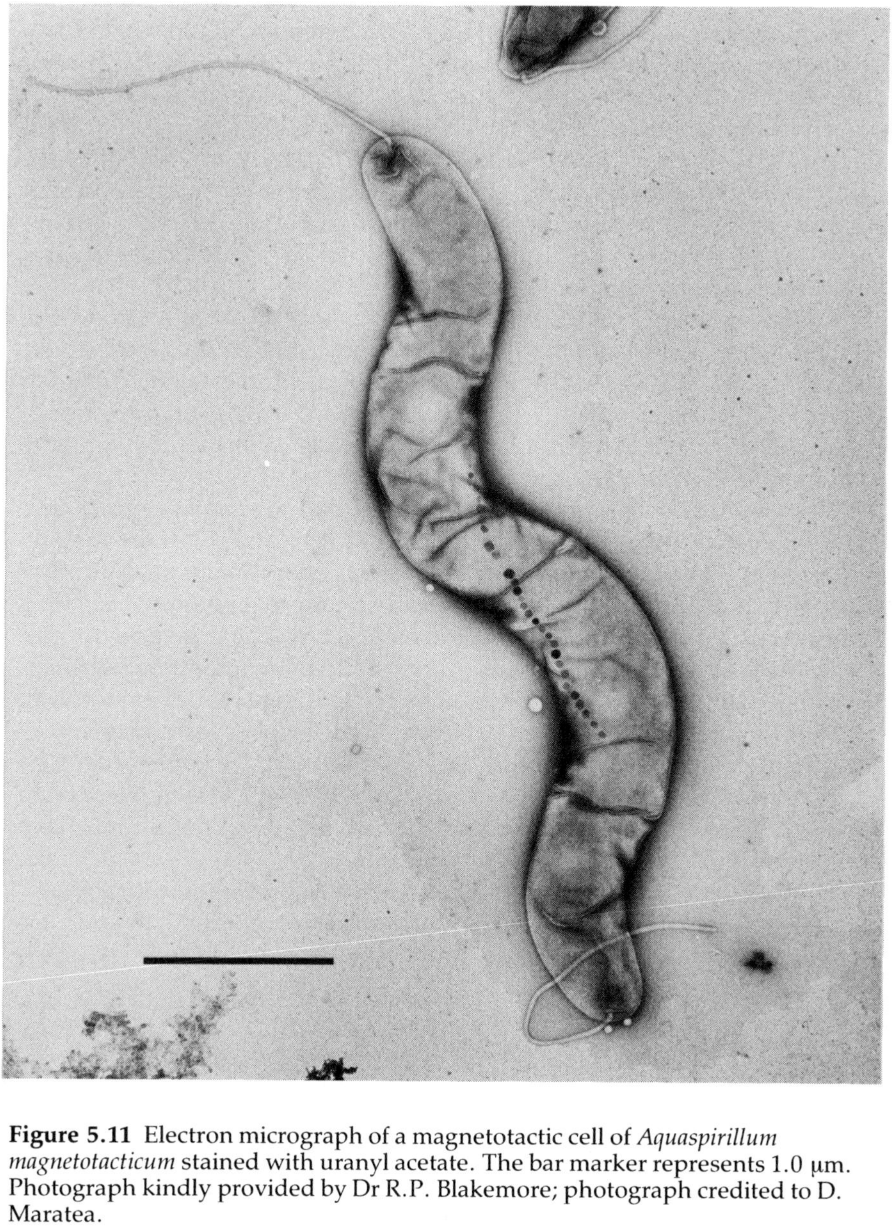

Figure 5.11 Electron micrograph of a magnetotactic cell of *Aquaspirillum magnetotacticum* stained with uranyl acetate. The bar marker represents 1.0 μm. Photograph kindly provided by Dr R.P. Blakemore; photograph credited to D. Maratea.

Magnetotactic cells accumulate iron approximately 20 000- to 40 000-fold over its extracellular concentration. Siderophores (Chapter 3) have not been detected and in view of the relatively high iron concentration in the environment of these organisms ($\approx$1 mg l^{-1}), such high-affinity systems may not be required. Consideration of the mechanism of Fe_3O_4 synthesis from ferrous iron (the form likely in the environment) suggests that cells might derive some energy from this oxidation, but the energy yield is such that magnetite should not be regarded simply as a byproduct of iron oxidation.

Mössbauer spectra of *A. magnetotacticum* show, in addition to the typical six-line spectrum of magnetite, a central pair of lines attributable to ferritin. This is a significant finding since (i) ferritin was not known to be present in prokaryotes (although functional equivalents are recognized; Section 5.5), and (ii) ferritin is an intermediate in magnetite biomineralization in a marine chiton mollusc.

Other physical techniques have been applied to magnetosomes and have confirmed the presence of Fe_3O_4 (14–79% by weight of the magnetosome) but the occurrence of alternative magnetic minerals such as other oxides of iron or cubic iron sulphides in certain magnetotactic bacteria cannot be ruled out. Mössbauer spectroscopy of isolated magnetosomes shows high-spin Fe(III) in oxygen coordination. In some coccoid magnetotactic bacteria, X-ray analysis has revealed substantial amounts of phosphorus, localized in large spherical bodies with potassium.

A magnetotactic *Chlamydomonas* sp. has also been discovered (cited by Blakemore, 1982) containing a permanent magnetic dipole about ten times greater than those of magnetotactic bacteria. Magnetosomes have not been reported.

The occurrence of magnetite in micro-organisms has important consequences for the biogeochemical cycling of metals, since death and lysis of the organisms will result in sedimentation of fine magnetic particles.

5.10 ROLES FOR COPPER

Copper has an essential role in many enzymes, especially those involved in electron transfer and in the transport of O_2 and the catalysis of its reactions.

5.10.1 Copper in cytochrome oxidases

The complement of metals in mitochondrial cytochrome *c* oxidase is generally taken to be two irons (haem *a* and *a*$_3$) and two coppers.

However, as described in Sections 5.2.2(b) and 5.2.2(c), a third copper is now thought to be present. One of the coppers (Cu_B) is in close proximity to haem a_3, possibly even sharing a ligand (Figure 5.2). It is to this reduced iron-copper couple that oxygen is bound and reduced to water in two 2-electron steps. An Fe(III)-O-O-Cu(II) μ-peroxo-bridged species may be an intermediate. The location of the two copper atoms on the various subunits of the enzyme is still unclear, but (i) the sequence homology of subunit II with other copper-containing polypeptides and (ii) the association of coppers with certain subunits after 'gentle' dissociation of the enzyme suggest that the metals are bound to subunits I and/or II. A current review (Steffens, Biewald and Buse, 1987) is that subunit I binds two copper atoms and both haems and thus constitutes the 'universal' oxidase entity of this type. Subunit II binds one copper and functions as an electron conductor. Many details of the structure and function of the coppers remain obscure.

The copper in the purified two-subunit *P. denitrificans* enzyme gives an ESR spectrum very similar to that of the mitochondrial enzyme but, from cells grown in the presence of manganese, the signal is largely obscured by the intense spectrum of tightly-bound Mn(II) (Seelig *et al.*, 1981). This metal probably has no catalytic role.

The optical absorbance properties of the two copper atoms have long been in doubt. It is now widely believed that the broad band observed in both the bacterial (Poole, 1983) and mitochondrial enzymes (Wikstrom, Krab and Saraste, 1981) at 820–840 nm is largely due to Cu_A, though Chance and coworkers argue that both coppers contribute more equally. Interpretation of further broad bands at 740 nm and 655 nm is also controversial; none of these studies have been pursued extensively with microbial oxidases.

A promising approach to these problems that exploits the advantages of microbial systems is copper-limited growth in continuous culture. Such studies with copper-limited yeast and *Neurospora crassa* have demonstrated the loss of visible absorption bands of the *a*-type haems but retention of the oxidase apoprotein and porphyrin *a* (Keyhani and Keyhani, 1980). It remains to be seen whether controlled copper deficiency can be used to yield an oxidase defective only in copper – a powerful tool in studies of the properties and spectroscopic features of the coppers – or whether copper deficiency inevitably results in diminution of haem *a* synthesis as well. The few bacterial studies are more encouraging; membranes from various cyanobacteria grown in copper-depleted medium showed severely depressed rates of cytochrome *c* oxidation but unchanged levels of haems *a* and a_3 (Peschek *et al.*, 1982).

Copper-limited growth of *Paracoccus denitrificans* in a chemostat (Hubbard, Poole and Hughes, 1986) results in loss of the near infra-red optical signal generally attributed to Cu_A as well as diminution of the haems *a* and a_3. Copper may also be removed *in vitro* by dialysis of the purified enzyme against cyanide or treatment with bathocuproine sulphonate. This is an undesirably harsh treatment, however, since extremes of pH may be required and cyanide is a ligand for haem a_3. These methods seem to result in rearrangement of the axial ligands to iron in cytochrome a_3 (Weintraub, Muhoberac and Wharton, 1982); the relation of such a situation to environmentally induced copper depletion is unknown.

5.10.2 Copper in oxidases that reduce oxygen to peroxide

In this category copper is more common than iron. Little is known of the mechanism of O_2 activation (Malmström, 1982). In galactose oxidases, a 'non-blue' oxidase, galactose binds to Cu(II) but cannot reduce it. The role of the metal seems to be to interact with the substrate which then reacts directly with O_2 without any role for the metal in electron transfer.

5.10.3 Copper in blue oxidases

The blue copper oxidases in bacteria, algae, fungi (and higher plants) contain copper in an unusual coordination site and are so-called because of the colour imparted by an unusually strong charge-transfer absorption band near 600 nm. They constitute a group that is complex with regard to the prosthetic groups, with four (in laccase) or more (in ascorbate oxidase and ceruloplasmin) copper ions in at least three coordination environments (Malmström, 1982).

Type 1 Cu(II) ions give a strong blue colour and a distinctive ESR spectrum. They are found in some small proteins such as azurin, the electron donor to cytochrome cd_1 in *Ps. aeruginosa*, plastocyanin (Section 5.10.4) and recently in a soluble *p*-phenylenediamine oxidase of unknown function from *Nitrosomonas* (Di Spirito *et al.*, 1985). Type 2 Cu(II) has ESR properties more typical of simple Cu(II) complexes, but an exceptionally high affinity for anions, such as F^-, which act as inhibitors. Type 3 copper ions constitute the O_2-reducing site in the blue oxidases (e.g. laccase) and consist of an antiferro-magnetically coupled Cu(II)-Cu(II) pair (ESR-undetectable) as in the monooxygenase, tyrosinase (Section 5.4.2(c)) and the O_2 carrier, haemocyanin.

Laccase, common in fungi, oxidizes organic substrates and contains four coppers, one Type 1, one Type 2 and two Type 3. The catalytic mechanism has been extensively studied in *Polyporus versicolor*. An arsenal of spectroscopic techniques is being used to describe the structure of the copper sites. The enzyme must be reduced by the organic substrate before reaction with O_2 occurs. Electrons are taken up one at a time by the Type 1 Cu(II) and Type 2 Cu(II). Reduction of the Type 3 copper pair occurs by an intramolecular transfer of two electrons from the reduced Type 1 and Type 2 sites. Reactions of the reduced enzyme with O_2 give a peroxy intermediate species, with the Type 3 coppers as the H_2O_2-binding site. Further reduction then proceeds in one-electron steps mediated by the Type 1 copper, leading to the formation of an oxygen radical intermediate at the Type 3 copper site. As in cytochrome oxidase (Section 5.2.2(b)), the thermodynamically unfavourable formation of O_2^- is avoided, in this case by reaction of O_2 with the Type 3 site, already reduced by two electrons, allowing direct reduction of the ligand to O_2^{2-}.

5.10.4 Azurin and other small copper-containing proteins with redox roles

A number of copper proteins of low M_r and generally positive redox potential, comprising single polypeptide chains with one copper atom per molecule, have been identified (Table 5.9). Azurins, for example, are 'blue' copper proteins whose solubility and relative simplicity have led to extensive studies of their electronic and magnetic properties and the kinetics of their reactions with inorganic model oxidants and reductants or with other proteins. The precise roles *in vivo* are not completely

Table 5.9 Small copper-containing redox proteins

Protein	Source	M_r	Reference
Azurin	*Ps. fluorescens*	14 000	Ambler and Brown (1967)
	Paracoccus denitrificans	13 790	Martinkus *et al.* (1980)
Plastocyanin	*Chlamydomonas reinhardii*	13 000	Gorman and Levine (1966)
Rusticyanin	*Thiobacillus ferrooxidans*	16 500	Cox and Boxer (1978)
Stellacyanin	*Rhus vernicifera* (Japanese lac tree)	20 000	Peisach, Levine and Blumberg (1967)

understood. In *Ps. aeruginosa*, azurin donates electrons to the cytochrome cd_1 complex (Section 5.2.2(c)); a facile electron transfer between these components is also observed *in vitro* and has been extensively studied (Poole, 1983).

Plastocyanins were first isolated from *Chlorella* and have since been described in many green plants, eukaryotic algae and cyanobacteria. The french bean (*Phaseolus*) plastocyanin, containing 100 amino acid residues, is the smallest copper protein known. Plastocyanins are involved in photosynthetic electron transfer from Photosystem II to Photosystem I and fulfil in certain algae the role of a cytochrome *c* in higher plants. Indeed, in *Anabaena, Scenedesmus* and *Chlamydomonas*, plastocyanin and a soluble cytochrome *c* can function interchangeably according to the copper content of the medium (Wood, 1978). Copper-limited growth of *Scenedesmus* results in continued synthesis of the metal-free apoprotein, which can be reconstituted with copper, a process competitively inhibited by silver ions (Bohner, Sandman and Böger, 1981). A redox role for plastocyanin is facilitated by the copper-binding site, whose geometry is a compromise between those preferred by Cu(II) (square planar) and Cu(I) (tetrahedral) (Chapter 1).

Rusticyanin is, as its name suggests (Cobley and Haddock, 1975), a protein isolated from the iron-oxidizing *Thiobacillus ferrooxidans*. The enzyme is stable at low pH, commensurate with its role in this acidophile, and spectrally resembles azurin and plastocyanin. It may be the initial acceptor of electrons during Fe(II) respiration (see Section 5.7).

5.10.5 Copper in the reduction of nitrite and nitrous oxide

The anaerobic reduction of nitrate provides a mechanism of energy generation for certain facultative and a few obligatory anaerobic bacteria. In some organisms, the nitrite thus formed is metabolized further in the process of denitrification, ultimately yielding nitrogen. Copper is known to be involved in the action of certain of the enzymes involved, although others are four-haem metalloproteins (such as *Pseudomonas* cytochrome cd_1) or sirohaem-containing enzymes (as in *E. coli*).

The intermediates proposed and their formal oxidation states are as follows:

$$NO_2^- \ (+3) \overset{(1)}{\rightarrow} NO \ (+2) \overset{(2)}{\rightarrow} N_2O \ (+1) \overset{(3)}{\rightarrow} N_2 \ (O)$$

The nitrite reductase (which catalyses reaction (1)) of *Achromobacter cycloclastes* has a M_r of 69 000 and contains two copper atoms per molecule (Matsubara, Frunzke and Zumft, 1982), whereas the *Alcaligenes faecalis* enzyme is a tetramer of four identical subunits each of M_r 30 000 and containing one Cu(II) (Kakutani *et al.*, 1981).

The dissimilatory nitrite reductase of *Alcaligenes* spp. (NCIB 11015) is unique in possessing only Type 1 copper (Masuko *et al.*, 1984).

Clues to the role of copper in N_2O reductase (reaction 3) come from the finding that copper-limited *Ps. perfectomarinus* and *Alcaligenes* spp. accumulated N_2O and that anaerobic growth on N_2O required copper (Iwasaki, Saigo and Matsubara, 1980; Matsubara, Frunzke and Zumft, 1982; Iwasaki and Terai, 1982). In several pseudomonads, N_2O respiration seems consistently associated with the presence of a large (M_r 120 000) copper-containing protein (Matsubara and Zumft, 1982). In *Ps. perfectomarinus*, this large protein contains nine coppers per mole and consists of two subunits of identical size. A small copper-containing protein (M_r 38 000) is also present but appears not to be required for N_2O reduction. However, it has been pointed out (Snyder and Hollocher, 1984) that the specific activity of the purified 120 000 M_r protein is very low and that it can be separated from the N_2O reductase activity (M_r 89 000). The role of the larger copper protein remains uncertain.

5.10.6 Other copper-containing oxidoreductases

There are a number of microbial enzymes that possess copper together with other metal centres. Recently reported examples include a copper-containing iron-sulphur flavoprotein, which has NADH oxidase activity in *Acholeplasma laidlawii* (Reinards, Kubicki and Ohlenbusch, 1981) and a copper- and haem-containing glycerol oxidase from *Aspergillus japonicus* (Uwajima, Shimizu and Terada, 1984). The roles of the various metals are not known at present.

5.11 SUPEROXIDE DISMUTASES

5.11.1 Function

These enzymes (superoxide: superoxide dismutase, EC 1.15.1.1, here abbreviated to SOD) are abundant metalloproteins (Fridovich, 1981), which play an important role in defence against the biological production of the unstable superoxide radical (O_2^-). They catalyse the disproportionation of the radical leading to the production of peroxide according to the equation:

$$2H^+ + 2O_2^- \rightarrow O_2 + H_2O_2$$

The rate constant is close to that of the diffusion-controlled limit. The peroxide is in turn scavenged by catalases and peroxidases (hydroperoxidases; Section 5.2.3).

Production of the highly reactive superoxide, peroxide and hydroxyl (·OH) radicals is generally avoided in cellular reduction of O_2 by ensuring complete, 4-electron reduction to water (Section 5.2.2). Nevertheless, cells contain reductants (e.g. ferredoxins, haemoglobin and thiols) that can interact directly with O_2 to give such intermediates. In addition, some enzymes directly catalyse superoxide generation; these include xanthine oxidase and certain dioxygenases. Superoxide may peroxidize lipids, inactivate enzymes and can kill bacteria. Indeed, activated phagocytic cells release superoxide as an antibacterial measure; as a counter-measure, *Mycobacterium lepraemurium*, which grows inside macrophages, has SOD at up to 70% of its extractable protein, whilst virulent strains of *Nocardia asteroides* appear to secrete FeSOD to combat the oxidative metabolic burst it induces in human polymorphonuclear phagocytes (Beaman *et al.*, 1983).

Such a stringent selection pressure has resulted in the appearance of only three classes of the enzyme, namely managanese superoxide dismutase (MnSOD), iron superoxide dismutase (FeSOD) and an enzyme containing both copper and zinc (CuZnSOD). Extensive amino acid sequence homology is exhibited by the first two. All SODs operate by a redox cycle in which the metal at the active site is reduced by one O_2^-, then reoxidized by another O_2^-. In the case of CuZnSOD, the copper oscillates between Cu(I) and Cu(II), while in the other types the metals cycle between the divalent and trivalent states.

5.11.2 The copper-zinc enzymes

Copper-zinc superoxide dismutase is characteristic of the cytosol of eukaryotes and has been isolated from *Neurospora* and yeast. An interesting exception is the symbiotic bacterium *Photobacterium leiognathi* which has acquired the enzyme probably as a result of gene transfer during a long-standing relationship with the host pony fish (Martin and Fridovich, 1981). More recently, a CuZnSOD (or bacteriocuprein) has been purified and characterized from the free-living bacterium *Caulo-*

bacter crescentus (Steinman, 1982). Analysis of tryptic peptides from the enzyme indicates that the enzyme is not as homologous to the eukaryotic CuZnSOD as is the enzyme in *Ph. leiognathi*.

The best-studied CuZnSOD, however, is that purified from bovine erythrocytes. It contains two identical subunits (M_r 16 000), each having 151 amino acid residues and bearing one copper and one zinc. X-ray crystallography shows each subunit to be essentially cylindrical. A pair of loops projects from the top and bottom of the cylinder enclosing and creating the active site. The copper is bound by four histidine imidazole ligands plus water. The ligands to the zinc, which probably has a role in formation and stabilization of the active site, are tetrahedrally arranged and one (His-61) bridges the copper to the zinc. During catalysis, the copper centre is reversibly oxidized and reduced by successive encounters with O_2^-, giving O_2 and H_2O_2.

$$\mathrm{ECu^{II}} + \mathrm{O_2^-} + \mathrm{H^+} \rightleftharpoons \mathrm{E'Cu^{I}} + \mathrm{O_2}$$
$$\mathrm{E'Cu^{I}} + \mathrm{O_2^-} + \mathrm{H^+} \rightleftharpoons \mathrm{ECu^{II}} + \mathrm{H_2O_2}$$

Recent studies of the yeast enzyme indicate that the active sites may be non-identical on the two subunits with respect to (i) the dissociation constants for the copper–protein and zinc–protein complexes (Dunbar, Johansen and Uchida, 1982), and (ii) titration with cobalt of the metal sites in the apoenzyme (Dunbar, Holmquist and Johansen, 1984).

5.11.3 Iron superoxide dismutases

The FeSODs and MnSODs are widespread amongst micro-organisms (Tables 5.10 and 5.11) and their similarities suggest evolution from a common ancestral protein. The iron-containing enzymes are found predominantly in prokaryotes but also in a few plants, *Euglena* and trypanosomatids (Le Trant *et al.*, 1983). The M_r values of the subunits generally lie in the range 18 000–22 000 and the enzymes are usually dimeric (see Table 5.10, however). One iron per subunit is found; lower values (e.g. for *E. coli*; Table 5.10) probably reflect some metal loss during isolation. In the resting state, the iron is present as Fe(III). The finding of Zn plus Fe in the *Methanobacterium bryantii* enzyme (Kirby, Lancaster and Fridovich, 1981) has been attributed to partial occupancy of the Fe sites by Zn or the adventitious binding of Zn to the protein.

Recently, the structure of the FeSOD from *Ps. ovalis* has been described at high (2.9×10^{-10} m) resolution (Ringe *et al.*, 1983). The enzyme (M_r =

Table 5.10 Iron superoxide dismutases in micro-organisms

Source	M_r	Subunits	Fe/mol	Reference
Bacteria				
Chromatium vinosum	82 000	2 × 41 000	2.0	Kanematsu and Asada (1978)
Desulfovibrio desulfuricans	43 000	2 × 21 500	1.6	Lumsden and Hall (1975)
Escherichia coli	39 000	2 × 18 000	1.0–1.8	Slykhouse and Fee (1976)
	37 000	(hybrid form SOD: see text)	0.8	Dougherty, Sadowski and Baker (1978)
Methanobacterium bryantii	91 000	4 × 24 000	2.7*	Kirby, Lancaster and Fridovich (1981)
Mycobacterium tuberculosis	88 000	4 × 22 000	4.0	Kusunose *et al.* (1976)
Cyanobacteria				
Anacystis nidulans	37 000	2 × 18 500	1.0	Cséke *et al.* (1979)
Plectonema boryanum	37 000	2 × 19 000	0.94	Misra and Keele (1975)
Spirulina platensis	37 000	2 × 18 400	1.0	Lumsden, Cammack and Hall (1976)
Protozoa				
Crithidia fasciculata	43 000	2 × 23 000	1.4	Le Trant *et al.* (1983)
Euglena gracilis	45 000	2 × 22 000	1.9	Kanematsu and Asada (1979)

* Plus 1.7 atoms Zn/mol, believed to be adventitiously bound.

Table 5.11 Examples of manganese superoxide dismutases in micro-organisms, and, for comparison, in higher animal cells

Source	M_r	Subunits	Mn/mol	Reference
Bacteria				
Bacillus subtilis	45 000	2 × ?	1.1	Tsukuda *et al.* (1983)
Bacillus stearothermophilus	40 000	2 × 20 000	4	Bridgen, Harris and Kolb (1976)
Nocardia asteroides	100 000	4 × 25 000	1–2*	Beaman *et al.* (1983)
Escherichia coli	40 000	2 × 20 000	1.2	Weisiger and Fridovich (1973)
Rhodococcus bronchialis	80 000	4 × 21 000	2.2	Ichihara *et al.* (1980)
Streptococcus mutans	42 000	2 × 19 000	1.2	Weisiger and Fridovich (1973)
Thermus aquaticus	80 000	4 × 21 000	2	Sato and Harris (1977)
Thermus thermophilus	80 000	4 × 20 000	2.0	Sato and Nakazawa (1978)
Eukaryotic algae				
Porphyridium cruentum	40 000	2 × 20 000	2.0	Misra and Fridovich (1977)
Yeast				
Saccharomyces cerevisiae	100 000	4 × 22 700–25 000	3.8	Ravindranath and Fridovich (1975), Ditlow *et al.* (1982)
Higher animals				
Chicken liver mitochondria	80 000	4 × 20 000	2.3	Weisiger and Fridovich (1973)

* Also contains Fe and Zn.

44 000) comprises two identical subunits roughly triangular in form (52 × 40 × 32 × 10^{-10}m). To each is bound a high-spin Fe(III) atom by four protein ligands. The FeSOD from *E. coli* is indistinguishable (Stallings *et al.*, 1983).

5.11.4 Manganese superoxide dismutases

Mn(III)SODs (Table 5.11) are found in prokaryotes and in the mitochondria or chloroplasts of eukaryotes, lending some support to the endosymbiont hypothesis of mitochondrial and chloroplastic evolution. The amino acid sequences of the yeast and bacterial enzymes show considerable homology (Ditlow *et al.*, 1982). The subunit M_r values, amino acid sequences and oligomeric structure also show strong similarities to the FeSODs. Indeed, *Propionibacterium shermanii* and *Streptococcus mutans* can synthesize either Fe or MnSOD, with an apparently identical protein moiety, depending on the metal supplied (Meier *et al.*, 1982; Martin *et al.*, 1984). However, the bacterial enzymes are often dimeric, whereas those from yeast and higher animals are tetrameric (Table 5.11); all contain 0.5–1.0 Mn atoms per subunit. In addition to the examples in Table 5.11, dimeric enzymes of M_r in the range of 40K–45K have been reported recently in *Mycobacterium lepraemurium, Serratia marcescens, Paracoccus denitrificans* and *Streptococcus faecalis*.

An FeSOD has been identified in *E. coli* (Dougherty, Sadowski and Baker, 1978; Clare, Blum and Fridovich, 1984) that consists of two subunits which appear to be a hybrid of MnSOD and FeSOD. This form can be generated by mixing the Mn and Fe-enzymes, removing the metals (with 8-hydroxyquinoline and 2.5 M guanidinium chloride) and dialysis against Mn(II) or Fe(II) salts. The apohybrid can be reconstituted to an active form with either metal, whereas only Mn(II) restores activity to the MnSOD (Ose and Fridovich, 1979) and iron competes with manganese for binding. It seems likely that most metals (e.g. Zn, Co and Ni) bind exclusively at the active site whereas Fe can bind to other sites as well. Interestingly, the iron (0.9 atom/mol) in the *Rhodococcus bronchialis* enzyme (Ichihara *et al.*, 1980) is tightly bound, together with Mn, throughout electrophoretic preparation of the enzyme; no function can yet be ascribed to the iron. Characteristics of both Mn and FeSOD are also exhibited by the unusual enzyme from *Nocardia asteroides*; Fe, Mn and Zn are contained in a single, electrophoretically distinct tetrameric enzyme (Beaman *et al.*, 1983).

Recently, Stallings *et al.* (1984) have elucidated the structure of a tetrameric MnSOD from *Thermus thermophilus*. In accord with partial amino acid sequences available, the highly helical fold of Fe dismutase is conserved in MnSOD.

5.11.5 Modulation of activity

Cyanide reversibly inhibits CuZnSOD by complexing to Cu(II) but does not inhibit the manganese or iron enzymes. Azide can also be used to distinguish between the enzyme types (Misra and Fridovich, 1978). The CuZnSOD and FeSOD are irreversibly inhibited by H_2O_2 but only at non-physiological concentrations and alkaline pH. Inactivation of the CuZnSOD also results from removal of copper by diethyldithiocarbamate.

The role of superoxide dismutase in protection against O_2^- toxicity has been demonstrated most convincingly in *E. coli* by Fridovich and others. Thus, when *E. coli* is grown under strictly anoxic conditions, only the FeSOD is formed; even the immeasurably low levels of O_2 in unstirred cultures open to the air suffice to induce the MnSOD and the induction correlates well with resistance to hyperbaric O_2. Secondly, paraquat and other compounds (e.g. pyocyanine, menadione) that increase the rate of intracellular O_2^- production all induce MnSOD, and result in increased resistance to hyperoxia. In *E. coli*, the FeSOD is constitutive, being maintained even during strictly anaerobic growth, whereas the Mn-containing enzyme appears to allow regulation of total SOD activity to match O_2^- generation. Superoxide itself, however, may not be the inducer of the MnSOD. The evidence comes from *E. coli* strains carrying the structural gene for the iron enzyme (*sodB*) on plasmids that allow overproduction of the enzyme; MnSOD induction by O_2 was quantitatively similar in cells differing 7-fold in their levels of FeSOD (Nettleton *et al.*, 1984).

Regulation of SOD levels can also be manipulated by the metal content of the medium. Thus, elevated manganese or iron levels also increase activities of the corresponding SOD, possibly by stimulating conversion of the apo to the holoenzyme (Pugh, Diguiseppi and Fridovich, 1984). Iron represses synthesis of the MnSOD; an iron-containing protein, but not FeSOD itself, has been implicated (Hassan and Moody, 1984; Moody and Hassan, 1984).

Certain organisms lack superoxide dismutase but contain instead millimolar levels of Mn(II) as a functional replacement (Section 5.12.2(b)). Iron in the magnetotactic bacteria (Section 5.9) may also assume such a role.

5.12 MANGANESE

5.12.1 Non-redox roles

Manganese serves both redox and non-redox roles in biological processes. The latter include activation of a number of enzymes *in vitro*, but the binding of the metal is often of low affinity and the situation *in vivo* is unclear. The best-studied high-affinity, non-redox Mn enzyme is pyruvate carboxylase, which is equally well activated by Mg(II). Other manganese-activated enzymes include DNA polymerase I and RNA polymerase. For the former, NMR studies have shown that the metal facilitates a rearrangement in the structure of the substrate (thymidine triphosphate) that is necessary for insertion into the double helix. In RNA polymerase, the Mn(II) is on the elongation side of the enzyme, close to ATP in the enzyme–Mn(II)–substrate complex, thus providing a conformation suitable for base pairing with the DNA template (Eichhorn, 1980). The α-isopropylmalate synthetase (EC 4.1.3.12) from *Alcaligenes eutrophus*, the first enzyme of the biosynthetic pathway to leucine, appears to be an Mn(II)-requiring protein (Wiegel, 1978; Section 4.3.5).

5.12.2 Redox roles

(a) In photosynthesis

Although managanese has long been implicated as an essential component of Photosystem II, its precise role remains uncertain (Amesz, 1983; Hipkins, 1983). Early experiments with photosynthetic microorganisms established that manganese was involved in photosynthesis. Manganese-deficient algae showed lower rates and quantum efficiencies of photosynthesis yet, in the case of *Ankistrodesmus braunii*, retention of anaerobic light-induced hydrogen uptake (driven by Photosystem I), pointed to a role for manganese in O_2 evolution (Amesz, 1983).

Isolated chloroplasts contain about 27–29 Mn/400 chlorophylls (400 chlorophylls corresponding to one photosynthetic unit). The manganese can be resolved by washing into loosely-bound metal and bound metal. Only some of the latter, sometimes referred to as the 'very strongly bound manganese' (possibly 3–6 Mn/400 chlorophylls) is believed to be associated with the oxidation of water. Perhaps the best evidence for a direct involvement of Mn in water oxidation comes from higher plants: ESR studies show that the quantity of heat-releasable Mn from spinach chloroplasts after flash illumination shows a cyclic oscillation with a period of four. This distinctive pattern is similar to that of flash-induced oxygen evolution, which has often been attributed to the storing of charge by Mn until two molecules of water are oxidized to O_2 by four oxidizing equivalents.

There have been several reports of the isolation of Mn–protein complexes from diatoms, green algae and higher plants. These preparations exhibit various metal contents and sometimes include copper. Recent claims for the isolation of manganese-containing proteins, which appear to reconstitute O_2 evolution when incorporated in liposomes containing extracted chloroplast material, have been equivocal (Barber, 1984). The development of methods for preparing purified, actively O_2-evolving Photosystem II preparations from blue-green algae (and spinach chloroplasts) and which retain the strongly-bound manganese appears to hold future promise (Amesz, 1983).

(b) Manganese (II) as O_2^- scavenger

The role of manganese (III) in certain superoxide dismutases was discussed in Section 5.11. However, a few O_2-tolerant micro-organisms lack any superoxide dismutase. These include several mycoplasmas, two disseminating strains of *Neisseria gonorrhoeae* and *Lactobacillus plantarum* (Archibald and Fridovich, 1981a, b). In the last organism, high (20–25 millimolar) intracellular levels of Mn(II) take the place of SOD in scavenging O_2^-, and endow a capacity for detoxification that is comparable to that exhibited in aerobically grown *E. coli*. The activity in *L. plantarum* extracts is heat-stable, dialysable and inhibited by EDTA. Mn(II) deprivation results in loss of O_2-tolerance. This substitution of millimolar concentrations of Mn(II) for micromolar SOD seems to be common in the lactic acid bacteria (Archibald and Fridovich, 1981a, b) but two strains have been found that lack both mechanisms and are consequently very O_2-sensitive. The Mn(II) is primarily associated with a large complex (>0.45 μm) believed to be a polyphosphate-protein aggregate (Archibald and Fridovich, 1982a). It is thought that Mn(II) is oxidized by O_2^- to give Mn(III), a strong oxidant which could react with a reductant such as NAD(P)H. Recent studies *in vitro* (Archibald and Fridovich, 1982b) show that although Mn(II) is always oxidized to Mn(III) with the production of H_2O_2, the fate of the Mn(III) is dictated by the experimental medium. Interestingly, in pyrophosphate buffer, Mn(III) is re-reduced by a second O_2^-, making the reaction a true metal-catalysed dismutation, like that catalysed by SOD (Section 5.11). The finding that *Methanobacterium bryantii*, a representative of the primitive Archaebacteria, possesses SOD (Kirby, Lancaster and Fridovich, 1981) suggests that high intracellular Mn(II) concentrations are probably not the ancestors of SOD activity but are rather an adaptation to life in manganese-rich environments, such as decaying plant material.

(c) Pseudocatalase

Catalases from higher organisms are haem proteins and many micro-organisms possess similar enzymes (Section 5.2.3). However, some bacteria are unable to synthesize haem but can produce catalases that are cyanide and azide-insensitive and have been called pseudocatalases. The possession of such an enzyme by *Lactobacillus plantarum* is one of several peculiarities in its defences against oxygen toxicity imposed by, or adopted in response to, the independence of this organism on iron. Kono and Fridovich (1983a) have recently purified this pseudocatalase; the enzyme (M_r 172 000) is composed of six non-covalently associated subunits of M_r 28 300, each containing about one atom of Mn(III) but no Fe, Co, Cu or Zn. The enzyme is pink in concentrated solution, the optical spectrum resembling MnSOD.

The functional significance of the enzyme has been demonstrated using a strain lacking the Mn catalase which, in the stationary phase of growth, rapidly loses viability when compared with a catalase-positive strain (Kono and Fridovich, 1983b). Growth in the presence of haemin, which allows catalase synthesis, protects the deficient strain from this loss of viability.

5.13 IRON AND COBALT IN NUCLEOTIDE AND NUCLEIC ACID BIOSYNTHESIS AND DEGRADATION

In most organisms, deoxyribonucleotides are produced by direct reduction of the corresponding ribonucleotides. The enzyme involved (ribonucleotide reductase, EC 1.17.4) has been studied extensively in *E. coli* and *Lactobacillus leichmannii*, although many other bacteria and eukaryotic microbes contain an Fe-ribonucleotide reductase. The *E. coli* enzyme (Table 5.12) is a large molecule (M_r 240 000) with two subunits stabilized by a Mg^{2+} ion. The iron is essential for enzyme activity. Protein B2 of the *E. coli* enzyme contains two non-identical, high-spin Fe(III) atoms in an antiferromagnetically coupled binuclear complex. The presence of an ESR-detectable radical, assigned to a tyrosine residue, is closely linked to the presence of iron. The function of the metal seems to be to generate this tyrosine radical by catalysing a one-electron aerobic oxidation and

Table 5.12 Properties of purified ribonucleotide reductases

Organism	Cofactor	Other metal requirements	M_r	Reference
E. coli	2Fe/B2 subunit	Mg(II) (stabilizer)	240 000 1B1 (160 000): 1B2 (78 000)	Thelander and Reichard (1979)
Anabaena 7119	5′ deoxyadenosyl cobalamin	Ca(II)	72 000	Gleason and Frick (1980)
Lactobacillus leichmannii	5′ deoxyadenosyl cobalamin	None	76 000	Thelander and Reichard (1979)
Brevibacterium ammoniagenes *Micrococcus luteus*	Mn(II) or Mg(II)	–	–	Schimpff-Weiland, Follman and Auling (1981)
Euglena gracilis	5′ deoxyadenosyl cobalamin	Mg(II)	400 000 (tetramer?)	Hamilton (1974)

then stabilizing the radical for reduction of the –OH groups in the substrate (Thelander and Reichard, 1979).

The *Lactobacillus* enzyme (Table 5.12) is a smaller molecule (M_r 76 000) with no subunit structure or cation requirements and uses cobalt, as adenosylcobalamin, as cofactor. This presumably serves a similar role to iron in the *E. coli* enzyme. Such adenosylcobalamin-dependent reductases are found mostly in bacteria (Gleason and Hogenkamp, 1972) and cyanobacteria (Gleason and Frick, 1980; Table 5.12) but have also been reported in *Euglena* spp. and the fungus *Phitomyces chartarum* (Thelander and Reichard, 1979). The physiological electron donor for both the *E. coli* and *Lactobacillus* enzymes is the small sulphydryl protein, thioredoxin.

There is a preliminary report of a Mn^{2+} or Mg^{2+}-requiring enzyme in certain gram-positive bacteria (Table 5.12).

5.14 MOLYBDENUM

5.14.1 Introduction

Molybdenum is found in a number of 'open-sided' metal sites in biological processes. In such cases, a ligand or substrate is allowed close approach or co-ordination to the metal and, in the case of molybdenum-containing enzymes, these sites are coupled to electron transfer. Thus, the additional electron transport groups that occur in molybdenum (Table 5.13) are iron-sulphur centres (as in nitrogenase, xanthine oxidase and respiratory nitrate reductase), FAD (xanthine oxidase) or haem (in sulphite oxidase).

The molybdenum-containing cofactors (Moco) from a range of molybdoproteins (e.g. in Table 5.13) are functionally interchangeable and may be identical. Indeed, the factor is normally assayed (Hawkes and Bray, 1984) by *in vitro* complementation of the *Neurospora crassa* mutant *nit-1*, which lacks co-factor but synthesizes the apoprotein for nitrate reductase. Bacterial nitrogenase is the single exception, since the cofactor here also contains iron (FeMoco; see Section 5.14.2(a)). The chemical structure of Mocos, where known, is that of a molybdenum–pterin complex (Figure 5.12). The cofactors in *E. coli* and *Neurospora crassa* are synthesized continuously even in the absence of molybdenum. In *E. coli*, molybdenum cofactor is soluble but associated with a protective carrier molecule (≈40 000 Da) from which it is easily removable (Amy and Rajagopalan, 1979). The synthesis of molybdoenzymes such as nitrate reductase, trimethylamine-*N*-oxide reductase and formate dehydrogenase is dramatically reduced in the presence of tungstate (Giordano

Table 5.13 Some molybdenum-containing enzymes‡

Enzyme	Source	Other redox centres present	Reference
*Biotin-sulphoxide reductase	*E. coli*	Activity requires thioredoxin-like protein	del Campillo-Campbell and Campbell (1982)
Carbon monoxide oxidase	Carboxydotrophs	FAD	Meyer (1985)
†Chlorate and tetrathionate reductases	*Proteus mirabilis*	?	Oltmann *et al*. (1979)
Formate dehydrogenase	*E. coli*	FeS, haem, Se	Ingledew and Poole (1984)
Nicotinic acid hydroxylase	*Clostridium barkeri*	FeS, FAD, Se	Dilworth (1983)
Nitrate reductase (assimilatory)	*Chlorella vulgaris*	haem, FAD	Giri and Ramadoss (1979)
*Nitrate reductase (respiratory)	*E. coli*	FeS, cytochrome *b*	Ingledew and Poole (1984)
Nitrogenase (MoFe protein)	Various N_2-fixing prokaryotes	FeS	see text
Sulphite oxidase	*Thiobacillus novellus*	haem	Toghrol and Southerland (1983)
Tryptophan side-chain oxygenase	*Pseudomonas* spp.	protohaem IX + haem-like chromophore	Takai *et al*. (1982)
Xanthine dehydrogenase (= purine hydroxylase) I and II	*Aspergillus nidulans*	FeS, FAD	Arst, Tollervey and Sealy-Lewis (1982)
Xanthine dehydrogenase	*Clostridium acidiurici*	FeS, W, Se	Wagner, Cammack and Andreesen (1984)
Xanthine oxidase		FAD, FeS	Hughes (1987)
Trimethylamine-*N*-oxide reductase	*Escherichia coli*		Ingledew and Poole (1984)

* Certain *bis* (for biotin sulphoxide reductase) and *chl* genes (for nitrate reductase) may be identical, the enzymes sharing a common Moco.
† Both catalyse the reduction of chlorate to chlorite.
‡ EXAFS data have also been reported for several molybdenum-containing enzymes, in both oxidized and reduced forms (Bordas *et al*., 1980; Cramer *et al*., 1978, 1981, 1984).

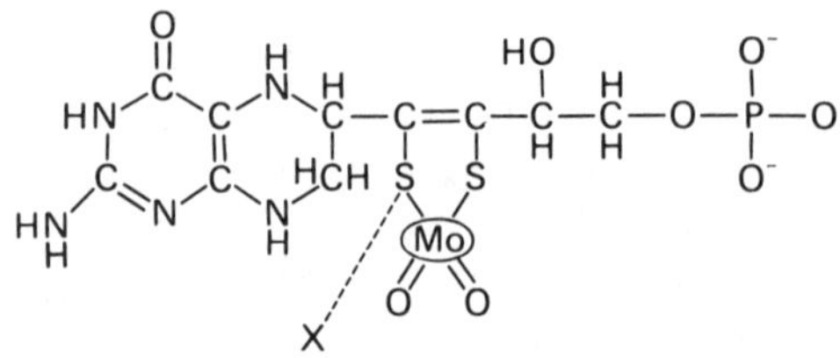

Figure 5.12 Proposed structure of the molybdenum-containing cofactor of nitrate reductase (reproduced with permission from Ingledew and Poole, 1984).

et al., 1984), which competes with molybdenum for the same site. Tungsten-containing (or metal-free) cofactor complements nitrate reductase in the *nit-1* assay forming an intact, but inactive, enzyme (Amy and Rajagopalan, 1979).

The redox chemistry of molybdenum in biological systems is less well understood than that of iron or copper. Its role in nitrogenase, however, has attracted great interest and it is this example that is treated here. Respiratory nitrate reductase was discussed in Section 5.3.5(c).

5.14.2 Nitrogenases

(a) Function

Dinitrogen is a particularly stable, triply-bonded molecule but can be reduced to ammonia by an exclusively prokaryotic enzyme, nitrogenase. The agricultural and biogeochemical significance of this nitrogen fixation is well known (McCormick, 1988) and need not be reiterated here. Nitrogenase catalyses the ATP-dependent reduction of N_2 to NH_3 as well as the reduction of certain other small unsaturated molecules, notably acetylene (Postgate, 1978). Writing the reaction catalysed by nitrogenase as

$$N_2 + 8H^+ + 8e^- + nATP \rightarrow 2NH_3 + H_2 + nADP + nP$$

emphasizes the uncertainty of the ATP involvement and the fact that H_2 evolution is an intrinsic part of the N_2 reduction process (for a discussion of the latter point, see Robson and Postgate, 1980).

The existence of the V-nitrogenase has been demonstrated genetically by deleting the structural genes encoding the well-known Mo-nitrogenase from *Azotobacter*. Such strains still fix N_2, albeit poorly as measured by acetylene reduction, and contain a nitrogenase comprising stoichiometric amounts of vanadium and only traces of molydenum (Robson *et al.*, 1986). Very recently (for a conference report, see Cammack, 1988), Bishop, Eady and collaborators have identified, genetically and biochemically, another nitrogenase which lacks both these metals and contains only iron (Fe-nitrogenase). All three nitrogenases comprise similar large and small subunits, and it is the larger subunit, sometimes called dinitrogenase, at which N_2 is actually fixed, and which bears the respective active cofactor, namely Fe-Mo-co, Fe-Va-co (or putatively Fe-Fe-co). The smaller subunit in each case is an iron-sulphur protein.

Both proteins of Mo-nitrogenase are irreversibly damaged by dioxygen (Robson and Postage, 1980). The smaller, more O_2-sensitive Fe protein binds MgATP and transfers electrons from a suitable donor to the larger protein (Mortenson and Thorneley, 1979). Figure 5.13 is a simplified scheme to illustrate the interaction of the two proteins in the Mo-containing enzyme, which is the subject of subsequent sections.

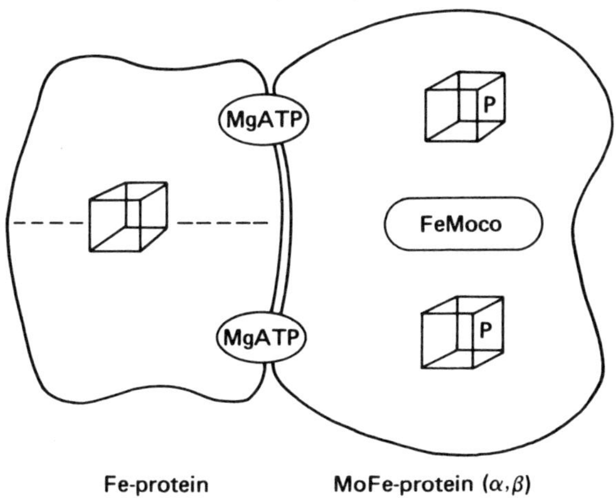

Figure 5.13 This shows the arrangement of subunits and clusters in nitrogenase. Each α, β half of the tetrameric MoFe protein contains two [4Fe-4S] clusters and one FeMoco. These [4Fe-4S] clusters differ from normal four-iron clusters, possibly in the identity of the non-sulphidic ligands, and are termed P clusters. The Fe protein, made up of two subunits, contains one [4Fe-4S] cluster. The reduced Fe protein binds 2 mol MgATP, and after undergoing a conformational change, binds to the MoFe protein. This is followed by hydrolysis of the MgATP and the transfer of one electron to the MoFe protein, and ultimately to the substrate. The Fe protein is thought to protect the N_2-reducing site on the MoFe protein from protons, which will react preferentially with molybdenum-bound hydrides. The two MgATP molecules may act as bridging groups.

(b) The Fe protein

This ('component II') comprises two identical subunits whose size varies somewhat between species. In *Cl. pasteurianum* and *Az. vinelandii*, for example, the M_r values are 57 674 and 64 000, respectively. The four Fe atoms are arranged as a [4Fe-4S] cluster. This has been demonstrated by the extrusion technique, during which the four cysteine thiols liganded to the four Fe atoms are replaced quantitatively by the four thiol ligands of added phenylthiol. The extruded product is easily identified, since the properties of synthetic [4Fe-4S] (phenylthiol) are well established. Two cysteine thiols from each subunit (Cys-97 and Cys-132 for *Az. vinelandii* nitrogenase) are liganded to the [4Fe-4S] cluster, which is bound symmetrically between the subunits.

The reduced iron protein undergoes a conformational change on binding MgATP, which results in a decrease in $E_o \approx 100\,mV$, while the ESR spectrum becomes more axial and there are changes in the MCD spectrum. The complex of (reduced iron protein.MgATP) binds to the MoFe protein and transfers one electron to it. The MgATP is only hydrolysed in the adduct with the MoFe protein.

(c) The MoFe protein

The MoFe protein ('component I') is isolated as an $\alpha_2\,\beta_2$ dimer with total molecular masses for the *Cl. pasteurianum* and *Az. vinelandii* enzymes respectively of 220 and 250 kDa. The protein contains two complex clusters, each containing both iron and molybdenum. An FeMoco has been isolated, containing 6–8 Fe and 4–6 S^{2-} per molybdenum. A more recent estimate of the metal and sulphur composition, with the sulphur determined radiochemically, is $MoFe_6S_{8\text{ or }9}$ (Nelson, Levy and Orme-Johnson, 1983) and is more in line with recent EXAFS studies of the FeMoco. Although this factor has not been shown to catalyse the reduction of N_2 to NH_3, it has been widely studied, compared with various model (i.e. synthetic) FeMo clusters and demonstrated to give fully active protein on addition to a Mo-free MoFe protein.

The most active preparation studied (based on the rate of reduction of the artificial substrate, acetylene) is the MoFe protein from *Klebsiella pneumoniae* (Kp1). This contains, per mol, two Mo atoms and equal amounts (33 ± 3 mol) of Fe and acid-labile sulphur, indicating the presence of four [4Fe-4S] clusters and two iron-molybdenum clusters (FeMoco), each containing six iron atoms (Smith *et al.*, 1980; Kurtz *et al.*, 1979). Core extrusion techniques have confirmed the presence of these clusters in the MoFe protein (Averill, 1983). The [4Fe-4S] clusters differ from normal clusters with this stoichiometry, having two different types of iron atom. Each cluster, designated a 'P' cluster, contains three iron atoms of one type (D atoms) which differ from the fourth. An unusual aspect of the P clusters is that all four iron atoms are Fe(II), giving a fully reduced cluster. As all four iron centres are in the same oxidation state, the difference between them most probably resides in the identity of the non-sulphide ligands.

Other physical techniques are being exploited to explore the structure of the nitrogenase (Shah *et al.*, 1984). Thus, EXAFS has provided some structural information on the local environment of the molybdenum (Cramer, Wahl and Rajagopalan, 1981; Mortenson and Thorneley, 1979) but not of the iron. Low-temperature MCD at high magnetic fields is providing data on the individual paramagnetic centres. Recently, single crystals of the MoFe proteins of *Cl. pasteurianum* and *Az. vinelandii* have been grown and used in preliminary X–ray diffraction studies, in an attempt to reveal ultimately the secondary and tertiary structures of the proteins and the organization of the metal clusters (Weininger and Mortenson, 1982). The potential cysteinyl ligands for the Fe-S and MoFeS centres have been identified in the *Az. vinelandii* protein (Lundell and Howard, 1981). One tryptic peptide of each of the α and β subunits shows considerable homology with bacterial ferredoxins.

Genetic studies may be expected to shed more light on the structure and regulation of the MoFe protein. In *Klebsiella pneumoniae,* four of the fifteen genes in the *nif* cluster are believed to be involved in the biosynthesis of the FeMoco (Imperial *et al.*, 1984). Unlike the Moco of other molybdoenzymes, the FeMoco of nitrogenase is not synthesized constitutively but is repressable, e.g. by ammonia (Pienkos, Shah and Brill, 1977). The use of mutants defective in FeMoco (Ugalde *et al.*, 1984) has shown that under derepressed conditions Mo is scavenged from the medium only if FeMoco can be synthesized, but that FeMoco can be synthesized in the absence of component I.

(d) The mechanism of nitrogenase

It is now accepted that the reduced iron protein in nitrogenase binds two mol MgATP, followed by a change in conformation and a decrease in redox potential. The $(MgATP)_2(Fe)^-$ complex binds to the MoFe protein, leading to hydrolysis of the ATP, transfer of one electron to the MoFe protein, and dissociation of the $(MgADP)_2(Fe)$ species from the $MoFe^-$ protein. This dissociation step is rate-limiting in the catalytic cycle (Thorneley and Lowe, 1983), which is therefore independent of substrate concentration.

In the absence of substrate, transfer of electrons to the MoFe protein leads to the formation of molybdenum-bound hydrido groups by reduction of protons. There is evidence for a trihydridomolybdenum site. The evolution of dihydrogen from nitrogenase in the absence of N_2 is due to the reaction of protons with these hydrido groups. Binding of dinitrogen to nitrogenase also results in the evolution of dihydrogen due to the displacement of two of the three hydrido groups.

Detailed kinetic studies on the mechanism of nitrogenase from *K. pneumoniae* have been reported by Lowe and Thorneley (1984) and Thorneley and Lowe (1984). Protons are required at the active site to protonate reduced dinitrogen species, but their access must be controlled to prevent the protonation of hydrides to give dihydrogen competing with nitrogen fixation. The simplest explanation is that the binding of the iron protein to the MoFe protein prevents access to H^+. The scheme in Figure 5.13 suggests that the two MgATP complexes serve to bridge the Fe protein and the MoFe protein, to give this protected region where substrate reduction and electron transfer occurs. The fact that the oxidized iron protein has to be dissociated from the MoFe protein before it can accept another electron from the physiological electron donor suggests that there is only one electron transfer site on the iron protein. The dissociation of the iron protein will also allow limited access by protons to the active site. The slow turnover rate of nitrogenase reflects the need to have eight electron-transfer steps, each of which will involve dissociation of the iron protein from the MoFe protein.

The sequence of chemical events occurring during the reduction of dinitrogen has been much studied. Electron transfer in nitrogenase is complicated in view of the number of redox centres available to receive and transfer electrons. Binding of dinitrogen displaces dihydrogen, with a transfer of two electrons to N_2. The reduction of dinitrogen probably involves the formation of an $Mo{=}N{-}NH_2$ species, while N–N bond cleavage to give the first molecule of ammonia probably leads to the formation of Mo=N or Mo=NH intermediates, which are then reduced to give a second molecule of ammonia.

5.15 NICKEL

5.15.1 Roles

Since 1979, nickel has become firmly established as an essential metal for several enzyme-catalysed reactions in micro-organisms (Thauer, Diekert and Schönheit, 1980; Lancaster, 1988). It thus joins other metals in the latter half of the first transition metal series (Mn, Fe, Co, Cu) that are indispensable for many biological processes. Previously, only urease in plants and rumen bacteria (Schneider and Kaltwasser, 1984) provided examples of a biological function for nickel. Most of the nickel-requiring microbial enzymes are involved in some aspect of gas metabolism (Poole and Dow, 1985) and include, for example, hydrogenase, carbon monoxide dehydrogenases, and the methyl cofactor M reductase of methanogens. In addition, *Bacillus subtilis* forms a nickel-containing intracellular pigment, and nickel has been implicated in endospore formation (Thauer, Diekert and Schönheit, 1980) and in the ureases of some algae (Rees and Bekheet, 1982), several heterotrophic bacteria (Hausinger, 1987) and purple sulphur bacteria (Bast, 1988).

5.15.2 Factor F-430 of methanogenic bacteria

Just as the antibacterial effects of certain platinum compounds were accidentally discovered by platinum contamination (from electrodes) of bacterial cultures (Chapter 6), so the requirement for nickel by methanogens was long-overlooked, due to the use of stainless steel needles in anaerobic culture techniques, thereby providing enough dissolved nickel to sustain growth. Stainless steel fermenters might mask nickel requirements of other organisms.

A yellow, (λ_{max} 430 nm) low M_r nickel-containing compound was subsequently isolated from strictly anaerobic methanogens and called factor F-430. It has been shown to occur in a protein-bound form (Moura *et al.*, 1983) and be the prosthetic group of methyl coenzyme reductase, which catalyses the conversion of methyl coenzyme M into methane (Ellefson, Whitman and Wolfe, 1982). All methanogenic bacteria studied require nickel for growth and synthesize this factor (Diekert *et al.*, 1980, 1981). Studies of the uptake of precursors of tetrapyrrole synthesis (Diekert *et al.*, 1980; Gilles and Thauer, 1983) suggest that factor F-430 is a nickel tetrapyrrole. Nuclear magnetic resonance spectroscopy suggests an unusual tetrahydroporphyrin structure (Figure 5.14) for F-430M (the methanolysis product of F-430). The nickel porphinoid is a (tetrahydro) derivative of the corphin system combining structural elements of both porphyrins and corrins. The ring is substantially buckled, allowing stronger interaction between the Ni(II) and the nitrogen donor atoms of the corphin structure.

5.15.3 Carbon monoxide dehydrogenase

Certain acetogenic bacteria, such as *Acetobacterium woodii, Clostridium thermoaceticum* and *Cl. pasteurianum* oxidize carbon monoxide to CO_2

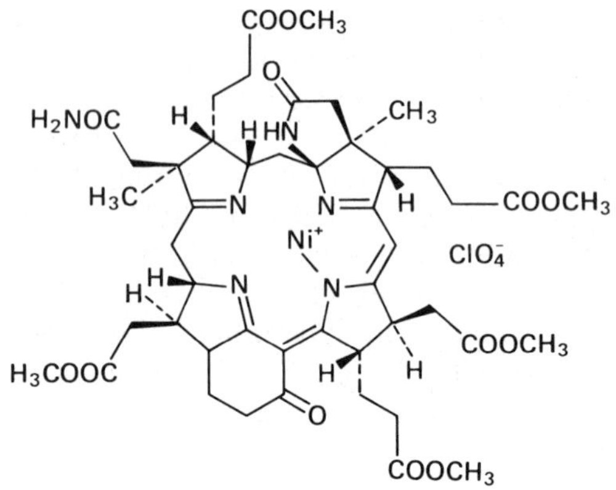

Figure 5.14 The structure of F-430M (reproduced from Hughes, 1985).

anaerobically, utilizing a nickel-containing, O_2-sensitive, carbon monoxide dehydrogenase. Table 5.14 shows molecular properties and metal contents of the CO dehydrogenases purified from several methanogens and acetogens. In *C. thermoaceticum* and *A. woodii*, the larger subunits appear to be identical, but the small ones differ. Quantification of the metals suggests two mol of nickel per mol of dimeric enzyme and 2 or 3

Table 5.14 Some carbon monoxide dehydrogenases*

Source	M_r	Subunits	Metal/hexamer			Labile sulphide
			Ni	Fe	Zn	
Methanogens						
Methanosarcina barkeri	232 000 ($\alpha_2\beta_2$)	92 000 18 000	Yes			
Methanobrevibacter arboriphilicus	140 000		Yes			
Acetogens						
Clostridium thermoaceticum	440 000 ($(\alpha\beta)_3$)	77 600 70 900	6	32	3	42
Acetobacterium woodii	480 000 ($(\alpha\beta)_3$)	80 000 68 000	5	27	varies	36
Clostridium pasteurianum			Yes			

* From Hughes (1987) with permission, which should be consulted for references.

[4Fe-4S] clusters per dimer. The *A. woodii* enzyme may have a 3Fe centre.

Ragsdale *et al.* (1982, 1983) suggest that the nickel might form part of a corrinoid structure. The nickel is thought from EPR data to bind CO and cycle between Ni(II) and Ni(III), which is a little surprising from a chemical point of view. Ni(III) complexes are well known but nickel is not commonly regarded as a redox metal.

5.15.4 Hydrogenases

Nickel is a component of some, but not all, hydrogenases (Moura *et al.*, 1984). A role for nickel has been inferred from the effects of the metal and chelating agents on hydrogenase activity (Pedrosa and Yates, 1983). These enzymes provided the first example of a catalytic involvement of nickel in redox changes. The nickel is protein-bound and thought to cycle between Ni(II) and Ni(III) during reaction with hydrogen, the substrate. The original proposal for such a role was based on ESR spectra, in which all g values were above 2.0 (indicating a d^7 system) and with no hyperfine structure. Lancaster (1980, 1982) suggested that the paramagnetic centre was low-spin Ni(III) in a rhombically distorted octahedral environment. Strong support has come from studies on hydrogenase isolated from *Methanobacterium thermoaceticum* and from *Desulfovibrio gigas* grown on media enriched in ^{61}Ni(II), and isotope with a nuclear spin of 3/2, the enzyme exhibiting the predicted hyperfine splitting in the EPR spectra (Albracht, Graf and Thauer, 1982; Moura *et al.*, 1982). Table 5.15 summarizes the properties of some nickel-containing hydrogenases. All contain Fe-S clusters or a flavin as additional cofactors.

It now appears that the Ni(III) signal observed originally arises from an inactive form of the enzyme. Thus, incubation of the *D. gigas* enzyme with the substrate (H_2) results in loss of the Ni(III) EPR signal and appearance of a transient radical type of signal, which is subsequently but transiently replaced by a different Ni(III) signal. In the hydrogenase from a *M. thermoautotrophicum* strain, removal of H_2 from the enzyme by argon results in the recovery of a Ni(III) signal, confirming the participation of the nickel centre in electron transport between the various redox centres of the enzyme (Kojima *et al.*, 1983). There is some evidence that Ni(III) and a $[4Fe\text{-}4S]^{3+}$ cluster interact magnetically in the hydrogenase from *Chromatium vinosum* (Albracht, Kalkman and Slater, 1983).

Several unanswered questions remain. The first two relate to the redox potential of the nickel. The mid-point reduction potentials measured for the Ni(III)/Ni(II) couple of the *D. gigas* hydrogenase lie in the

Table 5.15 Some nickel-containing hydrogenases*

Source	M_r	Ni (mol)	Fe (mol)	FeS clusters	Others
D. gigas	89 000	1	11/12	[3Fe-xS] 2 [4Fe-4S]	
Vibrio succinogenes	≈100 000	1	2	?	
M. thermoautotrophicum					
F420-reducing	170 000	2.5–3	33–43		2–3 FAD
Methylviologen-reducing		1	15/20		
D. desulfuricans		0.7	11	[3Fe-4S] 2 [4Fe-4S]	
Chromatium vinosum		+	7	[3Fe-xS] [4Fe-4S]	
A. eutrophus		1–2		2 [4Fe-4S] 2 [2Fe-2S]	FMN
Methanosarcina barkeri		0.5		2 [4Fe-4S]	FMN
Nocardia opaca					
	large dimer 64 000, 31 000	none	9.3 }	[4Fe-4S]	FMN
	small dimer 56 000, 27 000	2	3.9 }	[2Fe-2S]	

* From Hughes (1987) with permission, which, with Schneider, Cammack and Schlegel (1984) should be consulted for references.

range −145 mV to −220 mV (Teixeira *et al.*, 1983), values much lower than those of model peptide complexes of Ni(III), with E_m values of +790 to +960 mV. Presumably, the hydrogenase nickel is bound by different amino acid residues; one cysteine residue has been implicated in the coordination environment (Sugiura, Kuwahara and Suzuki, 1983). Second, the measured potential of the Ni(III)/Ni(II) couple in hydrogenase is higher than the $2H^+/H_2$ couple by about 200–280 mV, so that there are mechanistic difficulties in suggesting that the nickel activates dihydrogen. Further measurements are required, as is validation that the potentials apply to mechanistically relevant species of the metal. Finally, only about half of the total nickel is present as Ni(III), perhaps due to the existence of a further EPR-undetectable nickel. This, or an EPR-detectable cluster, could be the site of hydrogen activation.

5.16 ZINC

Zinc does not function directly in redox processes. Its role in the enzymes listed in Table 5.16 is often that of a strong Lewis acid, in which substrates are coordinated, polarized and activated. In other cases zinc may have a structural or template role (Sigel, 1983), but the precise role of the metal is not always understood. This is in part because zinc is 'silent' so that electronic and nuclear spectroscopy cannot be used to give direct information on site symmetry and ligand groups around the native metal. Sometimes, metal substitution studies have yielded useful data on these problems. A common feature, however, seems to be the ability of the zinc centre to accommodate the different geometrical requirements of a succession of intermediates, involving changes in the coordination number of the zinc.

5.16.1 Carbonic anhydrase

Carbonic anhydrase is a most effective catalyst of the reversible hydration of CO_2 and dehydration of HCO_3^-:

$$OH^- + CO_2 \rightleftharpoons HCO_3^-$$

Although the enzyme is present in micro-organisms, plants and animals, the form studied most intensively comes from bovine or human red blood cells (Hughes, 1987; Lindskog, 1982). It contains one zinc per 30 000 M_r, which sits near the bottom of a cleft in the enzyme molecule. The zinc is bound by the imidazole groups of three histidine residues.

Table 5.16 Some zinc-containing enzymes of micro-organisms

Enzyme	Common source	Role of zinc	Reference
Alcohol dehydrogenase	Yeast	Polarization of C–O bond	Prince (1979) Hughes (1987)
Alkaline phosphatase	*E. coli*	See text	Coleman and Gettins (1983) Hughes (1987)
Aspartate transcarbamylase	*E. coli*	See text	Hughes (1987)
Carbonic anhydrase	Many microbes, animals, plants	See text	Hughes (1987)
Cyclic nucleotide phosphodiesterase	*S. cerevisiae*	?	Londesborough and Suoranta (1983)
RNA polymerase	*E. coli*	Structural and catalytic	Solaiman and Wu (1984)
F_1-ATPase	*Micrococcus lysodeikticus*	?	Mollinedo *et al.* (1981)
Glyoxalase	Yeast	Polarization of substrate group	Hughes (1987)
β-lactamase II	*Bacillus cereus*	?	Hughes (1987)
Metalloproteinases	Various fungi and bacteria	Conformational integrity?	Murao *et al.* (1982 a,b) Matsumato, Maeda and Okamura (1984)
Ornithine transcarbamylase	*E. coli*	Allosteric cofactor	Kuo, Lipscomb and Kantrowitz (1982)
Phenylalanine-tRNA synthetase	*E. coli*, yeast	Effector	Plateau, Mayaux and Blanquet (1981)
Phospholipase C	*Bacillus cereus*	Stabilization of protein structure	Little *et al.* (1982)

The fourth coordination site is occupied by water, giving a distorted tetrahedral geometry. The zinc-bound water molecule is probably involved at the active site in controlling catalytic activity, the deprotonated form being the active species. The zinc can be removed from carbonic anhydrase by dialysis against certain chelating agents, and the apoenzyme reconstituted by one of a number of divalent cations. The Co(II) enzyme exhibits about 50% of the activity of the zinc enzyme. The metal appears to be present in a five-coordination geometry.

5.16.2 Carboxypeptidases

The carboxypeptidases are exemplified by bovine carboxypeptidase A. It contains one mole of zinc per protein of M_r 34 500. These enzymes cleave the C-terminal amino acid residue from certain peptides and proteins. The zinc is bound in a protein cleft to give a five coordinate environment around the metal. The mechanism of peptide hydrolysis involves either the attack of a zinc-bound hydroxide group or coordination and polarization of the carbonyl group (Hughes, 1987).

5.16.3 Aspartate transcarbamylase

This is an allosteric enzyme that requires zinc solely for structural stability (Hughes, 1987). A product of the reaction of carbamyl phosphate with L-aspartate is carbamyl-L-aspartate, a key precursor in pyrimidine biosynthesis. The *E. coli* enzyme comprises five subunits, three of which (R or regulatory subunits) bind two Zn each. The zinc-binding site involves four cysteinyl residues, with tetrahedral geometry and this domain is the major site of interaction between R and C (catalytic) subunits. Under zinc-deficient conditions, the enzyme dissociates into subunits. If *E. coli* is grown in a cadmium-containing, zinc-deficient medium, the enzyme retains activity but contains 6 Cd(II) per molecule, also bound by cysteinyl ligands.

5.16.4 Alkaline phosphatase

This non-specific phosphomonoesterase has been most studied in *E. coli*. It contains up to 4 Zn^{2+} and 2 Mg^{2+} per molecule. Each of the two subunits has three metal-binding sites, two of which must be fully occupied by Zn(II) for maximum activity (i.e. 4 Zn(II) in total). The A site metal coordinates the phosphate group of the substrate, whilst the function of the B site metal is to enhance transfer of the phosphate to Ser-102 by charge neutralization. Product phosphate is then dissociated. Metal substitution experiments have made a great impact on understanding metal binding to the three sites, particularly through NMR studies (Hughes, 1987).

5.16.5 'Zinc fingers'

Zinc fingers are zinc-stabilized structural domains in proteins believed to be important in protein binding to DNA and thus gene expression (Klug and Rhodes, 1987). These domains, now identified in a number of proteins or sequences in *E. coli*, yeast and higher animal cells, are rich in histidine and cysteine residues, common ligands for Zn. In the well-characterized transcription factor IIIa of *Xenopus*, a domain consists of 30 amino acids looped around a zinc atom which has a purely structural role in maintaining protein folding. Metals such as Co or Cd have not yet been shown to substitute for Zn.

5.17 COBALT

The main requirement for cobalt by micro-organisms is as vitamin B_{12}, which has important functions in two classes of enzymes. The first catalyses C atom transfer reactions and is exemplified by the methyltransferase involved in methanogens and by the transmethylase involved in the production of acetate from pyruvate in *Cl. thermoaceticum* (Wood *et al.*, 1982). In the second class are enzymes containing 5′ deoxyadenosylcobalamin, such as ribonucleotide reductases (Section 5.13).

Naturally-occurring cobalt compounds involve the corrin ring structure shown in Figure 1.6. Cobalamin is a 5-coordinate complex having four nitrogen donor atoms from the corrin ring and an axial position filled by an N atom of the 5′6-dimethyl benzimidazole ring, which is itself linked to the corrin. In cyanocobalamin, more commonly known as vitamin B_{12}, cyanide is present in the other axial position.

In adenosylcobalamin, this axial position is filled instead by the 5′-C of 5′-deoxyadenosine, to give a remarkably stable organocobalt compound. Adenosylcobalamin is an essential coenzyme for several enzymes especially mutases (e.g. glutamate mutase) and dehydrases (e.g. glycerol dehydrase). These enzymes catalyse various arrangements in which the Co–C bond of adenosylcobalamin undergoes cleavage. The adenosyl radical produced abstracts a hydrogen atom from the substrate.

Adenosylcobalamin is also a coenzyme for ribonucleotide reductase (Section 5.13).

In methylcobalamin, the upper axial position is occupied by a methyl group. Methylcobalamin is an especially important cofactor in the generation of methane from CO_2 by methanogenic bacteria. The final step involves coenzyme M. This reacts with methylcobalamin, and 2-mercaptoethanesulphonic acid to give methylcoenzyme M, which is then reduced to methane. The reductase involved has a nickel-containing prosthetic group, factor F-430 (Section 5.15.2).

5.18 SUMMARY

The transition metals and zinc have many diverse roles in microbial physiology and biochemistry. This results from their ability to: (i) bind and activate small molecules and larger substrates, and to control the electron distribution within these coordinated molecules, (ii) take up disordered geometries, imposed by the protein, sometimes for a series of intermediates, (iii) provide redox potentials finely tuned for interaction with other redox species, and (iv) to bridge ligand groups and to stabilize structures. All these properties depend upon the cooperation of the protein, or other metal-binding group, which enforces certain geometries upon the metal and provides the donor groups and other interactions that control the redox potential and ensure vacant coordination sites.

Transition metalloenzymes are essential catalysts of redox reactions. The change in oxidation state of the metal is linked to the oxidation or reduction of substrates, often coordinated to the metal, or to the transfer of an electron. Iron is the best studied and most well-understood of the transition metals in biology, particularly in haem compounds and the iron-sulphur proteins. Both these classes of compound are important electron transfer catalysts. In the former case, the redox potential is controlled by the choice of the axial ligand and by subtle interactions with the protein: mid-point potentials lie between −300 and +700 mV, showing the flexibility of the haem system in this context. The iron-sulphur proteins include the rubredoxins which contain FeS_4 groups, with cysteine sulphur donors, and those with iron-sulphur clusters, which may have two, three or four iron atoms bridged by sulphide and with the FeS_4 environment completed by cysteine residues. These clusters all transfer one electron but are associated with a wide range of redox potentials in iron-sulphur proteins, probably due to cycling between alternative overall oxidation states in the cluster. Iron haem enzymes are also involved in the catalysis of the reactions of hydrogen peroxide in the catalases and peroxidases. These enzymes involve common high oxidation state intermediates with $Fe^{IV}=O$ (oxyferryl) groups. The most important users of dioxygen are the cytochrome oxidases, the terminal electron acceptors in the respiratory chain, which transfer four electrons to reduce the oxygen molecule to water. At least five classes of cytochrome oxidase are known in bacteria, cytochromes aa_3, a_1, d, cd_1, and o. The aa_3 oxidase is similar to the mitochondrial enzyme but has a much simpler subunit structure. The aa_3 oxidase contains four metal centres, with two copper centres in addition to the a and a_3 cytochromes, to ensure rapid reduction of dioxygen without the build up of free, toxic oxygen species. The peroxidases and superoxide dismutases are part of the defence mechanism against these toxic species. Superoxide dismutase may contain Cu and Zn, Fe, or Mn(III) depending upon its origin. Manganese(III) also has a redox role in the oxygen-evolving reaction in photosynthesis, where the water is bound to pairs of manganese centres for oxidation, and also in pseudocatalase.

The oxygen atoms of dioxygen can be incorporated into substrates by mono or dioxygenases. Monooxygenases include copper proteins (such as tyrosinase), an unusual combination of a 2Fe-2S cluster plus mononuclear non-haem iron (for example, putidamonooxin), and the haem cytochrome P-450. Dioxygenases usually contain iron, as haem or iron-sulphur.

It is not surprising that iron-storage proteins are now being discovered in bacteria; *Azotobacter* has a ferritin-like storage system. Iron is also present as magnetite crystals in magnetotactic bacteria. This causes these bacteria to swim North or South, depending upon which hemisphere they originated from. This allows them to swim down into sediments.

Copper in the blue copper proteins is well known as a catalyst of electron transfer. Some blue oxidases are also known; these contain four copper centres, distributed among three types of copper, Type 1 (blue), Type 2 (normal Cu(II)) and Type 3 (a pair of antiferromagnetically coupled Cu(II) ions). Copper is involved in some of the enzymes associated with denitrification, namely nitrite reductase and N_2O reductase.

Molybdenum is the only essential second row element. It is present in a small number of enzymes (nitrate reductase, sulphite oxidase, xanthine oxidase, for example) which are characterized by having a free site on the molybdenum for the coordination of substrate, which is followed by transfer of an oxo group between the molybdenum and the substrate. A molybdenum-containing cofactor isolated from these enzymes appears to be similar in all cases and functionally interchangeable. Molybdenum in nitrogenase is an exception, both in the character of the reaction being catalysed, and in the presence of a different molybdenum cofactor, an iron-molybdenum-sulphide cluster. It is now known that vanadium-iron nitrogenase and iron nitrogenase also exist, and this may suggest that molybdenum is not the binding site for N_2.

Cobalt is well known for its role in vitamin B_{12} coenzymes, where it is bound in the corrinoid ring structure. The role of nickel in micro-organisms has been appreciated more recently, but it is now firmly implicated in the activity of some hydrogenases, carbon monoxide dehydrogenase and methanogenic bacteria. In the first two examples, nickel is bound to protein and is thought to cycle between Ni(II) and Ni(III), while in methanogens, it is present in factor 430M from methyl coenzyme reductase. F-430 is a tetra-pyrrole combining structural features of porphyrins and corrins.

Zinc has different roles from the transition metals. It behaves as a strong Lewis acid in hydrolytic reactions, for example in carboxypeptidase and carbonic anhydrase. Zinc can act either by coordination and polarization of a substrate, with enhancement of its reactivity towards water, or by increasing the acidity of a coordinated water molecule to give a zinc-bound hydroxide group which shows high reactivity to substrates. Aspartate transcarbamylase is a rare example of a zinc enzyme where the metal has a structural role rather than a catalytic one. Zinc also appears to stabilize structural domains (zinc fingers) in certain proteins which are implicated in the binding of the protein to DNA.

6

Metal toxicity

Mad as a hatter

Lewis Carroll
From: Alice in Wonderland

Every beauty, when out of its place, is a beauty no longer

Voltaire
Philosophical Dictionary (1764)

6.1 INTRODUCTION

Metals play indispensable roles in cell growth and maintenance of metabolic functions. However, when a metal with no known biological function competes with or replaces a functional metal, toxicity results. Metal toxicity to man is an increasingly important environmental concern. A recent analysis (Nriagu and Pacyna, 1988) shows that the annual total toxicity of metals mobilized by human activities now exceeds the combined total toxicity of all radioactive and organic wastes generated each year. Although toxicity is generally viewed as a property of soft heavy metals, detrimental effects can also result from a surplus (or deficiency, Chapter 2) of essential metals. Indeed, a number of metals, at first investigated because of concern over their toxic nature, were subsequently found to have essential biological roles. Thus, zinc was studied because of its toxic fumes long before discovery of its role in carbonic anhydrase (Vallee and Ulmer, 1972). Iron and cobalt are examples of the reverse situation; their biological roles were appreciated prior to the discovery of toxic effects. This chapter describes the binding of toxic metals to cell surfaces, metal transport in and out of cells, selected cases of toxicity and the sometimes-elaborate mechanisms used by microbes to acquire resistance to, or tolerance of, heavy metals. These mechanisms include minimizing the accumulation by cells of the metal (Section 6.2), biotransformation of the toxic metal to a chemically distinct, often volatile, form (Section 6.3), and binding of the metal to render it innocuous and prevent entry of the metal into the cell (Sections 6.2 and 6.5). Highly specific mechanisms for resistance to ions such as Hg^{2+} (and organomercurials), compounds of arsenic and organoarsenicals, Cd^{2+}, Pb^{2+} and Ag^{+} are frequently determined by genes on plasmids or transposons. The identity and function of genes responsible for resistance to metal ions is described in Section 6.4. Finally, we examine man's use of metals and metal-containing compounds in controlling cancer (Section 6.7.6) and microbial infections (Section 6.8). Although the fact that micro-organisms display such resistance to potentially toxic metals is exploited implicitly in a number of biotechnological applications of metal–microbe interactions (Chapter 7), our increased knowledge of the biochemistry and genetics of resistance has thus far made little impact on the applied field. Great hope exists, however, in (i) the use of improved strains for metal leaching, accumulation of toxic metals and radionuclides, and (ii) designing improved metal-based biocides.

6.1.1 Factors that modulate metal toxicity in laboratory culture

There is an enormous literature on the toxic effects of heavy metals on micro-organisms; reduction of growth rates, extension of the lag phase and perturbations in morphology and physiology are well-documented. A comprehensive review of these studies is impossible here and probably of limited value, especially since many studies in this field fail to recognize the importance of metal speciation. References to original work can be found in the reviews of Gadd and Griffiths (1978) and Sterrit and Lester (1980a,b). We draw attention in this section to the caution required in assessing data on metal toxicity.

Microbiological studies on heavy metals are generally performed in synthetic growth media, the constituents and properties of which can greatly influence the free concentrations, and thus toxicity, of metals.

Complex organic medium constituents such as yeast extract, proteose peptone and amino acids (especially cysteine and glutamic acid) all bind large amounts of Hg^{2+}, Pb^{2+}, Ag^{+} and Cu^{2+}, though less Cd^{2+} (Ramamoorthy and Kushner, 1975). This affinity of cations for organic ligands follows the Irving-Willams series and explains, for example, why addition of yeast extract to cultures of *Aerobacter aerogenes* protects the bacteria from the otherwise toxic effects of the Cu^{2+} found in some laboratory distilled water, alleviates silver toxicity to iron oxidation by *Thiobacillus ferrooxidans* (Tuovinen *et al.*, 1985) and why *Tetrahymena pyriformis* tolerates 100-fold higher concentrations of copper in a rich organic medium than in a defined one (Nilsson, 1981). Similarly, in Nelson's medium, which is particularly rich in amino acids, Ramamoorthy and Kushner (1975) were unable to detect free Hg^{2+} using a solid-state ion-specific electrode until 160 ppm of the metal was added, though total amounts of 10 ppm were toxic. Humic and fulvic acids and chlorophyll effectively reduce the toxicity of nickel to fungi (Babich and Stotzky, 1982), probably by chelation or complexation of the toxic divalent cation. Equally, of course, organic complex-forming substances may decrease the free concentrations of *essential* metal ions and thus influence the outcome of competition between essential and toxic metals. Living or dead cells may also constitute surfaces that bind metals and alleviate the toxicity to other cells in a population.

Certain metal cations are readily lost from solution through adsorption to glass surfaces. This phenomenon makes measurement of metal concentrations at micromolar levels problematic. In a few cases, notably Hg^0, volatile metal species may be lost to the atmosphere.

Although wholly defined media, containing the minimum of organic constituents, are frequently recommended for metal toxicity studies, inorganic salts (particularly the anionic species) can also affect heavy metal toxicity dramatically. For example, phosphate protects against copper toxicity, phosphate or carbonate against lead, fluoride against aluminium, and sodium chloride against cadmium. The lethal effects of Co and Zn are decreased in media prepared with hard water (Wong, 1980). In contrast, the toxicity of zinc to fungi, bacteria and coliphage are generally increased by NaCl, due to the formation of a soluble zinc–chloro complex. However, one unusually cadmium-tolerant *Pseudomonas* species has been reported (Onishi *et al.*, 1984) for which 2–4 M NaCl provides protection against $CdCl_2$ at 2.5 mg ml^{-1}. Nitrate and chloride attenuate the toxicity of inorganic tin compounds in solid media (Hallas, Thayer and Cooney, 1982). The presence of inorganic salts probably contributes to the influence of natural aqueous media on metal toxicity. Nickel and mercury, for example, exhibit enhanced toxicity in lake water compared to sea water (Babich, Schiffenbauer and Stotzky, 1983), probably due to the high Mg^{2+} content of sea water and its alkaline pH. There are several reports that Mg and Zn alleviate the toxicity of Ni to micro-organisms (Babich and Stotzky, 1982), possibly by competition between these cations, of similar non-hydrated ionic radii, for common sites on the cell surface. Calcium and magnesium alleviate the toxicity of cadium to *A. niger*.

Other metal combinations are synergistic in their toxic effects. For example, Ni and Cd were synergistic towards growth and survival of *K. pneumoniae* (Ainsworth, Tompsett and Dean, 1980) but antagonistic towards growth, photosynthesis and N_2 fixation by *Anabaena inequalis* (Stratton and Corke, 1979). For the latter organism, the effects of Ni and Hg were complex. When added at sublethal concentrations and when Hg was added first, or when both metals were added simultaneously, synergism was noted for growth rate and cell yield. When, however, Ni was added first, antagonism occurred. Such cooperative effects have received only scant attention.

The choice of medium pH and buffer is of obvious importance. Phosphate, citrate and tris buffers bind many metals, notably copper, whereas MES (2-(*N*-morpholino)ethanesulphonic acid) binds little metal (Blanchard, 1984) and has been reported to allow maximal expression of copper toxicity (Baldry and Dean, 1980). In general, metals exist as free cationic species at acid pH but frequently precipitate as hydroxides or oxides at alkaline pH. The effects of pH may be complex and depend not only on the metal but also on its redox state. Zinc, for example, precipitates as $Zn(OH)_2$ near pH 5 but at pH 8.5 and above forms the zincate ion which can be precipitated by calcium. In soft water at neutral pH, the non-charged aquamonocarbonato complexes of Cu(II) and Zn(II) prevail, and may be more toxic than the charged aqua-metal ions or dicarbonato complexes present at lower *or* higher pH (Stumm and Bilinski, 1972). Alkaline pH ($\approx$8.5) eliminates the toxicity of nickel to various fungi (Babich and Stotzky, 1982). The observed, strong pH-dependence of the adsorption of trace metals is largely due to the pH-dependent variation of metal species (Stumm and Bilinski, 1972).

Caution needs to be exercised in attributing to the metal *per se* the results of its addition to a culture, since toxicity can result quite indirectly. The apparent toxicity of lead to fungi and algae, for example, has been attributed to the depletion of phosphate by precipitation as $Pb_3(PO_4)_2$ (Ross, 1975; Schulze and Brand, 1978).

In solid media, the choice of solidifying agent is crucial, a fact well illustrated by the toxicity of inorganic tin compounds (Hallas and Cooney, 1981a). Thus, $SnCl_4.5H_2O$ precipitates in liquid media and increasing the amount added does not increase the free (dissolved) metal concentration. However, increasing the amount in the presence of agar significantly decreased viable counts, suggesting the formation of a soluble, toxic, tin–agar complex. Hallas, Thayer and Cooney (1982) found tin toxicity to be reduced in media solidified with silica gel or gelatin; Cd^{2+}, Cu^{2+}, Pb^{2+}, Ni^{2+} and Zn^{2+} exhibited similar patterns of toxicity.

6.1.2 Metal toxicity in natural environments

In the natural environment, as in the laboratory, a host of factors interact to influence the toxicity of the metals present (Gadd and Griffiths, 1978; Babich and Stotzky, 1980). In 1979, the US Environmental Protection Agency (EPA) set new criteria for pollutants, including heavy metals, in aquatic environments. They noted that the 'toxicity of certain compounds may be less in some waters because of differences in acidity, temperature, water hardness and other factors. Conversely some natural water characteristics may increase the impact of certain pollutants.' In recognition of these factors, the criteria set for marine and fresh ecosystems were quite distinct (Babich and Stotzky, 1982; Abel and Barlocher, 1984). There is an urgent and long-standing need for further studies of the speciation of metals in natural waters, that is, data on the oxidation state of the metal, the type of complexes present, their stability and the rate of complex formation. As well as providing a firm basis for understanding the toxicity of metals in natural waters, such information would account for apparent anomalies in metal assays in natural systems. For example (Stumm and Bilinski, 1972), analysis for ferric iron often appears to show supersaturation of the water, probably explicable by the presence of soluble organic iron(III) complexes or stable colloidal suspensions of inorganic iron compounds.

The results of competition for a metal between microbial surfaces and sediment and other non-biological surfaces in natural environments are unpredictable at our present level of knowledge and further experiments are needed. Ramamoorthy, Springthorpe and Kushner (1977) showed that bacteria (albeit at a higher concentration than found in natural waters) compete quite effectively with sediment in accumulating mercury. However, in a study of the uptake of nitrosyl ruthenium compounds, Gibson *et al.* (1984a) demonstrated that the modest ability of marine bacteria to accumulate the metal compounds was masked by uptake of natural sediments, clays and, particularly, $Fe(OH)_3$.

Since natural populations of microbes appear to reflect metal levels accurately, it has been suggested that bacterial populations could be used to assess the extent of metal pollution of an environment. These natural communities consist not only of the tolerant bacteria themselves but also of less tolerant types whose survival in the potentially hostile environment is allowed by the protecting influence of the tolerant organisms. Likewise, the inhibition of microbe-mediated ecological processes, such as glucose degradation, by pollutants could be used to quantitate the sensitivity of natural ecosystems to heavy metals (Babich, Bewley and Stotzky, 1983).

6.2 BINDING AND TRANSPORT OF TOXIC METALS

The intracellular accumulation of metals generally comprises initial, rapid and reversible binding of the metal to the cell surface layers and, in some cases, a subsequent transport of the metal across the membrane to the interior. Transport in the reverse direction (efflux) sometimes plays a role in detoxification. Intracellular transport can be followed by accumulation of the metal in particular cell compartments, again often resulting in detoxification.

6.2.1 Binding of toxic metals to cell surfaces

Many metals bind with various degrees of tenacity to the largely anionic outer surface layers of microbial cells. This is sometimes a prelude to transport and may be exploited biotechnologically to detoxify effluents and concentrate metals of value (Chapter 7).

In a series of detailed studies, isolated cell walls from *Bacillus subtilis* were shown to bind numerous metals (Beveridge and Murray, 1976). Some metals (Mg^{2+}, Fe^{3+}, Cu^{2+}, Na^{+} and K^{+}) bound in amounts that made them visible by electron scattering in thin sections. Relatively small amounts of the toxic metals Hg^{2+}, Sr^{2+}, Pb^{2+} and Ag^{+} were bound, however (Table 6.1). The metals shown are arranged in order of the extent of binding and compared with respect to their ionic radius, heat of hydration and the number of water molecules in their hydration shells, but there appear to be no obvious correlations between these properties and binding. The soft heavy metals show relatively little adsorption.

In such cases, binding results largely from the many anionic sites on the wall, especially (i) the phosphodiester groups of teichoic acids, (ii) free carboxyl groups of the peptidoglycan, (iii) the sugar hydroxyl groups of both major classes of wall polymers, and (iv) the amide groups of peptide chains. The peptidoglycan, however, is especially significant, binding metal ions and forming nucleation sites for secondary deposition.

Table 6.1 The extent of uptake of metals by walls from *Bacillus subtilis* and lack of correlation with chemical properties of the elements (by courtesy of Beveridge and Murray, 1976)

Characteristic	Large uptake								→	Small uptake			→	No uptake			
	Mg^{2+}	Fe^{3+}*	Cu^{2+}	Na^{+}	K^{+}	Mn^{2+}	Zn^{2+}	Ca^{2+}	Ni^{2+}	Hg^{2+}	Ag^{+}	Sr^{2+}	Pb^{2+}	Li^{+}	Ba^{2+}	Co^{2+}	Al^{3+}
Ionic radius ($\times 10^{-10}$ m)	0.78	0.67	0.72	0.95	1.33	0.91	0.83	0.99	0.78	1.12	1.13	1.27	0.80	0.78	1.43	0.82	0.57
Heat of hydration (kcal mol^{-1} at 25°C)	464	1072	507	100	79	445	492	382	507	441	116	350	505	125	316	497	1122
Number of water molecules in hydration shell	6	6	4	4	4 or 6	6	6	6	6					4		6	6

* Complex water chemistry.

Walls from *E. coli* and most other gram-negative bacteria have a lower charge capacity than gram-positive walls, but have a complex three-layered structure that also binds and immobilizes many metal ions, including Pb^{2+} and Hg^{+} (Beveridge and Koval, 1981). Gram-negative bacteria generally exhibit greater metal tolerance than gram-positives.

The formation of a complex between a metal and an outermost layer of the microbial wall envelope, in particular with an extracellular or peripheral polymer, is an important means of metal removal from waste waters. For example, the potential of *Zoogloea ramigera* for removal of cadmium and mercury (Chapter 7) has been evaluated and strains producing a gelatinous matrix were found to bind significantly more metal than strains lacking such a layer. Metal binding by polymers from *Klebsiella aerogenes, Pseudomonas* spp. and other bacteria has been demonstrated (Sterritt and Lester, 1980a). Similar binding phenomena are apparent with wall constituents of eukaryotic microbes, the bound metal sometimes reaching a substantial proportion of the wall weight. Growth of *Saccharomyces cerevisiae* with mercuric chloride, for example, results in extensive metal binding to the wall; isolated walls can bind their own weight of mercury to high-affinity sites (Murray and Kidby, 1975). Hyphal walls from *Neurospora crassa* grown in the presence of high copper concentrations become blue and contain 12% copper by weight; iron and zinc may also be bound, probably to chitin or chitosan (Subramanyam, Venkateswerlu and Rao, 1983). Melanin binds Cu^{2+} and Cd^{2+} avidly, and, in the fungus *Aureobasidium pullulans*, the presence of these metals induces melanization (Mowll and Gadd, 1984).

Although normally regarded as a prelude to transport, surface binding of metals may persist when the transmembrane accumulation mechanism is defective. Thus an *E. coli* mutant with tenfold-lowered sensitivity to thallium still bound the metal just as avidly as did the wild-type (Norris *et al.*, 1976).

6.2.2 Membrane transport

(a) Introduction

Many toxic metals are transported by systems that normally translocate an essential, chemically related metal. Thus cadmium shares an uptake system with Mn^{2+} (Section 6.2.2(c)) and Tl^{+} is a substrate for certain K^{+} transport systems in bacteria (Section 3.3.2(e)) and eukaryotic microbes (Norris *et al.*, 1976). In the case of the arsenate and vanadate anions

(Sections 6.2.2(b) and 6.2.2(d)), phosphate transport systems are responsible for uptake.

For other metal ions, information is sparse and incomplete. Copper and cadmium ions are rapidly accumulated by several fungi (Ross, 1975) following initial binding to the cell surface. Copper uptake may be energy-dependent in *S. cerevisiae* (Ross, 1977; Norris and Kelly, 1979) and the uptake causes release of K^+. In the same organism, Mg^{2+} reverses the inhibitory effect of Mn^{2+} by competitively inhibiting uptake of the latter. Curiously, Mn^{2+}-resistant mutants take up *more* Mn^{2+} than do sensitive wild-type strains, the increased resistance residing not in transport processes but in the ability to sequester and render the Mn^{2+} insoluble (Bianchi *et al.*, 1981).

Cells that feed by phagocytosis may ingest the metal directly. For example, lead acetate (0.1–0.2%), in a rich organic growth medium forms a precipitate which is ingested by *Tetrahymena*, but lead also enters directly through the cell membrane (Nilsson, 1979).

Toxic metal ions may inhibit their own transport or that of other metals. The former appears to be the case for vanadate and may be due to the inhibition by vanadate of the plasma membrane ATPase responsible for generating an electrochemical gradient. An example of the latter situation is the inhibition of zinc uptake by silver and mercury ions, thus killing the cells (Failla, Benedict and Weinberg, 1976).

For other metals, very little passes through the cell membrane, providing some degree of resistance to the toxic metal. Metals that appear to be excluded by various metal-tolerant microbes include lead, mercury copper, cobalt, zinc, cadmium and manganese (Foster, 1977; Gadd and Griffiths, 1978, 1980; Pan-Hou, Nishimoto and Imura, 1981). An extreme example is afforded by *Penicillium ochro-chloron*, which grows at saturating concentrations of $CuSO_4$, yet accumulates little copper (Gadd *et al.*, 1984). In organotin-resistant bacteria, up to 99% of the applied metal complex remains associated with the wall–membrane complex (Blair *et al.*, 1982). Similarly, most of the uranium removed from solution by *Thiobacillus ferrooxidans* is associated with the wall and membrane. Cells poisoned with cyanide, however, showed an 8 to 11-fold increase in the cytoplasmic uranium concentration, due perhaps to an increase in membrane permeability or the inhibition by cyanide of an energy-dependent efflux pathway for the metal (DiSpirito, Talnagi and Tuovinen, 1983), a situation resembling that of cadmium transport (Section 6.2.2(c)). An increase in nickel efflux explains the increased resistance of spontaneous mutants of *Alcaligenes eutrophus* (Lohmeyer and Friedrich, 1987).

The means by which surface layers bind metals without screening negatively-charged binding sites and reducing the effective surface potential required for transport of essential ions is unclear. However, the transport mechanisms for certain toxic metal species and their role in metal resistance are well described and are covered in the following section. The transport system for mercuric ions is part of a complex, plasmid-encoded detoxification mechanism described in Sections 6.3.2(c) and 6.4.1.

(b) Arsenate and arsenite

Arsenate is transported into bacteria by two phosphate transport systems (Figure 6.1). Arsenate resistance results from reduced net accumulation but not from alteration of these transport systems. Thus, plasmid R773, one of several plasmids that confers resistance to metals and certain antibiotics (Silver, 1983; Section 6.4) does not alter the kinetic parameters of the phosphate transport system or the K_i for arsenate. Instead, reduced net accumulation is due to rapid arsenate efflux (Figure 6.1), which is not driven chemiosmotically (Silver and Keach, 1982; Mobley and Rosen, 1982; Rosen and Borbolla, 1984) but directly by ATP. In this

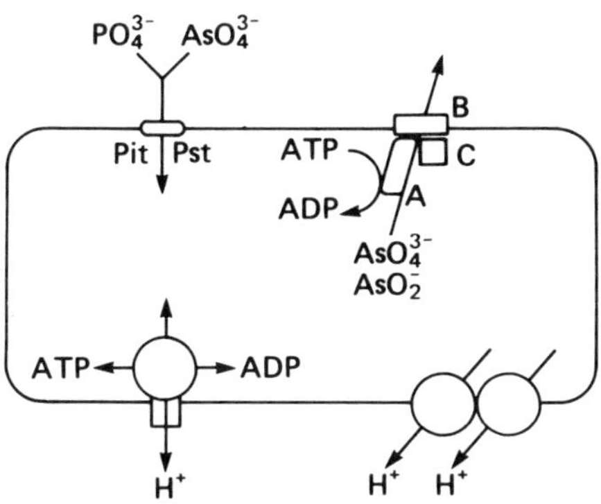

Figure 6.1 The influx and efflux systems for arsenate in *Escherichia coli*. The constitutive Pit (P_i transport) and inducible Pst (phosphate specific transport) systems both transport arsenate with a K_i (for arsenate as competitive inhibitor of phosphate transport) of 25 μM. Proceeding clockwise, are shown the arsenate efflux mechanism, proton translocation by the respiratory chain and the proton-translocating F_1F_0-ATPase. Arsenate efflux is via an anion pump comprising the polypeptides Ars A, Ars B and Ars C (Section 6.4.2); the last is not required for arsenite transport. The pump is energized directly by ATP hydrolysis and not by coupling to the proton motive force generated by respiration or F_1F_0-ATPase activity. Modified from Silver (1983) and Silver, Rosen and Misra (1986).

respect, the pathway resembles the Na^+ efflux system of *Streptococcus faecalis* (Section 3.12.7). The current concept is of a three polypeptide complex. The ArsA peptide contains two nucleotide-binding domains and is probably the catalytic, arsenite-translocating ATPase. The ArsB protein is a transmembrane channel for arsenical compounds, the specificity of which may be governed by a third subunit.

(c) Cadmium

Cadmium transport occurs via the Mn^{2+} active transport system in *Staphylococcus aureus* (Figure 6.2; Tynecka, Gos and Zajac, 1981a,b; Perry and Silver, 1982), *Bacillus subtilis* (Laddaga, Bessen and Silver, 1985) and *Lactobacillus plantarum* (Archibald and Duong, 1984). Thus, Cd and Mn are competitive inhibitors of Mn and Cd uptake, respectively. In *Escherichia coli*, however, Zn^{2+}, but not Mn^{2+}, inhibits Ca^{2+} uptake (Laddaga and Silver, 1985). The enigma is that despite the presence of an active Cd^{2+} transport system in *E. coli* and another gram-negative organism, *Pseudomonas aeruginosa*, and the consequent accumulation of high intracellular Cd^{2+} levels, gram-negative bacteria are generally more resistant to the metal in the medium. This resistance may be related to the presence of metallothionein-like proteins in such bacteria (Section 6.6; Table 6.3).

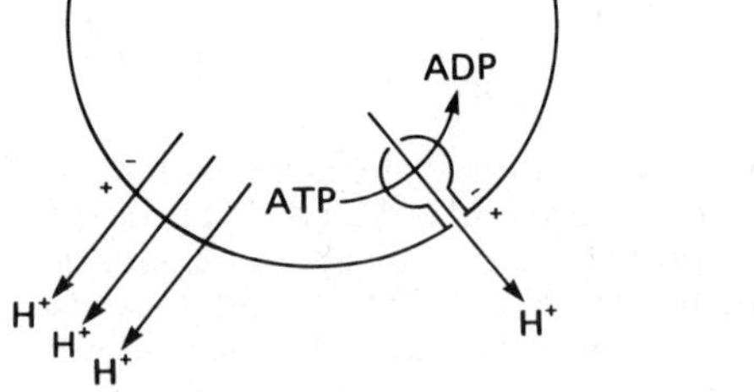

Figure 6.2 Models for cadmium (II) uptake and efflux in staphylococci. Cadmium(II) and manganese(II) are accumulated by a membrane transport system that utilizes the transmembrane electrical potential (positive outside). A plasmid-encoded efflux system rapidly excretes Cd(II) (and possibly Zn (II)), probably via a $Cd^{2+}/2H^+$ antiport. Reproduced with permission from Silver (1983).

Resistance to Cd^{2+} in staphylococci results from a constitutive reduction in net accumulation of the metal, and is due to a plasmid-located gene *cadA*, conferring high (typically 100-fold) resistance. Resistant strains rapidly excrete Cd^{2+} by an energy-dependent pathway. It is possible that the *cadA* gene product might also cause Zn^{2+} efflux, although this has not been directly demonstrated. However, the presence of the *cadA*$^+$ gene does not affect the transport of Mn^{2+}, another related metal. A second plasmid gene, *cadB*, has also been identified; this confers a low level of Cd^{2+}-resistance that is masked by the presence of *cadA*$^+$. Its function is unknown, but clues come from the Zn^{2+}-resistance that accompanies *cadB* gene function and from the synthesis of a metallothionein-like protein governed by this gene. Thus, the *cadB* gene product may be an inducible or amplifiable metallothionein-like protein (Section 6.6), binding both Cd^{2+} and Zn^{2+}.

In *Bacillus subtilis*, a chromosomal, rather than plasmid, gene has been implicated in Cd^{2+} resistance (Laddaga, Bessen and Silver, 1985), attributable to reduced Cd^{2+} uptake.

Certain eukaryotic micro-organisms also accumulate cadmium. *Saccharomyces cerevisiae* accumulates over 300 nmol Cd^{2+} (mg dry wt)$^{-1}$ in 1 h from a 0.5 mM $CdSO_4$ solution (Norris and Kelly, 1977; Gadd and Mowll, 1983). In this organism and in *Aureobasidium pullulans*, calcium competes with cadmium for uptake sites, suggesting the transport of cadmium via a calcium uptake system (Mowll and Gadd, 1984). Indeed, the two metals have similar ionic radii (0.97 and 0.99 $\times 10^{-10}$m respectively). In *A. pullulans*, Cd^{2+} uptake is balanced by a stoichiometric exchange for K^+ ions that maintains appropriate ionic balance across the cell membrane. A Cd^{2+}/K^+ antiporter, however, seems unlikely (Mowll and Gadd, 1984).

(d) Vanadate

Vanadate is a potent inhibitor of plasma membrane ATPases (Section 6.7.4). In *Neurospora crassa*, vanadate uptake occurs via one of the two phosphate transport systems (Figure 6.1). This system (II) is derepressed during phosphorus starvation and has a K_m of 3 μM for phosphate and 8.2 μM for vanadate. The latter value is close to the K_i for vanadate inhibition of phosphate transport. Mutants able to grow with vanadate (*van*) have been isolated and characterized. They are unable to take up vanadate from the medium and are also deficient in the phosphate transport system. The K_m for phosphate transport is elevated 35-fold in the *van* mutants, suggesting that the *van* locus may code for a structural component of the transport system (Bowman, Allen and Slayman, 1983).

6.3 BACTERIAL TRANSFORMATIONS OF TOXIC METALS AND METALLOID ELEMENTS

6.3.1 Introduction

The transformations of metals and metal compounds by micro-organisms fall into two broad categories: (i) the reduction or oxidation of inorganic forms, and (ii) conversions from inorganic to organic forms and vice versa, namely methylation and demethylation. In this section, we deal largely with transformations, in both categories, of the toxic metal cations mercury, cadmium, lead, tin and thallium, and, briefly, the possible biotransformation of the metalloids arsenic, selenium and tellurium. The redox transformations and their roles in leaching of metals are discussed in Chapter 7, while the special roles of iron as oxidant and reductant in microbial energetics are covered in Chapters 2 and 5. These topics have been reviewed by Jernelöv and Martin (1975), Konetzka (1977), Summers and Silver (1978), Thayer and Brinckman (1982) and Thayer (1984).

Considerable interest attaches to these transformations, not least at the ecological level since microbe-mediated transformations modify (generally increasing) the toxicity of metal species in the environment. Alkylmetal toxicity depends greatly on the number of alkyl groups present and on the nature of these groups. To a large extent, the toxicity reflects the lipophilicity of the species. Even a single methyl group introduced onto a metal or metalloid will elicit dramatic changes in volatility and lipid solubility. In the extreme case, fully methylated derivatives of the metal are gases or volatile liquids at ambient temperatures and are virtually insoluble in water. In this state, they escape into the atmosphere and may be widely dispersed, perhaps from anthropogenic sources to susceptible living organisms.

6.3.2 Mercury

(a) Speciation and microbial involvement

The so-called mercurous cation is not a simple monomeric Hg^+ ion, but a dimeric species, $^+Hg{-}Hg^+$ with a strong metal–metal bond. This species is unstable towards disproportionation to metallic mercury (Hg^0) and the toxic mercuric ion (Hg^{2+}). Of particular interest is methyl mercury, which is 50 to 100 times more toxic than inorganic Hg^{2+} and

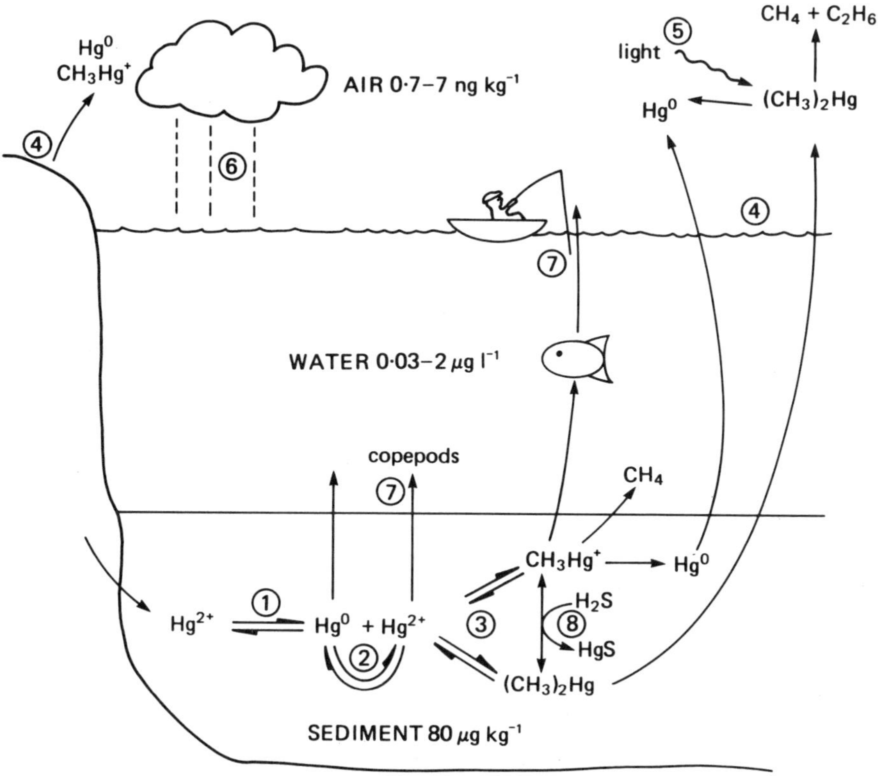

Figure 6.3 Mercury mobilization in the environment. The interconversions of mercury, many catalysed by bacteria, set up a dynamic system of reversible reactions. A steady state of toxic methylmercury is established in sediments. The figures for typical mercury contents of air, water and sediment are shown. The most significant transformations and movements are (1) disproportionation; (2) reduction (and vaporization); (3) methylation and demethylation; (4) vaporization; (5) photolysis; (6) washing from the atmosphere by rainfall; (7) ingestion; (8) disproportionation, catalysed by S^{2-} of microbial origin. Based on Wood (1974), Summers and Silver (1978) and incorporating data from Berk and Colwell (1981) and Wood (1984).

yet, due to its ready diffusion through biological membranes (Wood, 1984), is the most significant form of mercury in both salt and fresh waters (Figure 6.3). It was responsible for the human poisonings at Minamata and Niigata in Japan in the 1950s caused by ingestion of seafood containing methylmercuric compounds, and later in Iraq, due to the use of flour treated with alkylmercury fungicides (Konetzka, 1977). Mercury (Hg^0) has a high vapour pressure compared to other metals, such that the atmospheric concentration can rise to 2 ppm. However, methyl and dimethyl mercury are even more volatile and allow escape of the metal from aqueous media. Microbial activity is partly responsible for the escape of Hg^0 and organomercurials (Imura *et al.*, 1977), particularly at sewage works, perhaps at coal and metal ore deposits (Olson, Iverson and Brinckman, 1981), and for the methylation of mercury that accumulates in certain fish (Figure 6.3). Microorganisms, particularly bacteria and ciliated protozoa, have also been implicated in the transfer of inorganic mercury in marine food chains (Berk and Colwell, 1981). It should be noted that although many bacteria oxidize Hg^0 to Hg^{2+} (as do red blood cells), these bacteria do not generally methylate mercury.

(b) Mechanisms of methylation

The primary route for methylation, and thus mobilization, of mercury in bottom sediments and sludge of polluted waters occurs via the biogenic extracellular methyl donor methylcobalamin (methylcob(III)alamin; CH_3B_{12}). The mechanism of this non-enzymatic methylation is well documented (Ridley, Dizikes and Wood, 1977; Thayer and Brinckman, 1982), partly because of the commerical availability of the methyl donor and the many cobalt compounds with similar chemistry. There are still uncertainties about some mechanistic details, however, and in particular whether monomethyl or dimethyl mercury is the initial product. The cobalt does not change its oxidation state during the reaction, indicating that the methyl group is transferred as a carbanion (CH_3^-). Other potential methylating agents (*S*-adenosylmethionine and N^5-methyltetrahydrofolate) donate carbonium methyl groups (CH_3^+).

Dimethyl mercury is synthesized by an identical mechanism except that CH_3Hg^+ is the reacting species (instead of Hg^{2+}) and that the rate is about 6000 times slower than that for methyl mercury production (Wood, 1974).

Evidence that methylcobalamin-dependent methylation serves as a detoxification process comes from recent experiments with a vitamin B_{12}-requiring mutant of *Clostridium cochlearium*. The auxotroph, lacking the mercury-methylating activity, showed higher sensitivity to inorganic mercury, while both strains (the auxotroph and the wild-type) were similarly sensitive to methylmercury (Pan-Hou and Imura, 1982a,b).

Certain bacteria, such as *E. coli* K12, do not produce methylcobalamin but do methylate mercury. It is possible that such organisms utilize traces of the vitamin found in the medium or environment. A second possibility is that enzymatic transfer of methyl groups from alternative methyl donors occurs intracellularly. Wood (1974) has argued that these donors cannot be *S*-adenosylmethione or 5-methyl-tetrahydrofolate, which transfer CH_3^+.

The methylation of mercury in the fungus *Neurospora crassa* involves the complexing of mercuric ion with homocysteine or cysteine. A methyl donor such as choline or betaine, but not methylcobalamin, is required plus a transmethylating enzyme (Landner, 1971).

(c) Detoxification of organomercury compounds

The most common mechanism of microbial resistance to Hg^{2+} and organomercurials is an inducible system for conversion to volatile Hg^0, although for *p*-hydroxymercuribenzoate there is no evidence for hydrolysis or degradation and resistance may be attributed to the existence of a 'permeability barrier'. In this section, we describe the pathways and enzymes of detoxification. The plasmids that confer resistance, and the detailed molecular information available from sequencing studies, are covered in Section 6.4.

The organomercurials RHg^+ where R is phenyl, methyl or ethyl are cleaved to give benzene, methane and ethane respectively (Figure 6.4) which, to varying degrees, are volatile. The Hg^{2+} is then reduced to highly volatile Hg^0 (Robinson and Tuovinen, 1984).

(i) Mercuric reductase The enzyme responsible for reduction of Hg^{2+} to Hg^0 has been studied in plasmid-containing strains of *Pseudomonas* spp., *E. coli* and *S. aureus*. A general picture has emerged of a soluble, cytoplasmic enzyme. The enzyme cofactor is generally NADPH, but some mercuric reductases (e.g. in a *Streptomyces* strain) can use NADH or NADPH. The *E. coli* enzyme has been extensively purified and shown to have a M_r of 170 000–180 000, consisting of three ($M_r \approx 63\,000$) polypeptide subunits (Schottel, 1978). More recently, a dimeric enzyme of M_r 110 000 has been purified from *E. coli* containing the cloned mercury resistance genes from plasmid NR1. The enzyme is specific for mercury and contains redox-active cysteine residues at the active site (Rinderle, Booth and Williams, 1983). In a soil *Pseudomonas* K62, NADH can substitute for NADPH and the M_r of the active (monomeric) enzyme appears to be only 67 000.

No other enzyme activity has been detected with other metals as substrates, except that induced resistant cells have been reported to produce colloidal Ag^0 and Au^0 from Ag^+ and Au^{3+}, respectively. It has been speculated (Williams and Silver, 1984) that genetic engineering of mercuric reductases could therefore result in an enzyme capable

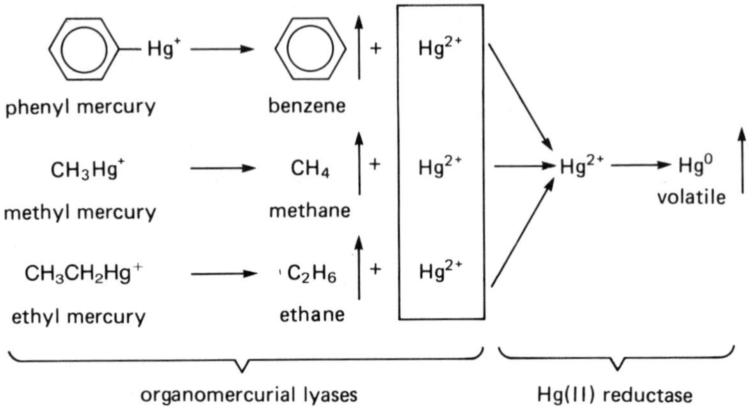

(a)

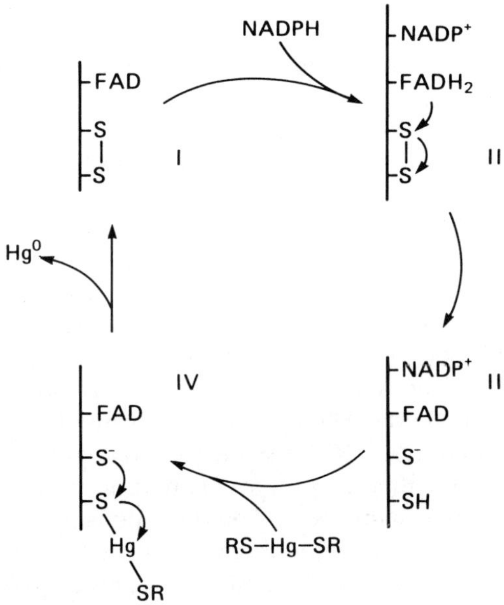

(b)

Figure 6.4 (a) The detoxification of Hg (II) and organomercurials (Summers and Silver, 1978). Reproducd with permission from the *Annual Review of Microbiology*, vol. 32, copyright 1978 by Annual Reviews Inc. (b) Proposed mechanism of mercuric reductase. Reproduced with permission from Williams and Silver (1984).

of extracting these precious metals from low-level soluble material (Chapter 7).

Recent studies, reviewed by Williams and Silver (1984), have suggested a mode of action for the reductase. All the enzymes studied are flavoproteins, having one bound FAD moiety (Figure 6.4(b)). However, anaerobic titrations of the enzyme suggest an additional electron acceptor and, on the basis of spectral similarities between the spectra of the mercuric reductase and of lipoamide dehydrogenase, a redox-active disulphide has been implicated (Figure 6.4(b)). The mechanism probably involves electron transfer from NADPH to FAD, then to the active site disulphide and eventually to the chelated Hg^{2+}. The 540 nm absorbance of species III (Figure 6.4(b)) the 2-electron reduced form, is attributed to a charge-transfer complex between one reduced thiolate at the active site and flavin.

Structural homologies between mercuric reductase, glutathione reductase and lipoamide dehydrogenase are evident from the DNA sequence analysis of mercuric reductase genes (Section 6.4.1(a)). The redox potentials of the three proteins are also similar for the addition of either two or four reducing equivalents.

(ii) Organomercurial lyases The most complete information comes from the soil *Pseudomonas*, isolate K62. Two enzymes have been identified with somewhat different properties and substrate ranges. Both have M_r values of about 19 000 and require a reduced thiol, preferably L-cysteine, for activity. In *E. coli*, there is kinetic evidence for two enzymes which have a larger M_r than those in *Pseudomonas* (Schottel, 1978). DNA sequence information on the *merB* gene (Section 6.4.1(a)) has recently suggested that the *Serratia* lyase may have a M_r of about 23 000.

(iii) The Hg^{2+} transport system The resistance mechanism also involves, perhaps surprisingly, a specific plasmid-encoded transport of Hg^{2+} *into* the cell, presumably to prevent this highly toxic ion reacting with sensitive sulphydryl groups in the membrane. Operation of such a system is indicated by the fact that mutants with inactivated mercuric reductase are hypersensitive or supersensitive (Hg^{ss}) to Hg^{2+}.

(d) Mercury transformations by eukaryotes

These mechanisms have been less well studied (Jeffries, 1982), but in yeast two modes of resistance to mercury have been identified. One involves reduction and volatilization and the other involves some constituent of methionine synthesis. The yeast *Cryptococcus* and the alga

Chlamydomonas have been shown to be capable of reducing Hg^{2+} to Hg^0, whilst in *S. cerevisiae* methionine auxotrophy is associated with the acquisition of methylmercury-resistance. The basis of the resistance is not understood.

6.3.3 Tin

(a) Methylation and dealkylation of tin compounds

It is now well established that tin can be methylated in the environment (Wood, 1974), contributing to a biological cycle for tin in which the metal species are biotransformed and mobilized (Ridley, Dizikes and Wood, 1977). Micro-organisms, especially *Pseudomonas* spp., have been isolated from, for example, Chesapeake Bay, Plastic Lake (Ontario) and San Francisco Bay and shown to generate methyltin, dimethyltin, trimethyltin and tetramethyltin (Hallas, Means and Cooney, 1982; Thayer and Brinckman, 1982). It is likely, however, that abiotic formation of methylated tin compounds (Guard, Cobet and Coleman, 1981) and anthropogenic effects dominate the flux of the metal through the biosphere (Byrd and Andreate, 1982).

The widespread use of organotin compounds as fungicides (Section 6.3.3(b)) and their high phytotoxicity has encouraged studies of their microbial degradation. For example, triphenyltin (fentin) acetate is degraded by bacteria and filamentous fungi in soil as well as photochemically (Barnes, Bull and Poller, 1983). Bis(tributyltin)oxide can be slowly degraded by pure cultures of bacteria and fungi to the less toxic monobutyl tin and small amounts of dibutyltin derivatives, which were further metabolized under the conditions tested (Barug, 1981). Attempts to isolate micro-organisms that could use bis(tributyltin)oxide as sole carbon source failed.

A number of bacteria (*Pseudomonas* spp.) resistant to tributyltin and inorganic tin have been isolated from an industrially polluted region of the Chesapeake Bay and shown to accumulate tributyltin to substantial levels, probably by passive adsorption to the cell envelope (Blair *et al.*, 1982). In this case, no evidence could be obtained for possible transformation products of tributyltin.

(b) Toxicity of organotin compounds

In addition to the relatively small-scale methylation of tin in the environment by microbes, there is a massive industrial production of organotins. In the USA, the total weight of such compounds produced from 1965 to 1976 was about 84 × 10^6 kg; the forecasted growth rate from 1976 to 1986 would have resulted in an annual production in 1986 of about 8.4, 4.5, 3.6 and 4.1 × 10^6 kg, respectively, of butylisooctylmercaptotin acetate, methylisooctylmercaptotin acetate, dibutyltin dilaurate and bis(tributyltin)oxide. Over two-thirds of the world's annual production of organotins is used in the thermal stabilization of polyvinyl chloride (PVC) plastics. The other main uses are (i) as catalysts in the production of rigid polyurethane foam and for room temperature vulcanization of silicone elastomers, (ii) in biocides, particularly in marine anti-fouling paints, and (iii) in anthelminthic applications (Thayer, 1984).

All such compounds eventually reach the natural environment where, in view of their low vapour pressure, there is little tendency to escape to the atmosphere. Chemical and biological degradation (dealkylation) contribute to the eventual fate of the compounds thereafter, and so the toxicity of these compounds has attracted much interest. A feature of organotin metabolism, distinguishing it from that of organic compounds of arsenic and mercury, is that the most toxic organotins are orders of magnitude *more* toxic than inorganic compounds of tin.

In mammals, tetraethyltin is successively destannylated by a microsomal, cytochrome P-450-mediated system to more toxic species. Whereas mammalian toxicity is maximal with the ethyltin compounds, insects are most affected by the trimethyltins and fungi and bacteria are generally more susceptible to the tri-n-propyl (Hallas and Cooney, 1981a,b) and butyltins, though there are significant variations in sensitivity patterns. This selectivity allows effective use of this family of compounds as partly specific biocides.

The biochemical basis of the toxicity of organotins in best understood in mammals, although similar mechanisms appear to operate in all cell types studied. The principal inhibitory site is the energy-transducing ATPase of the inner mitochondrial membrane, where organotins block proton flux and thus ATP synthesis in the intrinsic F_0 portion in an oligomycin-like fashion. In yeast, triethyltin is not covalently bound to the binding site, which genetic studies show to be distinct from that for oligomycin and venturicidin (Cain and Griffiths, 1977). However, organotins also inhibit the Na^+, K^+-ATPase of cell membranes, the Ca^{2+}-translocating ATPase of sarcoplasmic reticulum, and act as ionophores, facilitating OH^-/halide exchanges across biological and artificial membranes (Aldridge, 1976; Selwyn, 1976). Inhibition of membrane-bound ATPase activity and mediation of an OH^-/halide antiport has also been reported for *E. coli* (Singh and Bragg, 1979a,b; Yamada and Watanabe, 1979a,b).

Recently, a further trialkyltin derivative, dibutylchloromethyltin chloride (DBCT), has been examined for biological activity. Unlike trialkyltin, DBCT binds in an apparently irreversible manner and appears to behave as a covalent inhibitor of mitochondrial ATPase and oxidative phosphorylation. It has been used as a specific affinity label for a component of the ATP synthetase. The suggestion that this component is lipoic acid and represents a key factor in oxidative phosphorylation (Cain, Partis and Griffiths, 1977) was not supported by studies with bacteria (Singh and Bragg, 1978). In addition, DBCT is not as specific as once supposed, and also inhibits electron transport in plant mitochondria (Moore, Linnett and Beechey, 1980).

Tripropyltin chloride is also adsorbed by membrane phospholipids, causing leakage of cell constituents and inhibiting certain transport systems. The bacteriostatic and bactericidal effectiveness of bis(tributyltin)oxide (TBTO)-containing formulations with *Legionella pneumophila* appears to result partly from the inhibition by TBTO of energy metabolism, suggesting the same mode of action as in other cell types.

(c) Volatilization of tin as stannane

A further pathway for volatilization of tin and its loss from aqueous phases to the atmosphere has recently been described. The macroalga *Enteromorpha* produces tetramethyltin under aerobic conditions; however decaying algal material in sea water induces, anaerobically, the volatilization of tin as stannane, SnH_4 (Donard and Weber, 1988). The chemical similarities between tin and carbon, and evidence for the presence of methanogens in the test material, suggest that tin might substitute for carbon in the microbial pathway for methane formation.

6.3.4 Lead

(a) Evidence for biomethylation of lead

There is still controversy as to whether lead compounds can be biomethylated. There have been several reports that sediment microorganisms can form tetramethyllead using either inorganic lead nitrate or trimethyllead acetate as substrates. Several gram-negative aerobic rods have been implicated, including species of *Pseudomonas, Acinetobacter, Flavobacterium* and *Aeromonas* (Summers and Silver, 1978; Thayer and Brinckman, 1982). A study of sites around Lancaster in England (Harrison and Laxen, 1978) suggested that the amounts of tetraalkyllead in air could not be explained by vehicle emission alone and were more likely to arise by airborne transport from biogenic sources in mud flats in the nearby Morecambe Bay estuary. In contrast to these claims are the results of attempts to produce alkyl leads in anaerobic and aerobic sediments (Reisinger, Stoeppler and Nürnberg, 1981). Labelled and unlabelled substrates were used as methyl donors and labelled and unlabelled lead compounds as acceptors. No biomethylation could be detected, but earlier findings that tetramethyl and trimethyllead are the products of a chemical sulphide-induced conversion were confirmed. Another possible route to tetraalkyllead compounds (Thayer and Brinckman, 1982) is via a crossover mechanism involving biogenic methyl iodide (CH_3I, which occurs in appreciable quantities in marine waters) and abiotic lead. Toxic methyllead compounds are formed.

(b) Toxicity of organolead compounds

Alkyllead compounds, particularly tetraethyllead, are still used in certain countries on a massive scale as 'antiknock' agents in petrol. The combustion of petrol provides the single largest technological source of lead in the atmosphere and terrestrial environments (9000 and 3000 tonnes yr^{-1}, respectively). Organolead compounds are more toxic than inorganic lead compounds due to their greater liposolubility and ready absorption through biological membranes. Tetraethyllead is dealkylated in the liver to form the triethyllead cation which is believed to be the principal toxic form *in vivo*.

There are conflicting reports on which of the organolead compounds are most toxic to micro-organisms. The tripropyllead compound is more liposoluble than the triethyl analogue and was more toxic in our studies on *E. coli* (Gibson *et al.*, 1980). A conflicting finding (Huber and Kirchmann, 1978) is confused by the used of culture respiration rates as a measure of growth. The choice was unfortunate since, in Cl^--containing media, triethyltin compounds *stimulate* respiration and the mode of action of both classes of triethylmetal compounds appears to be similar (Gibson *et al.*, 1980). Tetramethyllead is also toxic, illustrated by its effect on the alga *Scenedesmus* (Silverberg, Wong and Chau, 1977).

Diverse effects of organolead compounds have been described. In *E. coli*, trialkylleads inhibit growth, respiration and ATPase activity and, significantly, at only 1 μM, mediate a halide (Cl^- or Br^-)/OH^- exchange (Figure 6.5). Trimethyllead acetate-treated *Aeromonas* accumulates the compound in the membrane and wall, possibly the sites of methylation to the tetramethyl derivative, or reflecting operation of a pathway for exclusion of the lead compound from the cell. There are several reports

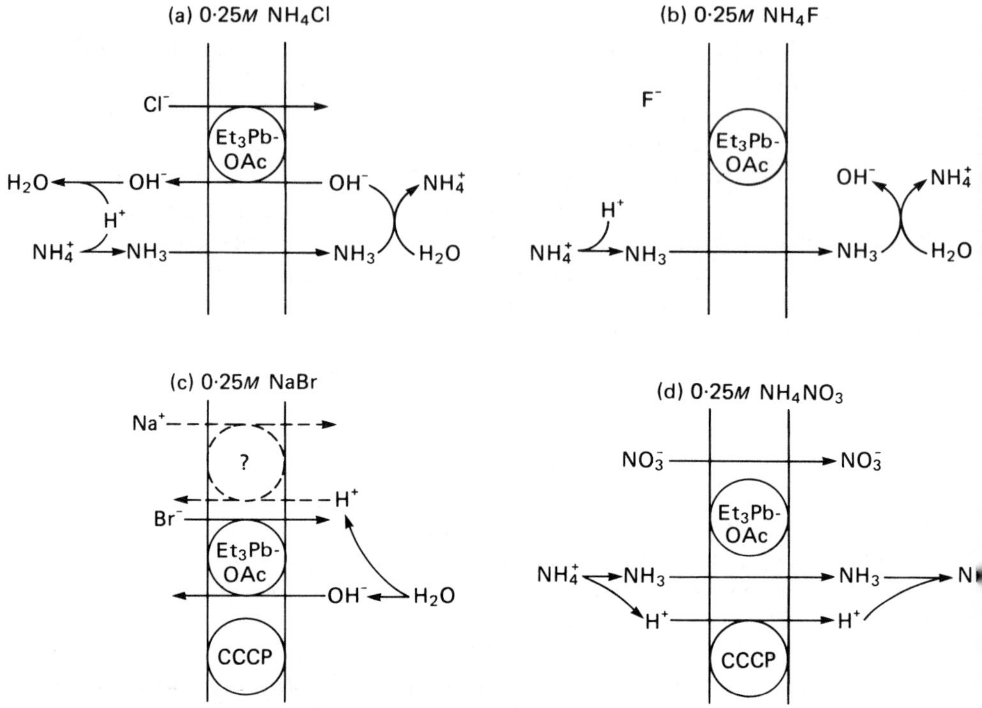

Figure 6.5 Mode of action of triethyllead acetate in facilitating ion movements across the cytoplasmic membrane of *E. coli*, based on experiments with spheroplasts suspended in iso-osmotic solutions. Spheroplasts are stable, and thus do not swell, in solutions iso-osmotic with 0.25 M sucrose (not shown). Spheroplast swelling occurs, detected spectrophotometrically by light scattering changes, when the overall transport of ions from the suspending medium is electrically neutral and produces no net movement of H^+ or OH^- across the membrane. In each scheme (a to d), the pair of vertical lines represents the cytoplasmic (spheroplast) membrane, with the cytoplasm on the right and extracellular space on the left. Large circles in the membrane represent specific pathways provided by triethyllead acetate (Et_3PbOAc) for Cl^-/OH^- exchange or Br^-/OH^- exchange and by CCCP for H^+ movement. Passive diffusion is shown for NO_3^- and NH_3. These schemes explain entry of solutes as follows. In (a), entry of NH_4Cl (or NH_4Br) and spheroplast swelling is shown. In (b), the ineffectiveness of Et_3PbOAc in facilitating a F^-/OH^- exchange does not lead to swelling. In (c), charge imbalance prevents swelling despite the presence of the Et_3PbOAc-mediated Br^-/OH^- exchange; operation of the Na^+/H^+ antiport shown by the broken lines would result in swelling. In (d), NH_4NO_3 enters and causes swelling in the presence or absence of Et_3PbOAc, provided that CCCP is present. Reproduced with permission from Gibson *et al.* (1980).

of inhibition of cell division by tri- and tetraethylleads, resulting, for example, in giant cells in the flagellate *Poterioochromonas* (Röderer and Schnepf, 1977).

6.3.5 Arsenic-containing compounds

(a) History and speciation

The activities of micro-organisms played a seminal role in the discovery of biomethylation. Research in this area has its origins in the 'arsenic rooms' of the nineteenth century, in which people succumbed to arsenic poisoning by breathing a volatile compound exuded by fungi growing on wallpaper. Gosio showed that the compound ('Gosio-gas') contained arsenic. Subsequently, Challenger and others (i) identified it as trimethyl arsine ($(CH_3)_3As$), (ii) demonstrated its formation from arsenic trioxide by the fungi, (iii) showed that similar compounds could be formed from the oxides of selenium and tellurium, and (iv) identified *S*-adenosyl-methionine as the biogenic methyl donor (Thayer, 1984; Thayer and Brinckman, 1982).

The toxicity of compounds containing the metalloid element arsenic is related to the oxidation state of the arsenic atom. The least toxic inorganic form is arsenate, $As(O)(OH)_3$, with the metalloid in the +5 oxidation state. More toxic by a factor of 100–200 is arsenite, $As(OH)_3$ (+3 oxidation state), probably by virtue of its binding to sulphydryl groups. Arsenate (+5) is an analogue of inorganic phosphate, both in membrane transport systems (Section 6.2.2(b)) and in the reactions catalysed by enzymes such as kinases. Arsenylated sugars thus formed hydrolyse spontaneously.

Three types of biotransformation occur, oxidation, reduction and alkylation (Figure 6.6).

(b) Oxidation and reduction

Arsenite can be oxidized to the less toxic arsenate by several bacteria, such as *Achromobacter*, *Xanthomonas* and *Pseudomonas* spp., although the mechanism is not understood in detail (Konetzka, 1977). A soluble enzyme (arsenite dehydrogenase) has been identified that appears to be coupled to a membrane-bound, cytochrome-mediated electron transfer system to oxygen (Konetzka, 1977; Summers and Silver, 1978). *Thiobacillus ferrooxidans* can oxidize the arsenic in minerals such as arsenopyrite ($FeS_2.FeAS_2$), energite ($3Cu_2S.As_2S_5$) and orpiment (As_2S_3).

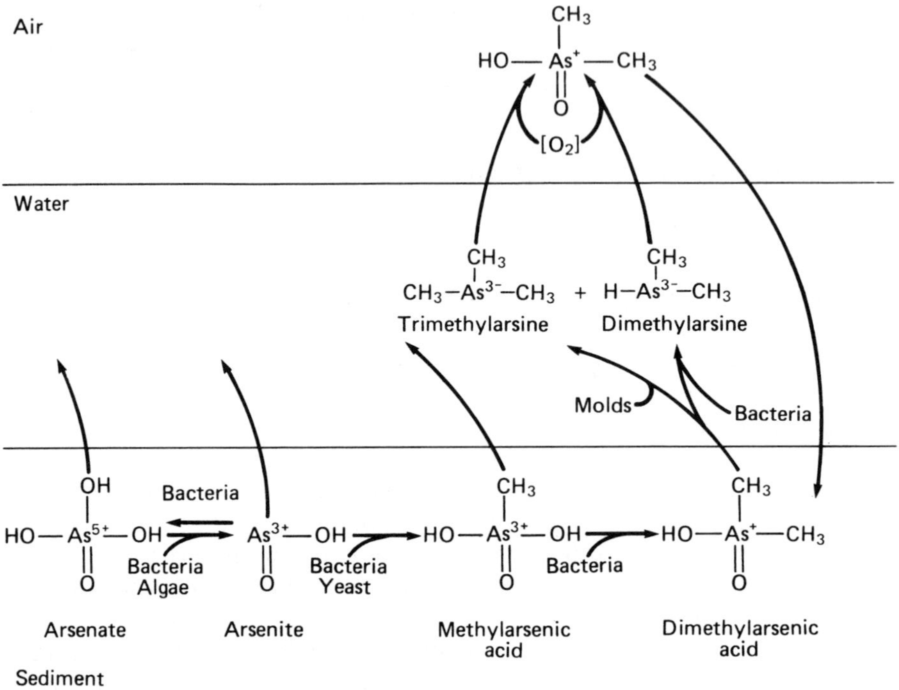

Figure 6.6 The biological cycle for arsenic (Wood, 1974). Arsenate is reduced to the more toxic arsenite by algae and bacteria; the reverse reaction is catalysed by several bacteria. The organoarsenicals are synthesized by various micro-organisms. Reproduced with slight modification and permission from *Science* 183, 1049–52, copyright 1974 by the AAAS.

Recently a high affinity arsenite oxidase has been partially purified from *Alcaligenes faecalis*. The enzyme appears not to contain haem or flavin but there may be FeS centres. Oxygen and cytochrome *c* are amongst the electron acceptors identified (Williams and Silver, 1984).

Reduction of arsenate to arsenite has been reported in *Chlorella* and *Micrococcus lysodeikticus*. The reduction of arsenate to the most reduced state, arsine, is well documented for a variety of fungi and *Methanobacterium*.

(c) Methylation

In the oceans, phosphate is so depleted by photosynthesis that arsenate concentrations may approach those of phosphate, leading to uptake of the metalloid by effective competition with phosphate. Marine algae and higher aquatic plants detoxify and excrete the arsenic as dimethyl-5-deoxyribosylarsine oxides. Conversion of arsenous acids to phospholipid derivatives of these oxides may be a mechanism for membrane-mediated excretion of arsenic by marine algae (Knowles and Benson, 1983).

Other methylated arsenic-containing compounds are known to be synthesized by micro-organisms. *Methanobacterium* methylates arsenate in a side reaction of methane biosynthesis (McBride and Wolfe, 1971), using coenzyme M (2,2′-dithiodiethanesulphonic acid):

$$\underset{\text{arsenate}}{O{=}As(OH)_3} \rightarrow \underset{\text{dimethylarsine}}{(CH_3)_2As}$$

Methylarsine (CH_3AsH) and dimethylarsine are produced by sewage bacteria, probably by mechanisms related to methane biosynthesis rather than using vitamin B_{12}-mediated methylation. *S*-Adenosyl methionine is involved in arsenic methylation by fungi (Summers and Silver, 1978). Other microbes identified as capable of methylating arsenite are *Serratia marinorubra*, *Rhodotorula rubra* and *Pseudomonas* spp. (Thayer and Brinckman, 1982). There are also scattered reports of arsenic biomethylation in higher organisms, including man (Thayer and Brinckman, 1982) but it is uncertain whether the reactions are catalysed by mammalian body processes or by microflora of the gastrointestinal (Wood, 1974) tract.

The transformations described thus constitute a biological cycle for arsenic (Wood, 1974; Figure 6.6), in which the reduced forms of arsenic may be utilized as an energy source by bacteria, methylated to give dimethylarsine and trimethylarsine of extreme toxicity, and then readily oxidized to give less toxic products such as cacodylic acid.

6.3.6 Tellurium-containing compounds

The oxoanions of the metalloid elements tellurium and selenium are toxic at appropriate concentrations to most micro-organisms. In general, gram-negative bacteria are more sensitive to tellurite than are gram-positives. The resistance to tellurite displayed by certain pathogenic species (e.g. *Corynebacterium diphtheriae* and group D *Streptococcus*) is not understood at a mechanistic level. A black deposit of metallic tellurium is formed by certain bacteria and the yeast *Schizosaccharomyces pombe* as a result of tellurite reduction, but it is not clear whether the resistant organisms have a tellurite reductase or whether they are impermeable to the compound (Smith, 1974).

Some pseudomonads methylate tellurium and selenium (Konetzka, 1977). *Pseudomonas aeruginosa* growing in the presence of potassium tellurite produces a strong garlic odour indicative of volatile methyl- or dimethyltellurium. *Candida humicola* generates an unidentified gas, probably dimethyltellurium, from tellurate.

6.3.7 Selenium-containing compounds

Selenium is unusual because of the narrow range of concentrations in which it exhibits biological activity as an essential nutrient or as a toxic element. Resistance to selenite is characteristic of *Salmonella* and *Shigella* species. Perhaps the most striking microbial transformation of selenium is the reduction of selenate or selenite to red amorphous selenium, which accumulates in cells and gives them a brick-red appearance. The reduction occurs in various organisms, including bacteria, cyanobacteria, yeasts and fungi (Konetzka, 1977). Further reduction to selenide has been reported. The mechanisms of reduction and cellular localization of the selenium granules are uncertain; in *E. coli*, the metalloid seems to be deposited in the wall and membrane, but in other bacteria accumulation occurs in the cytoplasm.

Recently, the oxidation of elemental selenium to selenite has been observed in a soil *Bacillus megaterium* (Sarathchandra and Watkinson, 1981). This has important consequences for mobilization of the element in soil, preventing selenium deficiency in ruminants, and constituting a step in the proposed biogeochemical cycle for selenium (Figure 6.7).

Several microbes, especially fungi and bacteria, methylate selenium in selenate or selenite producing, for example, dimethylselenide $(CH_3)_2Se$, dimethylselenite $(CH_3)_2Se_2$ and dimethylselenone $(CH_3)_2SeO_2$ (Reamer and Zoller, 1980). Production of these compounds (by a yet-unknown biochemical pathway) may represent a detoxification route and constitute a further step in the selenium cycle (Figure 6.7).

6.3.8 Cadmium

A heavy metal-tolerant strain of *Pseudomonas*, capable of generating metallic mercury from Hg^{2+} and making methylmercury, was also able to volatilize cadmium in the presence of cobalamin. Exposure of the volatile cadmium compound to Hg^{2+} gave methylmercury, suggesting that the volatile cadmium compound was methylcadmium (Summers and Silver, 1978).

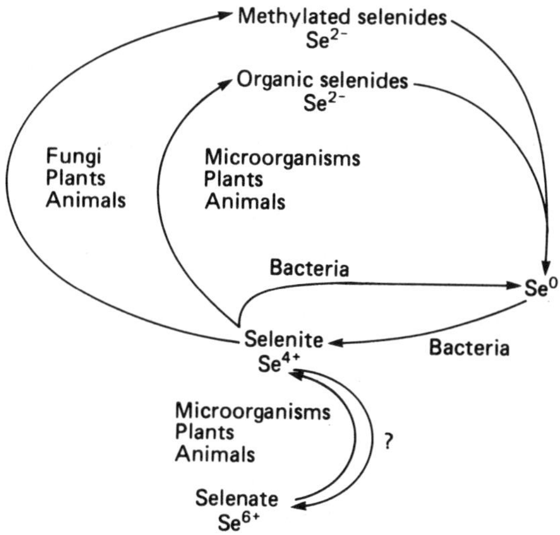

Figure 6.7 A biogeochemical cycle for selenium. Selenate or selenite are reduced to amorphous selenium by various micro-organisms; the reverse (oxidation) has been recently reported. Methylation gives volatile selenium compounds. Reproduced with slight modification from Konetzka (1977) by courtesy of Marcel Dekker Inc.

6.3.9 Chromate

Chromates (CrO_4^{2-}) and dichromates (CrO_7^{2-}) containing Cr(VI) are much more toxic and mutagenic than Cr(III) compounds. Enzymatic reduction of K_2CrO_4 has been reported in a chromate-resistant *Ps. fluorescens* strain. A membrane-associated metalloenzyme, using NADH as reductant, has been implicated but little further is known (Bopp, Chakrabarty and Ehrlich, 1983).

6.3.10 Antimony

Stibiobacter senarmontii is capable of oxidizing Sb^{3+} to Sb^{5+}. The reaction is probably catalysed by a specific oxidase (Lyalikova, Vedenina and Romonova, 1976).

6.3.11 Silver

A pseudomonad isolated from domestic sludge was resistant to 300 ppm silver nitrate in nutrient broth or agar and produced a reduced environment that might have caused the deposition of the less toxic metallic form (Belly and Kydd, 1982).

6.3.12 Methylation of other metals and metalloids

Thallium (Tl^{+}) is isoelectronic with Pb^{2+} and so the possibility arises that inorganic thallium compounds also undergo oxidative biomethylations. (Note that biomethylation of compounds of mercury (the other neighbour of thallium in the Periodic Table) proceeds without change in the oxidation state.) Thallium methylation by sediments under anaerobic conditions has been demonstrated in the laboratory (Huber and Kirchmann, 1978), but there appear to be no reports of methylthallium compounds in natural waters. Unlike the other metals discussed in this section, the dimethyl form of thallium is the only species likely to persist; CH_3TlX_2 compounds are unstable in aqueous solutions, whereas $(CH_3)_3Tl$ decomposes instantaneously to give $(CH_3)_2Tl^{+}$ and methanol.

Attempts to detect methylgermanium and methylantimony compounds in natural environments have proved negative, although the possibility of biogeochemical cycling of methylantimony has been discussed (Thayer and Brinckman, 1982).

No biological significance can yet be attached to the observed methylation of platinum complexes by methylcobalamin (Fanchiang, Ridley and Wood, 1979).

6.4 PLASMID-DETERMINED RESISTANCE MECHANISMS

Many microbial metal resistance mechanisms, like most clinically significant antibiotic resistance, are determined by genes located on plasmids, autonomously-replicating circular DNA molecules. Transposons are DNA elements which can insert at random into plasmids or the bacterial chromosome independently of the cell's recombination system. They have the ability to transpose from one DNA molecule to another and, with plasmids, allow rapid movement of the resistance genes between DNA molecules and between cells in a population. They greatly facilitate analysis of the resistance mechanisms by molecular genetic techniques. Plasmid-determined resistance to toxic metal ions (and to antibiotics) has been reviewed by Foster (1983, 1987) and Silver, Laddaga and Misra (1989).

The three best understood systems are those conferring resistance to mercuric ions and organomercurials, and to AsO_4^{3-} and Cd^{2+}. In the first case, volatile mercury is formed; the enzymology of this process has been described above (Section 6.3.2). The mechanisms of AsO_4^{3-} and Cd^{2+} resistances involve separate efflux transport systems (Section 6.2.2) for the toxic ion. In this section, emphasis is placed on the plasmid-encoding of these mechanisms of resistance and plasmid-determined strategies for resisting other toxic metals.

6.4.1 Mercury and mercurial resistances

(a) Organomercurial lyase and reductase synthesis

The presence of a heavy metal in an environment tends to select for bacteria carrying resistance plasmids (Summers *et al.*, 1978); in the case of mercury compounds, this correlation is well-established (Jeffries, 1982; Silver, 1981; Foster, 1983, 1987). The reduction of the mercuric ion (Hg^{2+}) to Hg^0 and, in the case of organomercurials, their prior hydrolysis releasing mercuric ion, is catalysed by enzymes which are encoded by genes on plasmids and transporons. These genes and mechanisms are the best understood of all plasmid-determined resistances to toxic metals.

The patterns of resistance to mercury compounds exhibited by various genera can be rationalized by our understanding of the underlying enzymology of the detoxification, namely the requirement for both an organomercurial lyase and a mercuric reductase. Thus, two resistance phenotypes have been recognized in gram-negative bacteria. Narrow spectrum resistance is found in strains that express the reductase alone and broad spectrum resistance reflects the presence of both a reductase *and* a lyase. There are complications, however. Thus, in *E. coli*, over 90% of the plasmids conferring resistance to mercuric ions also confer resistance to the organomercurials merbromin and fluoroscein mercuric acetate, but without hydrolysis or mercuric reduction. A 'permeability barrier' has been postulated to account for such resistance. The remaining plasmids confer additional resistance to only phenylmercuric acetate and thiomerosol (ethylmercurithiosalicylate), although Hg^0 is also formed from *p*-hydroxymercuribenzoate (pHMB) and ethyl mercury. This is attributed to low lyase activities and the greater toxicity of ethyl- and methylmercury. The spectra of resistances in *Pseudomonas* and other bacteria are given by Foster (1983) and in references cited by Summers and Silver (1978), Summers (1984), and Foster (1987).

Important advances have been made in the last few years, stemming from DNA sequencing studies of the previously identified *mer* operon.

Three systems have been investigated (Silver, Rosen and Misra, 1986; Foster, 1987; Silver, Laddage and Misra, 1989): (i) plasmid R100; (ii) transposon Tn501; (iii) plasmid pI258 of *Staphylococcus aureus*. These regions show a number of common features, suggesting that they have evolved from a common mercury resistance genetic element. Only the results from R100 will be summarized here, unless explicity stated otherwise.

Figure 6.8 shows a model of the genes and gene products of the mercury toxification system. The regulatory gene *merR* provides both positive and negative regulation. A promotor-operator region (OP) precedes the first structural gene *merT*. The latter encodes a transport system whose role is to trap extracellular Hg^{2+} and deliver it to the intracellular mercuric reductase. It is the expression of this function in the absence of reductase activity that renders certain *mer* mutants hypersensitive (Section 6.3.2(c)) to mercurial compounds. The gene product appears to be a very hydrophobic membrane protein that might make three passes through the cytoplasmic membrane. Of special interest are cysteine pairs on each side of the membrane. These are conserved in R100 and Tn501 and have been suggested (Figure 6.8) to be directly involved in the mercury transport mechanism. The gene *merA* encodes the subunit of the dimeric reductase. The active site sequence is highly conserved in the mercuric reductases encoded on R100 and Tn501. Of 44 amino acids in a row identical in the two reductases, 16 are also identical

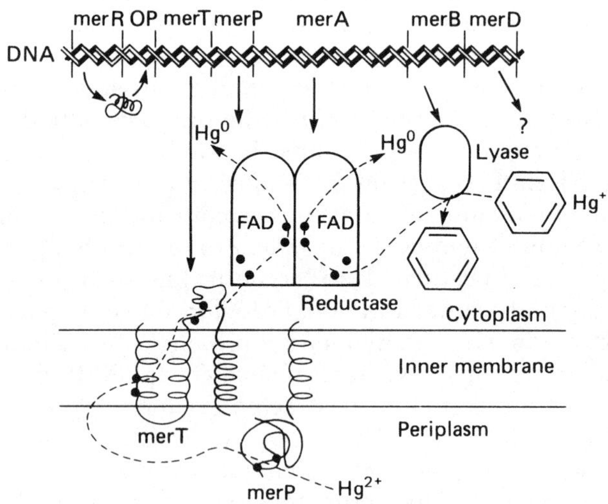

Figure 6.8 Current model for the mercuric detoxification system. Reproduced with permission from Silver, Rosen and Misra (1986) and Oxford University Press.

in human glutathione reductase. There are also striking sequence homologies for most of the lengths of the mercuric and glutathione reductases (Wood, 1984; Williams and Silver, 1984). The genes *merC* (in R100 only) and *merD* were subsequently identified by transposon mutagenesis but are of uncertain function.

MerB specifies the organomercurial lyase enzyme responsible for broad spectrum resistance. This gene is absent from R100 and Tn501 but has been studied on the plasmid pD41358 isolated from a hospital strain of *Serratia marcescens* (Griffin *et al.*, 1987). The lyase is thought to be a 212 amino acid polypeptide and shows considerable homology with the lyase from *Staphylococcus aureus* plasmid pI258 (Laddaga *et al.*, 1987). The mechanism of action of the enzyme is unknown but the *merB* sequence lacks a recognizable signal sequence, suggesting that the gene product remains cytoplasmic (Foster, 1987). Organomercurials presumably cross the membrane in a manner similar to Hg^{2+}.

Many of the questions raised by the model in Figure 6.8 are now amenable to direct experimentation using oligonucleotide-directed site-specific mutagenesis and should be answered in the near future.

Mercury sensitivity may be attributable to any one or more of the components of this complex resistance mechanism. Sensitive mutants have been described that retain the capacity for volatilization but show diminished transport of the heavy metal from the medium to the cytoplasmic mercuric reductase. There is some evidence (Barkay and Colwell, 1983) that transport processes other than those mediated by *merT* may be involved, emphasizing the complexity of bacterial resistance to mercurials.

(b) Mercury resistance in Clostridium cochlearium

The mercury resistance of this bacterium appears not to reside in mercury reduction and volatilization of the element, or in depression of binding or uptake. Instead, a single plasmid seems to determine demethylating activity as well as the ability to form H_2S (Pan-Hou and Imura, 1981). It is suggested that organomercurial degradation is followed by formation of insoluble mercuric sulphide. Plasmid-mediated formation of H_2S has been reported in other micro-organisms and may constitute an additional, widespread mechanism of resistance to heavy metals that react with H_2S to form insoluble sulphides.

6.4.2 Arsenate and arsenite resistances

Plasmid-mediated resistance to arsenicals is widely distributed in gram-negative and gram-positive bacteria (Summers, 1984). Resistance to arsenite and arsenate is achieved (Section 6.2.2(b)) by an inducible ATP-driven transport system that serves to extrude the toxic anion(s) from the cell (Mobley and Summers, 1987).

The arsenic resistance ATPase of gram-negative bacteria has been intensively studied at the physiological and more recently, the molecular levels. Cloning of fragments encoding components of the resistance mechanism and subsequent DNA sequence analysis has revealed four genes, *arsA*, *arsB*, *arsC* and *arsR* (Chen *et al.*, 1986; Silver *et al.*, 1986). The ArsA protein is a loosely membrane-bound arsenic-stimulated ATPase, the polypeptide having potential ATP-binding regions homologous to those in other ATPases. The ArsB protein is a transmembranous, hydrophobic protein which may contain the arsenic (and antimony) transmembrane channel, although the relevant residues have not yet been identified. In addition to these two large proteins, which constitute the efflux pump, there is a third, smaller protein ArsC, which appears to determine the substrate specificity of the pump. Deletion of the *ArsC* gene results in loss of arsenate, but not arsenite, resistance. The ArsR protein is a small DNA-binding regulatory protein that switches on the arsenic resistance mechanism.

Interestingly, the ArsA and ArsC polypeptides appear related to components of nitrogenase (Chapter 6). In particular, the ArsA protein has ATP-binding sites homologous with the nifH-determined iron-containing protein of nitrogenase. Since the latter enzyme requires metal oxyanions such as vanadate and molybdate, it is tempting to suppose that there may be ancestral relationships between nitrogenase proteins and the arsenite and arsenate recognition sequences of the efflux pump (Silver *et al.*, 1989).

6.4.3 Silver resistance

Silver-resistant plasmids have been found in isolates from clinical environments where silver salts are still used for treatments of burns (Section 6.8.1(b); Summers *et al.*, 1978). A recombinant between plasmid R1 and a Ag^+-resistance plasmid from *Citrobacter* has been introduced into *E. coli* K12 and extensively studied (Silver, 1981). Silver resistance is constitutive, but is markedly affected by the presence of halide ions with which silver ions readily precipitate. It is proposed that silver-sensitive cells are able to extract the metal from the metal halide, whereas cells with the resistance plasmid cannot. The metal in silver sulphadiazine is not precipitated in the presence of chloride, and, in support of the above hypothesis, the $AgNO_3$-resistant plasmid confers resistance to this compound, irrespective of the presence of chloride.

A silver-resistant strain of *Ps. stutzeri*, isolated from a silver mine, harboured a plasmid that specified silver resistance (Haefeli, Franklin and Hardy, 1984). The mechanism of plasmid-mediated silver resistance is unknown (Trevors, 1987a), but chromosomally encoded resistance results from a decrease in silver binding by a major outer membrane protein (Summers, 1984).

6.4.4 Other heavy metal resistances

There are numerous reports that other metal-resistance mechanisms are plasmid-determined, but relatively few details are available. Examples are given in Table 6.2.

6.5 RESISTANCE BY METAL DEPOSITION

Detoxification is an attractive explanation (of many) for the widespread ability of micro-organisms to deposit metals (Dexter-Dyer Grosovsky, 1982). It is likely, however, that detoxification by metal deposition is gratuitous and does not involve a 'custom-made' mechanism.

In higher organisms, there are several well-documented cases of a metal being deposited or precipitated in an innocuous form. For

Table 6.2 Examples of plasmid-determined resistance to heavy metals

Organism	Metals	Reference
H_2-utilizing *Pseudomonas*	Co, Zn	Mergeay, Gerits and Houba (1978)
Nostoc calcicola	Cd	Singh and Pandey (1982)
Marine pseudomonad	Mn*	Lidstrom, Engebrecht and Nealson (1983)
E. coli	Cu†	Tetaz and Luke (1983)
S. aureus	Cd	Tynecka, Gos and Zajac (1981 a,b) El Solh and Ehrlich (1982)
Various clinical and non-clinical isolates	Bi, Pb, Zn, Co, Ni, Te	Summers *et al.* (1978) Summers (1984)
Bacteroides spp.	Cd	Summers (1984)
Pseudomonas, E. coli ?	Cd	Summers (1984)
Pseudomonas stutzeri	Ag	Haefeli, Franklin and Hardy (1984)

* Mn oxidation (Summers and Silver, 1978), arguably not a detoxification mechanism.
† Other references to plasmid-encoded copper resistance are given by Trevors (1987b).

example, in the hepatopancreas of the snail, spherical, concentrically-layered deposits are found in which various metals, e.g. Zn, Cu and Pb, are present. Similarly, 'aurosomes' (gold-containing bodies) are found in lysosomes, bismuth-containing inclusion bodies are localized in nuclei, and spherical deposits of lead may be found in kidney cells. There are many apparently analogous, but incompletely described, situations in bacteria.

Hydrogen sulphide production is fairly widespread among micro-organisms (Gadd and Griffiths, 1978) and confers metal resistance by the precipitation of heavy metals as their insoluble sulphides. In *Clostridium cochlearium*, H_2S production is plasmid-determined (Section 6.4.1(c)). In other bacteria, such as *Desulfovibrio* spp., H_2S production may result from anaerobic sulphate reduction. Recently reported examples are the precipitation of sulphides of Cd, Hg and Pb by *Klebsiella aerogenes* (Aiking *et al.*, 1984; Aiking, Govers and van't Riet, 1985). In yeasts, metal tolerance has often been linked with H_2S production (Ross, 1975); colonies of such organisms appear black or dark brown in the presence of copper as a result of deposition of CuS in and around the wall. Other fungi produce copper-containing particles and deposit copper in the nuclei and chromosomes in a chemically undefined form (Quinn, Barker and Manchant, 1981). In a few cases, the metal may be deposited as another insoluble compound, for example as extracellular cadmium phosphate by *Citrobacter* spp. (Macaskie and Dean, 1984a) and *Klebsiella aerogenes* (Aiking *et al.*, 1984).

In certain fungi, organic substances are produced that reduce metal toxicity. Citric acid, for example, is produced by many fungi and chelates copper and other metal ions. Copper tolerance of other fungi, notably those isolated from copper-chrome-arsenic-treated wood, is attributable to precipitation of copper oxalate (Murphy and Levy, 1983). Mercury resistance in certain *S. cerevisiae* strains has been linked to a requirement for methionine, believed to be used in the synthesis of a 'simple diffusible substance' for detoxifying mercurials. Intracellular sulphydryl compounds have been observed in certain fungi and implicated in tolerance to mercury and copper (Gadd and Griffiths, 1978).

At high metal concentrations, intracellular precipitation may occur. In the oxidative detoxification of thallium by yeast, for example, the toxic metal is compartmentalized by precipitation within mitochondria as thallium oxide, and then discharged and excreted. There are numerous, incompletely documented studies of metal deposition within yeast and fungi (Gadd and Griffiths, 1978), often in apparent association with specific regions of the cell. Nickel-resistant mutants of *Synechococcus* trap the metal intracellularly by association with a granular polymer (Wood, 1984). Exposure of *Tetrahymena* to excess, but sublethal concentrations of, copper stimulates phagocytosis and the formation of small refractile granules, previously supposed to function in ion regulation but which can accumulate strontium, lead or copper (Nilsson, 1981). Toxic metals may be finally eliminated by fusion of these granules with vacuoles which defaecate other debris.

Metal deposition or retention by micro-organisms can reduce the toxicity of a metal to other species. For example, mycorrhizal infection of the ericaceous plant *Calluna* provides significant resistance for the latter to copper and zinc (Bradley, Burt and Read, 1981). Similarly, the accumulation of cadmium by tissues of the mussel *Mytilus edulis* is reduced when the animal also ingests metal-resistant bacteria (Flatau and Gauthier, 1983).

In *Thiobacillus ferrooxidans*, it has been suggested that resistance to silver is achieved, not by immobilization of the metal, but by its solubilization and *increased* mobility away from the bacterial membrane where it inhibits iron oxidation (Sugio, Tano and Imai, 1981). Reduced glutathione is thought to play a role in metal release.

6.6 METALLOTHIONEIN-LIKE PROTEINS

Perhaps the best understood of those metal resistance mechanisms that involve binding or sequestration of the metal is that exhibited by metallothionein (MT) and similar proteins in the cystosol of vertebrates, invertebrates, higher plants, algae, yeasts and other fungi (Kojima and Kagi, 1978; Hamer, 1986). Metallothioneins are small, cysteine-rich proteins that bind heavy metals such as Cd, Zn, Cu, Ag and Hg, and function in metal detoxification, homeostasis and possibly metal transfer to apometalloproteins. Their synthesis is regulated by heavy metals and, in animal cells, glucocorticoid hormones, stress conditions, and interferon. Two forms, designated MT-I and MT-II, have been purified from mouse liver and their sequences determined. Each form contains 61 amino acids, of which 20 are cysteine and involved in metal binding. The positions of these residues are highly conserved. Metallothionein completely lacks aromatic amino acids and, therefore, the typical protein absorption band near 280 nm.

Recently, small metal-binding proteins have been described in several eukaryotic microbes and in bacteria (Table 6.3). Striking analogies with the mammalian metallothioneins are evident. For example, the *N. crassa* protein (Lerch, 1980) is rich in cysteine residues and these occupy exactly the same positions as the first seven cysteinyl residues in the amino-terminal sequence region of the Zn-Cd metallothioneins. Only copper(I) is bound. The lack of free cysteine residues and disulphide bridges, the broad featureless absorption near 250 nm (Chapter 2) and

Table 6.3 Metal-binding, metallothionein-like proteins in eukaryotic and prokaryotic micro-organisms, and comparison with the mammalian enzyme

Organism	Protein	Metals bound (with g atoms mol^{-1})	M_r	Cysteine residues as % of total	Reference
Eukaryotes*					
Mammalian cells	Metallothionein	Cd, Zn, but variable (7)	6800	33	
Neurospora crassa	Metallothionein-like protein	Cu(I) (6)	2200	28	Lerch (1980)
S. cerevisiae	Copper-metallothionein ('copper-thionein', 'copper-chelatin')	Cu(I) (8) Zn (4)	6573	24	Prinz and Weser (1975) Weser *et al.* (1977)
Schiz. pombe	Cd-BP1 Cd-BP2 (= cadystin)	Cd (5.4) Cd (1.8)	4000 1800	43 43	Murasugi, Wada-Nakagawa and Hayashi (1985)
Tetrahymena pyriformis	Metallothionein-like	Cd (3.7) Zn (0.7) Cu (0.1)	11000	32	Nakamura *et al.* (1981)
	Cd-metallothionein	Cd, Zn	27000†	35	Piccinni *et al.* (1987)
Chlorella ellipsoidea	Ch-CdBP	Cd	9000	12	Nagano *et al.* (1984)
Prokaryotes					
Synechococcus spp.		Cd (1.3) Cu (0.6) Zn (0.3)	8100	15	Olafson, Abel and Sim (1979)
E. coli		Cd	3900		Khazadi and Mitra (1981); Mitra *et al.* (1975)
Ps. putida	Cd-BP1	Cd(4);Zn(1); Cu(2)	6700	13	Higham, Sadler and Scawen (1984)
	Cd-BP2	Cd(1);Zn(2); Cu(2)	6900	10	
	Cd-BP3	Cd(3);Cu(1.5)	3600	23	

* For a review of fungal metallothioneins, see Butt and Ecker (1987).
† ≈ 6000 M_r subunits.

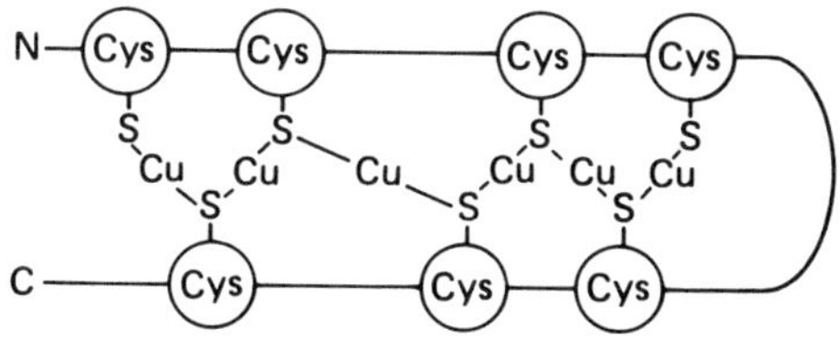

Figure 6.9 Proposed model for *Neurospora* copper metallothionein (Lerch, 1980). Absorption and ESR spectroscopy of the protein suggest the metal to be in the Cu(I) state. The lack of free cysteine residues and the u.v. absorption spectrum are indicative of the copper (I)-thiolate binding shown. The structure is supported by model studies of Cu(I) complexes with sulphur-containing ligands. (Reprinted with permission from *Nature*, vol. 284, 368–70. Copyright 1980 MacMillan Magazines Ltd.)

an orange luminescence suggest a Cu(I)-thiolate binding (Figure 6.9). The molecule probably serves several important physiological functions in addition to copper detoxification, such as copper storage and provision of copper for the copper-containing oxygenase (Chapter 5), tyrosinase (Beltramini and Lerch, 1982).

Increased resistance to copper in yeast results from elevated synthesis of copper-chelatin or copper MT, in turn arising from increased copy number of the *CUP1* gene (Fogel *et al.*, 1983; Butt and Ecker, 1987). Recent EXAFS studies (Bordas *et al.*, 1982) of the yeast copper-MT suggests a cubane structure strikingly similar to that of the 4Fe-4S clusters (Chapter 5). The power of DNA transformation techniques in yeast has recently allowed determination of the structure of the yeast protein by sequence analysis of the gene (Butt *et al.*, 1984). The 6573 Da protein differs in certain aspects from vertebrate metallothionein but, significantly, retains an arrangement of twelve cysteine residues characteristic of other metal-binding proteins. Transcriptional activation of the protein requires copper; however, zinc, cadmium and gold are ineffective in eliciting induction of the protein but this does not rule out the possibility that they may interact with it.

6.7 TOXIC EFFECTS OF SELECTED METALS AND THEIR COMPOUNDS

6.7.1 Cobalt and manganese

These metals have important biological roles but are toxic at high concentrations. Of interest is the observation that cobalt toxicity mimics in some respects iron deficiency, possibly resulting from inhibition by cobalt of haem synthesis. Resistant strains of fungi appear to have modified iron metabolism. Cobalt toxicity in *Klebsiella pneumoniae* is moderated under conditions of K^+-limitation but exacerbated with Mg^{2+}-limitation (Ainsworth, Tompsett and Dean, 1980). The latter effect may arise from increased effectiveness of the toxic metal in competing for the Mg^{2+} transport system.

Manganese and, to a lesser extent, cobalt are both capable of inducing cytoplasmically-inherited 'petite' or respiratory-deficient mutants in *Saccharomyces cerevisiae* – perhaps by interacting directly with mitochondrial DNA polymerase (Putrament *et al.*, 1977).

Manganese (and especially aluminium) are regarded as the elements that most effectively limit plant growth on acid coal spoils; both metals are also toxic to ectomycorrhizal fungi found on coal spoils (Thompson and Medve, 1984).

6.7.2 Zinc

The very diverse effects of zinc poisoning are probably attributable to interaction with sulphydryl groups in an equally wide range of proteins (Chapter 5). However, there seems to be some degree of specificity for electron transport systems (Singh and Bragg, 1974; Kleiner, 1978) although the reason for this is unclear. A histidine residue has been identified as the site of inhibition by zinc ions in cytochrome *o* purified from *E. coli* (Kita, Konishi and Anraku, 1984). Zinc also reversibly inhibits chloroplast electron transport at Photosystem II (Tripathy and Mohanty, 1980).

6.7.3 Cadmium

Cadmium is an element with no known biological function and is one of the most serious environmental pollutants, especially in the vicinity of smelters. It has a high affinity for –SH groups of proteins, for which it competes with Zn^{2+} (Vallee and Ulmer, 1972). It binds to the bases of DNA, with little apparent affinity for a specific nucleotide, and can induce single strand breaks. It is thus mutagenic and carcinogenic. It induces the synthesis of specific binding proteins in plants, animals and microbes (Section 6.6). Wide extremes of sensitivity to cadmium are displayed by micro-organisms (Babich and Stotzky, 1982). In general, actinomycetes are more tolerant than eubacteria, amongst which gram-positive bacteria are more sensitive. There are also wide variations within the fungi, with sporulation being especially sensitive.

6.7.4 Vanadium

Vanadate is required in trace quantities by many cell types (Chapter 2) but its precise biological function remains unclear. Until recently, no enzyme was known to contain vanadium but the recent discovery of vanadium nitrogenase (Chapter 5) illustrates a clear role. The toxicity of vanadium is in general poorly understood (Ramasarma and Crane, 1981). The burning of vanadium-containing fossil fuels and the mining and processing of the metal are contributing to escalating levels of vanadium in the biosphere. Both acute and chronic effects of occupational exposure have been documented. In eukaryotic cells and many micro-organisms, vanadates are potent inhibitors of certain enzyme activities, particularly of ATPases. Indeed, it was the finding that certain commercial preparations of ATP contained pentavalent vanadium and caused inhibition of ATPase activities in many laboratories that led to a resurgence of interest in vanadium and its compounds. Amongst the microbial enzymes that are vanadate-sensitive (with the K_i value in μM) are *E. coli* alkaline phosphatase (0.4), the dynein ATPase of *Tetrahymena* cilia (0.1) and the plasma membrane ATPase of *Neurospora crassa* (1.0) (Ramasarma and Crane, 1981).

6.7.5 Metal toxicity in microbiological mineral leaching

The use of microbial leaching for the recovery of metals from low grade ores has become increasingly important in recent years. The organisms involved and the feasibility of these processes are detailed in Chapter 7.

The strains of bacteria chosen for these operations must be tolerant to high concentrations of metal ions at acid pH. *Thiobacillus ferrooxidans* is selectively resistant to uranium and other mono and bivalent metal ions and adaptation to higher concentrations may be achieved by stepwise increases of the metal content in the culture medium or leach suspension. For example, a strain of *T. ferrooxidans* has been isolated which is active at uranium concentrations as high as 0.042 M and even higher concentrations of Cu (0.87 M), Ni (0.85–1.23 M) and Zn (1.7 M) have been achieved in certain bacterial leaching processes (Torma and Bosecker, 1982). Concentrations of other metals that can be tolerated are aluminium (0.37 M), zinc (0.15 M), cobalt (0.17 M), manganese (0.18 M) and chromium (0.1 M) (Lundgren and Silver, 1980). Tolerances are lower to silver (Hoffman and Hendrix, 1976; Tuovinen *et al.*, 1985) (10^{-8} M to 10^{-5} M Ag), mercury (0.05 M) and molybdenum (0.03 M) and the oxides of selenium, tellurium and arsenic are also inhibitory. There is some evidence that the mercuric ion specifically inhibits iron oxidation at the level of cytochrome oxidase (Imai *et al.*, 1975). In addition, high concentrations of ferrous and ferric ions cause a kinetically complex inhibition of ferrous iron oxidation.

Genetic selection for greater tolerance to toxic metals would be desirable. It is not yet clear whether acidophilic thiobacilli carry resistance factors for metals on plasmids, as occurs in heterotrophs but the potential for strain improvement exists and is now receiving experimental attention.

Similar problems of metal toxicity to a natural, mixed microbial population are evident in biological waste treatment processes. The toxic effects of metals on the bacterial sludge include deflocculation and a decrease in biological oxygen demand leading to poor effluent quality (Sterrit and Lester, 1980a).

6.7.6 Inhibition of microbial cell division by compounds of transition elements

A substantial literature has accumulated on the inhibitory effects of metal complexes on cell division. In bacteria inhibition of division, but not growth, may lead to the formation of filamentous cells. The most intensively studied compounds are those of the group VIIIB elements, i.e. platinum, rhodium and ruthenium, although other metal compounds such as organoleads may also lead to the formation of 'giant' cells (Section 6.3.4(b)). Interest in the phenomenon has been heightened by the use of such compounds in cancer chemotherapy (Rosenberg, 1978), although it is now clear that there is not necessarily a direct correlation between filamenting activity against bacteria and the chemotherapeutic value of a compound. Indeed, filamentation can be elicited by a number of remarkably varied agents.

(a) Platinum

The discovery of the antimicrobial activity of platinum complexes is a good example of scientific serendipity. In an investigation of the possible effects of an electric field on bacterial growth, Rosenberg, Van Camp and Krigas (1965) noted that *E. coli* formed long filaments, up to 300 times the length of a normal cell. The effective agent was subsequently found to be a platinum salt produced from the platinum electrode used. This salt, $[PtCl_6]^{2-}$, when present in media containing free ammonia and exposed to u.v. irradiation, formed three ligand-substituted species, one of which, the stable neutral species $[PtCl_4(NH_3)_2]^0$ elicited

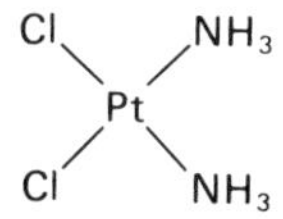

Figure 6.10 Structure of cisplatin.

marked filamentation. Subsequently, compounds of various group VIIIB transition metal ions, notably rhodium and ruthenium, were also found to inhibit bacterial cell division reversibly with relatively little effect on growth.

The most effective compound and the most intensively studied, of some 300 platinum drugs synthesized, is *cis*-dichlorodiammineplatinum (II) ('cisplatin', Figure 6.10). The compound penetrates the *E. coli* envelope and becomes associated with proteins and nucleic acids; a considerable affinity for sulphur ligands would be expected. Cisplatin is used as an antitumour agent (Rosenberg *et al.*, 1969) and may be currently the biggest selling anticancer drug in the US. It appears to inhibit DNA synthesis in eukaryotes and prokaryotes by generating intrastrand cross-links in DNA (Caradonna *et al.*, 1982). Ribonucleic acid produced from a cisplatin-complexed template is abnormally small (Eichhorn, 1980). Other metal complexes such as *cis*-diamminetetrachloroplatinum(IV) also induce filamentation in *E. coli* and are believed to have the same mode of action, but closer observation of the effects of such compounds suggests additional targets. For example, the phenotype of *E. coli* treated with *cis*-dichlorodiammineplatinum(II) is more like that of division mutants rather than of DNA mutants, as judged by growth in the form of long multinucleated filaments, not short filaments with few nucleoids. Interaction with a wall or membrane target has been suggested (Ferguson, Murray and Lancy, 1979).

Although the interaction with DNA is well-established, cisplatin does not significantly inhibit rates of DNA synthesis in growing *E. coli*, due to the bacterium's ability to remove interstrand cross-links. There is a good correlation between the mutagenic and antitumoral properties of platinum compounds and the induction of the recA protein, characteristic of the so-called 'SOS' functions which serve to repair damaged DNA (Little and Mount, 1982). The cisplatin-induced filamentation and the production of phage in lysogenic bacteria are both manifestations of this survival response.

Cisplatin and related drugs have been found to be mutagenic in tissue culture and carcinogenic in mice and rats. Direct mutagenicity has been established in the Ames test (Beck and Fisch, 1980) and shown to be correlated with the drugs' efficiency as anticancer agents.

(b) Ruthenium

As a result of the toxicity, and particularly nephrotoxicity, of cisplatin, compounds of other group VIII elements, especially the second row transition elements, i.e. rhodium, palladium and ruthenium, have been screened as potential anti-cancer agents. A number of ruthenium complexes induce filamentous growth in *E. coli* and some, such as *cis*-dichlorotetrakis(dimethylsulphoxide)ruthenium(II) and Ruthenium Red, exhibit anti-tumour activity. Certain ruthenium complexes are an order of magnitude less mutagenic and toxic than cisplatin (Gibson *et al.*, 1984b).

A large number of ruthenium-containing compounds have been studied in the authors' laboratories; dimeric, mixed-valence compounds of ruthenium [Ru(II), Ru(III)] have been found to be particularly effective in causing filament formation in growing cultures of *E. coli* K12. Three triply halide-bridged compounds of formula $Ru_2(NH_3)_6X_5.H_2O$ (where X is a halide) were tested (Gibson *et al.*, 1984b; Hughes *et al.*, 1987); in order of decreasing effectiveness (and with the concentration giving maximal effect) these were the bromo (10^{-5} M), chloro (10^{-4} to 10^{-5} M) and iodo (10^{-3} to 10^{-4} M) analogues. Filamentation elicited by the bromo and chloro compounds was at least as great as that produced by cisplatin but spontaneously reversible after 3–4 h, a phenomenon attributed to oxidation of the active mixed-valence form to inactive Ru(III) complexes. The compounds bound to mammalian DNA but had no measurable effect on the UV spectrum or content of DNA in *E. coli*, despite showing marked mutagenic activity in reverse mutation tests (Ames test) with *Salmonella typhimurium*. The mechanism of inhibition of division is not clear, but interaction with DNA, as for cisplatin, remains an attractive explanation. The dimeric complex could act as an inter-strand bridge, after two of the Ru-Cl-Ru bridges are broken, with each Ru bound to different strands. Oxidation to mononuclear Ru(III) centres will result in the cleavage of these inter-strand links.

In an attempt to identify those features of ruthenium complexes that determine their biological activity, we have studied the effects on *E. coli* of monomeric and dimeric complexes with ruthenium in the II or III oxidation states as well as mixed-valence complexes with ruthenium in the (II,III) and other oxidation states (Gibson *et al.*, 1985). In general, dimeric mixed-valence Ru(II,III) complexes were the most active class of compound. The pyrazine or bipyridyl-bridged complexes, however (Figure 6.11), which are known to involve strong metal–ligand interaction, stabilizing the Ru(II) oxidation state, were relatively inactive. Thus the instability of the mixed-valence state may be fundamental to activity;

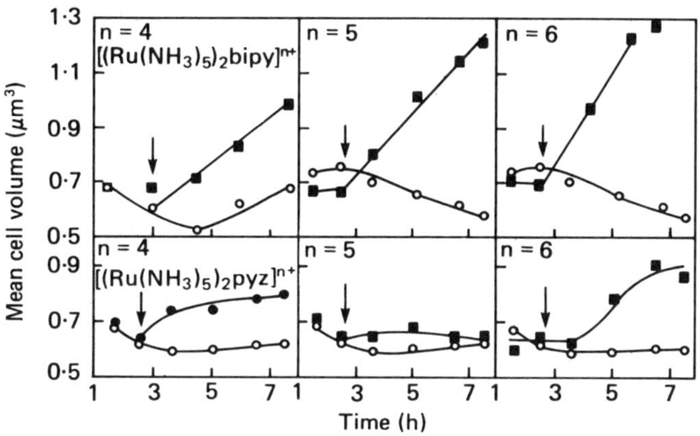

Figure 6.11 Effect on growth and cell division of *Escherichia coli* of complexes of the type $[(Ru(NH_3)_5)_2L]^{n+}$ where n=4, 5 or 6. In the top row, L= bipy (2,2′bipyridyl) and the compounds are $[(Ru(NH_3)_5)_2bipy]^{4+}$ $(PF_6)_4$, $[(Ru(NH_3)_5)_2bipy]$ $(ClO_4)_5$ and $[(Ru(NH_3)_5)_2bipy]$ $(ClO_4)_6$. In the bottom row are shown the analogous compounds where L = pyz (pyrazine). Compounds were added at the times shown by the arrows to concentrations of 0(○, control), 10^{-4} M (■) or 4×10^{-4} M (●). Mean cell volume (a measure of filamentation) was determined with a Coulter counter (reproduced with permission from Gibson *et al.*, 1985).

the complexes may have to be oxidized to the Ru(II,III) dimer prior to activity. Indeed, some Ru(III) complexes were significantly active in induction of filamentous growth in *E. coli*. One of these was $[Ru(NH_3)_5Cl]$ Cl_2, which did not inhibit electron transport, Mg^{2+}-ATPase activity or DNA synthesis in *E. coli*, but like $[Ru_2(NH_3)_6Br_3]Br_2.H_2O$ was a potent inhibitor of respiration-driven calcium transport in the organism. Calcium has been implicated in cell division (Chapter 4), so that the ability to interfere with calcium metabolism may explain the filamenting ability of the complexes.

(c) Rhodium

Much interest has been focused on rhodium as a potential anti-cancer agent. Half a century ago, it was found that $Cs_2[RhCl_6].H_2O$ showed activity against Ehrlich ascites carcinoma in mice. On the basis of a comparison with the properties of platinum anti-cancer compounds, Cleare (1974) suggested that Rh(III) compounds should be potentially useful. Rhodium(I) compounds, which could undergo oxidative addition under physiological conditions to the Rh(III) state, might also be considered.

Rosenberg van Camp and Krigas (1965) and Rosenberg, *et al.* (1967) found that $RhCl_3$ and $Rh(NH_3)_3Cl_3$, like a number of platinum complexes, caused elongation of cells of *E. coli* at 8 μg ml^{-1}. The complex $[Rh(NH_3)_5Cl]\ Cl_2$ was toxic, while *cis* and *trans*-$[Rh(en)_2Cl_2]NO_3$ (en = ethylenediamine) had no effect at this concentration. The compounds $(NH_4)_2\ [RhCl_6]$, $K_2Rh(NO_2)_6$, and $RhCl_3$ were also effective in filamenting cells, but less so than the platinum complexes. Induction of filamentation in *E. coli* B has also been reported in the presence of *trans*-$[Rh(NH_3)_4Cl_2]NO_3$, *mer*-$Rh(NH_3)_3Cl_3$, and *trans*-$[Rh(py)_4Cl_2]Cl.5H_2O$ (py = pyridine; Cleare, 1974).

Bromfield *et al.* (1969) investigated the effect of rhodium(III) complexes containing pyridine, 2,2′-bipyridyl, 1,10-phenanthroline or ethylenediamine on the growth of selected gram-positive and gram-negative bacteria. Only complexes of the type *trans*-$[RhL_4X_2]Y$ (L = substituted pyridine, X = Cl^-, Br^-; Y = Cl^-, Br^-, NO_3^-, ClO_4^-) were found to have significant antibacterial activity. In addition, $[Rh(py)_4Cl_2]$ Cl and [Rh(4-ethyl-py)$_4$Br$_2$]Br were found to induce filamentous growth of *E. coli* at 5.0 and 0.4 μg ml^{-1}, respectively; the filaments were 20–500 times the normal cell length. Inhibition of growth of *E. coli* by complexes of the type *trans*-$[Rh(L)_4Cl_2]Cl.nH_2O$ has been observed. Unlike platinum complexes, these complexes suppress the growth rate of the organism in the same concentration range as that required for filamentation. In the presence of *trans*-[Rh(4-ethyl-py)$_4$Cl$_2$]Cl.2H$_2$O, at a filament-inducing concentration of 7.4×10^{-6} M, DNA, RNA and protein syntheses are not appreciably altered in the filamenting cells, if the suppressed growth rate is taken into account (Seeman and Jack, 1980). Similar observations have been made for the *cis*-$(NH_3)_2PtCl_4$ system (Ferguson, Murray and Lancy, 1979); DNA and protein syntheses appear to proceed normally in the filamenting cells.

Monti-Bragadin *et al.* (1974) investigated the effects on bacteria of the Rh(I) complexes [RhbipyCOD]Cl and [RhphenCOD]Cl (where bipy = 2,2′ bipyridyl, phen = 1,10-phenanthroline, COD = *cis-cis*-1,5-cyclooctadiene): these have planar structures like cisplatin but are charged. Both rhodium complexes inhibited growth of *E. coli*, but only the bipyridyl complex had an inhibitory activity significantly greater than that of its free ligand. Modest filamentation was observed only at concentrations where protein and nucleic acid syntheses were also affected.

(d) Palladium

The palladium compounds Pd(en)Cl_2, Pd(tetramethylethylenediamine) Cl_2 and Pd(glycine)Cl_2 are toxic to *E. coli* at high concentrations and have no filamenting activity at low concentrations (Durig *et al.*, 1976).

However, two palladium(II) complexes of serine do induce filamentous growth of *E. coli* (Charlson, McArdle and Watton, 1981), and other palladium compounds exert moderate cytotoxic effects on mammalian cells, encouraging further studies.

(e) Cobalt

A number of octahedral cobalt(III) complexes such as chloropentaamminecobalt(III) chloride or nitropentaaminecobalt(III) chloride inhibit cell division in *E. coli*, forming long cells (Crawford, Talburt and Johnson, 1974). Most filamented cells were non-viable, however, the remainder dividing, after removal of the metal complexes, by formation of constrictions near the ends of the filamentous rods. The mechanism of action is unknown but the possibility of dysfunctional interaction in bridging purine dimers and inhibiting DNA replication is somewhat less likely for these cobalt complexes than for cisplatin.

6.8 THE USE OF METAL COMPOUNDS AS ANTIMICROBIAL AGENTS

A natural corollary to the toxicity to micro-organisms of metals and metal compounds is the use of such compounds by man to control infection in himself, his domesticated animals, and crops. In addition to the direct application of selectively toxic metals or compounds derived from them (e.g. the inorganic and organic derivatives of silver and mercury), metal-binding substances (Albert, 1979) have furnished many useful drugs in which chelation augments the toxic effect of a metal. Chelating agents may also be used as antidotes for metal poisoning, though these are outside the scope of the present book. An example of microbiological interest, however, is the use of desferrioxamine B, rhodotorulic acid (from yeast) and 2,3-dihydroxybenzoic acid (part of the enterochelin molecule) as antidotes for iron poisoning.

6.8.1 Bacteriostatic and bacteriocidal activities of metal-containing compounds

(a) Mercury

The toxic effects of mercury on micro-organisms have been recognized and exploited for more than a century. In 1880, Koch reported the inactivating effect of mercury bichlorate. Inorganic compounds such as red mercuric oxides, or preparations of chalk with mercury, were used orally for treatment of syphilis and other therapeutic uses, until they were abandoned because of their unfortunate side-effects. Indeed, the simile heading this chapter refers to the fate of hatters of the last century who used mercury(II) nitrate to treat felt.

More recently, organomercurials have been developed as antiseptics, disinfectants, fungicides and herbicides. They exert bacteriostatic effects by combining reversibly with sulphydryl groups of bacterial proteins. An extensive list is given by Foye (1977) and includes many of the compounds for which bacterial, plasmid-determined resistance mechanisms are known (Section 6.4).

(b) Silver

Silver nitrate ('lunar caustic') was introduced as a prophylactic against gonococcal ophthalmological infections in newborns in 1901. Such ionizable salts are caustic, however, and have resulted in the development of organic derivatives. Many of these are of unknown chemical composition (e.g. 'strong silver protein' and protargol), being prepared from proteins by precipitation with soluble silver salts. The silver is liberated when in contact with serous fluids and can react with bacterial proteins and nucleic acids. Silver nitrate releases the metal rapidly and frequent applications are required to maintain effective, local silver concentrations in wounds. Although organic silver compounds such as silver sulphadiazine release silver slowly, application is cumbersome and expensive; silver-coated nylon fabrics offer promise for continuous release of the metal (Deitch *et al.*, 1983). Silver nitrate is still used, however, for topical prophylaxis against *Pseudomonas* infections and for burns, but the occasional occurrence of resistant strains is a problem (Summers *et al.*, 1978).

(c) Copper and zinc

Basic cupric calcium sulphate (Bordeaux mixture) was introduced a century ago as an antifungal agent in crop spraying. Organic salts, however, have found wider application as mild antiseptics. Thus, copper citrate has been widely used in ointments for treatment of granulations and trachoma. A commerical mixture of zinc stearate, oleate and oxide is widely used in pharmaceutical powders and ointments. The organic compounds of these metals are more extensively used against fungi than against bacteria; Foye (1977) provides a list.

(d) Tin

Among the agents tested for inhibitory effects on dental plaque are fluorides, as the stannous or sodium salt. In 1956, however, it was shown that the usefulness was related not only to the fluoride content but to tin concentration. Recently, the uptake and strong binding of tin by oral plaque bacteria has been claimed (Attramadal and Svatun, 1980).

6.8.2 Antimicrobial activity of toxic chelates of heavy metals

(a) Oxine

Oxine (8-hydroxyquinoline) (Figure 6.12) has a similar antibacterial spectrum to penicillin. It exerts its toxic effect intracellularly by chelation of a metal, fortuitously present, and formation of a lethal complex – 'the cooperative effect'. In the absence of heavy metals, oxine, being liposoluble, enters bacterial and fungal cells but is non-toxic. The presence of Cu^{2+} and especially Fe^{2+} or Fe^{3+} renders the oxine toxic to gram-positive bacteria; oxine has only feeble activity against most gram-negative bacteria. The toxic species are the 1:1 or possibly 2:1 iron complexes; the saturated 3:1 (iron) complex (Figure 6.12) formed with excess oxine is not toxic. Cupric ions, but not iron, augment the antifungal activity.

Dimethyldithiocarbamic acid (DMDC) acts like oxine and is a potent fungicide as the 1:1 copper complex. Iron is ineffective. The mechanism

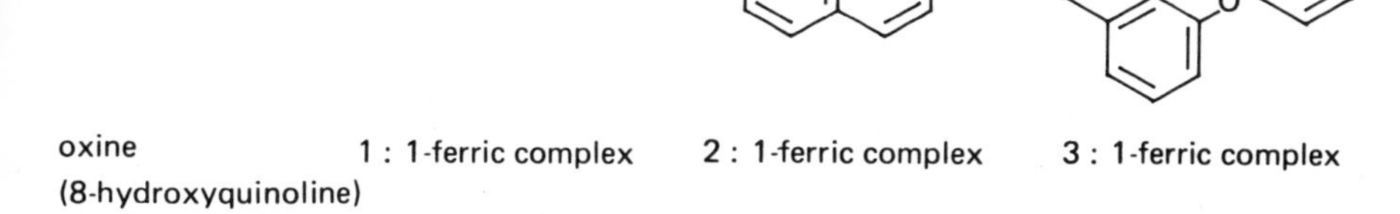

Figure 6.12 Equilibria between the oxine-ferric ion complexes. Only the 1:1 and possibly 2:1 complexes exhibit antibacterial action. Reprinted from Foye (1977), by courtesy of Marcel Dekker Inc.

of action appears to be catalysis of oxidation of –SH groups leading, in the presence of O_2, to formation of peroxide and superoxide. Cobalt provides protection against the copper complexes of both chelators, perhaps by moderating the chain nature of radical formation.

Several derivatives of oxine have been developed. Halogenation improves the performance against gram-negative bacteria; the most familiar example is 5-chloro-7-iodooxine (Vioform), widely used against travellers' diarrhoea and amoebiasis. Other substances, both chemically related and unrelated to oxine, but having similar modes of action, are described by Foye (1977) and Albert (1979).

(b) Tetracyclines

Tetracycline and its derivatives, chlorotetracycline and oxytetracycline, are chelating agents. Their selectivity rests on their being accumulated more readily by bacteria than by mammalian cells, possibly by forming a liposoluble complex with magnesium or another metal in the bacterial plasma membrane (Albert, 1979; Sigel, 1985). Their action is believed to occur on the ribosomes, also rich in magnesium.

(c) Isoniazid and other antitubercular drugs

Isoniazid (isonicotinic acid hydrazide) has metal-binding constants of the same order as those of amino acids. It is important as an antitubercular drug. Chelation with traces of copper renders the drug sufficiently liposoluble to penetrate the extremely lipophilic bacterial envelope. It is thought that inside the bacterial cell, isoniazid, the prodrug, is converted to isonicotinic acid which is then N-methylated to an analogue of NAD, which results in deranged lipid metabolism.

Thiacetazone is also believed to be absorbed more readily following chelation of the small amounts of copper in serum. It is used as an adjunct to isoniazid for controlling resistant strains of *Mycobacterium*.

Activation by Cu^{2+} and Co^{2+} ions has also been found for *p*-aminosalicylic acid, which acts as a metabolic analogue of *p*-aminobenzoic acid and salicylic acid, a constituent of *M. tuberculosis*.

(d) Other applications of chelation

A number of other antibacterial, antiviral, antifungal and antiprotozoal drugs such as salicylic acid, nalidixic acid, methisazone, tropolone and biallylamicol, respectively, bind metal ions, but it is not clear in all cases how or whether their efficiency depends on this ability (Foye, 1977; Albert, 1979).

Chelating agents can sometimes be used for countering drug resistance in bacteria. Thus, EDTA restores sensitivity in polymyxin-resistant strains of *Pseudomonas*, probably by removing calcium and magnesium from the plasma membrane. Oxytetracycline, a known chelating agent (Section 6.8.2(b)) renders penicillin-resistant *S. aureus* susceptible to the latter drug (Albert, 1979).

6.9 SUMMARY

Most metal ions, when present at sufficient concentrations, have the potential for being toxic to biological systems. Such metals may be essential at low concentrations (e.g. copper, zinc) or have no known biological function (e.g. cadmium, mercury) and compete with a functional metal. Metal speciation is a crucial, but frequently overlooked, factor in determining metal toxicity in both laboratory and environmental situations. Although the influence of organic chelating agents (which include most common medium constituents) in moderating metal toxicity is generally appreciated, pH, water hardness and buffers can influence observed toxicities.

Micro-organisms adopt diverse strategies in avoiding metal toxicity and many of these mechanisms are now understood at a molecular level. Metal resistance is frequently plasmid-encoded. For example, plasmid-encoded efflux pathways for arsenicals and cadmium are now well-established in certain bacteria and are quite distinct from the uptake pathways for phosphate and manganese, via which arsenate and cadmium, respectively, gain entry to cells. Many metal ions and compounds are subject to biotransformations that may increase or decrease their toxicity. Such transformations of mercury compounds are particularly well known because of their ecological impact, their well-established roles in cycling of the metal in the biosphere and the detailed information available on the genes and proteins involved. The primary route for methylation of mercury, and thus its mobilization in the environment, is via the methyl donor methylcobalamin. The resulting organomercurials may in turn be detoxified by plasmid-encoded mechanisms in certain bacteria, first by a lyase, which cleaves off the organic moiety, and then by a mercuric reductase which reduces Hg^{2+} to volatile Hg^0 which escapes. In contrast to the efflux pathways for metals such as cadmium, an essential feature of Hg^{2+} detoxification is the inward translocation of the metal ion to the intracellular reductase. The alkylation of compounds of tin and arsenic by prokaryotic and eukaryotic microbes is also well established and leads to increased potential toxicity of the metal. The toxicity of organotins and organoleads is of special concern in view of the widespread use of these compounds as biocides and as petrol additives, respectively. Microbial resistance to such compounds is relatively poorly understood. The biotransformations of these metals, and of compounds of arsenic, tellurium, selenium, cadmium, chromium and antimony play important roles in element cycling.

Metal toxicity may also be alleviated by deposition of the metal in an insoluble form. Precipitation as the metal sulphide is common in many bacteria and yeasts. Microbially-generated phosphate, citrate or oxalate may also result in extracellular deposition. Intracellularly, metals may be compartmentalized or bound by metallothioneins. These small, cysteine-rich proteins bind cadmium, zinc and other metals in higher eukaryotes and functionally and structurally analogous proteins are now recognized in many micro-organisms, including bacteria.

The toxicity of metal compounds has important implications in human and animal health, agriculture, ecological processes and biotechnology. Many metal-containing compounds are used as antimicrobial agents or serve important biological activity that can be assayed using microbial systems. Thus, cisplatin (*cis*-dichlorodiammineplatinum (II)) and related compounds of the group VIII elements, with potential anticancer properties, cause filamentation of bacteria by inhibiting cell division but not growth. Simple metal-containing compounds have long been used as antimicrobial agents. Inorganic and organic compounds of mercury have found use as antiseptics, disinfectants and biocides. Similarly, organic compounds of copper and zinc are used as mild antiseptics, and silver nitrate is still used for preventing infections of burns. Oxine, tetracyclines and isoniazid are all metal-binding agents whose antimicrobial action is augmented by metals. It seems likely that further understanding of microbial mechanisms of resistance to metal ions will make an impact on the development of improved antimicrobial agents and, on the other hand, enable the engineering of strains better suited to resist metal toxicity for technological applications.

7

Metals, micro-organisms and biotechnology

He personally analyzed the ashes with beggarly patience to extract from them all the valuable elements (and perhaps even less valuable) in a kind of ritual palingenesis.

Primo Levi
The Periodic Table (1984)

People find gold in fields, veins, river beds and pockets. Whichever, it takes work to get it out.

Art Linkletter
A Child's Garden of Misinformation (1965)

7.1 INTRODUCTION

The important role of micro-organisms in the solubilization and deposition of minerals in the environment has been stressed several times in this book. These naturally-occurring processes can also be exploited in the development of energy-saving, environmentally acceptable methods for the extraction of metals from ores and the recovery of high-value or toxic metals from industrial effluents and process streams.

The solubilization of metals by the bacterially-catalysed oxidation of minerals is exemplified by the well-known leaching of copper from dumps of low-grade ore and the in-place leaching of uranium, both of which are carried out commercially. Current interest centres on the leaching in vats, under controlled conditions, of high-value minerals such as gold-bearing pyrites. In this case, the pyrites is solubilized liberating the gold for full recovery on cyanidation.

Metals can be recovered from solution by several pathways, such as the binding of metal ions to microbial cell surfaces, the intracellular uptake of metals, or their precipitation through compound formation with microbially-produced ligands.

These processes sometimes lead to undesirable consequences in the environment, such as 'acid run-off' from mines and dumps, the corrosion of steel structures, the blockage of water pipes and cooling systems by deposits of microbial origin, and the possible bioaccumulation into the food chain of toxic metals, including radionuclides from nuclear waste.

7.2 RESERVES OF METALLIC ORES

The known reserves of certain essential minerals are small enough to make their exhaustion in the foreseeable future a matter of concern. The exact life-times of mineral reserves cannot be predicted accurately, as they are so dependent upon the global economic situation. Nevertheless, there is no doubt that long-term reserves of mercury, silver, uranium, lead, tin, zinc and copper, for example, are uncertain (Robinson, 1976). Uranium is a particular problem, in view of the importance of the nuclear power programme in satisfying long-term energy requirements.

Assessments of the life-times of mineral reserves are based upon the assumption that conventional extraction techniques will be applied to known reserves of ore. The situation would be alleviated considerably if new extraction processes were developed, which could be used economically to recover metals from low grade ores, dumps of slag or tailings, and effluents from inefficient extraction processes.

The sea also provides additional resources of raw materials (Cloud, 1969). Metal salts are present in seawater at very great dilution, but the total amount is enormous because the volume of seawater is so great (1.4×10^9 cubic kilometres). Examples are Zn, 1.52×10^{13}; U, 5.1×10^{12}; Cu, 5.1×10^{12}; Ag, 5.1×10^{11}; Hg, 5.1×10^{10} kg. The technical difficulties and energy costs involved in extracting and separating these elements are very great. Thus, while uranium is present in seawater at 3.6 g per 10^6 litres, the energy spent in recovering this is likely to exceed the energy obtained from using the uranium in a reactor. However, it is possible to use micro-organisms to recover metal ions from this level of dilution.

7.3 THE EXTRACTION OF METALS FROM LOW GRADE OR REFRACTORY ORES USING MINERAL-OXIDIZING BACTERIA

A common environmental problem associated with the mining industries is the presence of acidic drainage water. This results from the activity of bacteria which catalyse the oxidation of mineral sulphides to give dilute acidic solutions. The 'acid run-off' also contains metal ions that have been solubilized, notably iron and copper. This natural process has been exploited for over 2000 years for the recovery of copper, although the first large scale leaching operation was probably in the copper mines at Rio Tinto, Spain around 1700. Only in the 1940s was the essential role of bacteria in these processes discovered, with the isolation of *Thiobacillus ferrooxidans* from mine water.

Natural bacterial leaching is terminated by certain limiting conditions, which may be overcome by carrying out the process in a controlled environment. Bacterial leaching has been exploited in the large scale leaching of copper in dumps of low grade ore and also in the leaching of uranium. Current market factors have focused attention on the recovery of high value metals, particularly gold from refractory pyrite and arsenopyrite concentrates.

The bacteria responsible for the oxidation of sulphide and iron(II) occur naturally in sulphide-rich environments. They are autotrophic, obtaining energy by the oxidation of inorganic compounds and using carbon dioxide as the sole source of carbon. They are also acidophilic, preferring pH values in the range 2.5–1.5, and, in some cases, thermophilic.

Many reviews on bacterial leaching are available (Tuovinen and Kelly, 1972, 1974; Torma, 1977; Brierley, 1978, 1982; Curtin, 1983; Manchee, 1979; Kelly, Norris and Brierly, 1979; Lundgren and Silver, 1980; Silver, 1978; Torma and Bosecker, 1982; Bosecker, 1984; Johnson, 1985; Hutchins *et al.*, 1986) and the proceedings of a *Workship on Biotechnology for the Mining Industries* have been published (Ehrlich and Holmes, 1986).

7.3.1 Chemical aspects of microbial leaching

Metals in insoluble minerals can be solubilized either directly through the normal metabolic activities of micro-organisms or indirectly by chemical oxidation brought about by the products of metabolic activity, namely acidic solutions of Fe(III) (Silverman, 1967). The overall result of bacterial action upon pyrite is shown in Equation 7.1, while the individual reactions are given in Equations 7.2–7.5. These reactions occur chemically, but are catalysed by the bacteria by a factor of 10^6. Equation 7.4 describes oxidation of the pyrite by Fe(III), without participation of the bacteria, to produce elemental sulphur. The sulphur is subsequently oxidized by certain bacteria.

$$4FeS_2 + 15O_2 + 2H_2O \rightarrow 2Fe_2(SO_4)_3 + 2H_2SO_4 \quad (7.1)$$

$$2FeS_2 + 7O_2 + 2H_2O \xrightarrow{\text{bacteria}} 2FeSO_4 + 2H_2SO_4 \quad (7.2)$$

$$4FeSO_4 + O_2 + 2H_2SO_4 \xrightarrow{\text{bacteria}} 2Fe_2(SO_4)_3 + 2H_2O \quad (7.3)$$

$$FeS_2 + Fe_2(SO_4)_3 \rightarrow 3FeSO_4 + 2S \quad (7.4)$$

$$2S + 3O_2 + 2H_2O \xrightarrow{\text{bacteria}} 2H_2SO_4 \quad (7.5)$$

The sulphuric acid produced in these processes will lower the pH of the percolating solution, so encouraging the proliferation of the organisms, which are acidophilic. The percolating solution will leach out metal-soluble metal compounds, such as those of copper, so allowing the ultimate recovery of the metal. The breakdown of the sulphide lattice will also enhance recovery of metal trapped in the lattice.

The Fe(III) produced by bacterial oxidation of pyrite has an important role in the solubilization of metals, both in the oxidation of insoluble metal sulphides to soluble metal sulphates, and in the oxidation of insoluble uranium (IV) (UO_2) to the acid-soluble U(VI). Examples of oxidation of sulphides by Fe(III) are shown in Equations 7.4 and 7.6–7.9, while that of U(IV) is given in Equation 7.10. The chemical leaching of uranium by acidic solutions of Fe(III) has been carried out commercially. The microbiological method should be favoured, as the leaching components are continuously regenerated by microbial action.

Insoluble		*Soluble*		
$CuFeS_2$	$+ 2Fe_2(SO_4)_3 \rightarrow$	$CuSO_4$	$+ 5FeSO_4 + 2S$	(7.6)
Cu_2S	$+ 2Fe_2(SO_4)_3 \rightarrow$	$2CuSO_4$	$+ 4FeSO_4 + S$	(7.7)
CuS	$+ Fe_2(SO_4)_3 \rightarrow$	$CuSO_4$	$+ 2FeSO_4 + S$	(7.8)
ZnS	$+ Fe_2(SO_4)_3 \rightarrow$	$ZnSO_4$	$+ 2FeSO_4 + S$	(7.9)
UO_2	$+ Fe_2(SO_4)_3 \rightarrow$	UO_2SO_4	$+ 2FeSO_4$	(7.10)

It is difficult to distinguish between the direct and indirect mechanisms for the solubilization of mineral sulphides, as the Fe(III) will usually be present during the leaching of minerals. However, experiments conducted on synthetic iron-free cobalt and nickel sulphides, using washed cells of *Thiobacillus ferrooxidans* to ensure the absence of iron, showed a consumption of oxygen and the solubilization of the metal as the sulphate (Torma, 1971). This is good evidence for the direct attack of the microbial cells on the mineral. It should be noted that solubilization rates were appreciably enhanced by the addition of Fe(III), confirming, too, the existence of the indirect mechanism in the presence of Fe(III). It seems reasonable that both mechanisms should operate simultaneously. Other evidence for the direct attack of micro-organisms on the surface of the mineral comes from electron microscopy, which has shown that bacteria attach themselves to the surface, causing deep pitting and eventually the dissolution of the crystal (Bennett and Tributsch, 1978; Tributsch, 1976; Section 7.3.3).

7.3.2 The bacteria involved in the oxidation of minerals

A range of micro-organisms may be isolated from mineral-rich environments, whether these are natural ore deposits or manmade dumps. Coal dumps are also a useful source of mineral-oxidizing bacteria including thermophilic organisms. These autotrophic, chemolithotrophic or 'rock-eating bacteria' grow on inorganic nutrients present in the mineral, and use carbon dioxide as the carbon source. They obtain energy from the oxidation of inorganic compounds, particularly Fe(II) and reduced sulphur (usually sulphide). They may be grown in the laboratory either in a medium containing pyrite or other sulphide ores or in a medium containing Fe(II) and reduced sulphur species such as thiosulphate. The former option is preferable if mineral-oxidation studies are anticipated.

The growth of cultures cannot be followed directly when studying the bio-oxidation of minerals, as the mineral particles interfere with turbidity or Coulter counter measurements. It is also possible that the bulk of the cells are attached to the mineral surface. It is customary, therefore, to follow growth by measuring the solubilization of iron. Nevertheless it is important to obtain a direct measurement of biomass, including an assessment of bound cells and cells in the supernatant liquid. This can be attempted by analysing for molecules which must be of microbiological origin, such as protein, DNA or ATP. Standard methods are available for all these determinations, but they are difficult to apply to leaching organisms because of the low cell count and because of interferences from the bio-oxidized material. The usual approach is to heat samples with alkali to precipitate iron compounds and other inorganic material and then to apply the standard Lowry assay for protein (Bagdigian and Myerson, 1986; Pivovarova *et al.*, 1987). Cells are usually removed from the mineral surface by treatment with detergent. Other approaches have involved the determination of total organic carbon, total nitrogen and oxygen demand.

Methods have also been described for the extraction of DNA from sediments (Ogram, Sayler and Barkay, 1987) and for the enumeration of *Thiobacillus ferrooxidans* in a population of acidophilic bacteria by a combined immunofluorescence/DNA-fluorescence staining technique (Muyzer *et al.*, 1987). Epifluorescence microscopy, using certain nucleic acid dyes which fluoresce under blue light, has been employed to assay for *Thiobacillus ferrooxidans* attached to solid particles and free cells, and to show how the relative amounts of surface-bound and free bacteria change during the course of a pyrites-leaching experiment (Yeh *et al.*, 1987). The growth of free and attached *Thiobacillus ferrooxidans* in a suspension of a copper ore has been followed by direct counting techniques (Espejo and Ruiz, 1987).

A further difficulty is associated with the growth of these organisms on solid media, which is an important technique in the isolation and characterization of pure cultures. Recently, Johnson, MacVicar and Rolfe (1987) have reported a new solid medium which allows the isolation of *Thiobacillus ferrooxidans* and acidophilic heterotrophic bacteria. This paper should be consulted for an account of the alternative methods available in the literature.

Another major problem in studying these organisms and the detailed mechanisms of mineral oxidation is the poor understanding of their physiology. Most workers have been concerned with the isolation of new organisms and their mineral-oxidizing properties. Ingledew (1982) has reviewed the bioenergetics of *Thiobacillus ferrooxidans*.

(a) *Thiobacillus ferrooxidans* and other mesophiles

Thiobacillus ferrooxidans is the most-studied organism relevant to the leaching of metal sulphides (Tuovinen and Kelly, 1972; Torma, 1977). It is widely distributed in the environment. *T. ferrooxidans* is a rod shaped bacterium, approximately 2 μm long, and is most active in the pH range 1.5–3.5, with an optimum pH of 2.3, and a preferred temperature in the range 30–35°C. It is able to oxidize Fe(II) and reduced sulphur compounds. Many strains of *T. ferrooxidans* have been isolated, with different capacities to oxidize certain substrates, variations in DNA base composition, and with different resistances to metal ions and acidity (Pichuantes *et al.*, 1986). The 'thiobacilli' in dumps probably represent a mixed population of various strains, thus allowing growth and leaching activity over a wider range of conditions.

The organisms *Thiobacillus acidophilus* and *Thiobacillus organoparus* have similar morphology and growth properties to *Thiobacillus ferrooxidans*, but their role in metal leaching is uncertain. *Thiobacillus thiooxidans* differs in being unable to oxidize Fe(II), although it oxidizes reduced sulphur compounds effectively. It has enhanced acidophilic properties, generating pH values down to 0.65. It is probably unable to degrade sulphur compounds effectively. It has enhanced acidophilic properties, cultures as it could remove any sulphur deposited on the surfaces of minerals being leached by iron-oxidizing bacteria (Kelly, Norris and Brierley, 1979). This oxidation of sulphur may involve a bacterially-mediated reaction with Fe(III) (Brock and Gustafson, 1976; Kino and Usami, 1982). This reaction has also been shown for *Thiobacillus ferrooxidans* and *Sulfolobus acidocaldarius*, and may help to account for the production of Fe(II) from Fe(III) during growth of cultures on pyritic materials.

Leptospirillum ferrooxidans is less widespread than *Thiobacilli*, although this may reflect an inability to compete under the usual conditions of isolation. *Leptospirillum ferrooxidans* is unable to oxidize reduced sulphur but will oxidize Fe(II). Thus, it is sometimes found in mixed culture with *Thiobacillus thiooxidans*, when they will effect rapid dissolution of pyrite. *Leptospirillum ferrooxidans* grows optimally at 40°C and pH 1.2. It has a different morphology from *Thiobacillus ferrooxidans*, being vibrioid, and forming spirals of joined cells.

(b) Moderate thermophiles

A range of these organisms have been isolated from self-heating mineral and coal dumps (Brierley and Lockwood, 1977). Their optimal growth temperature lies between 45° and 50°C, and they are capable of oxidizing both Fe(II) and reduced sulphur (Marsh and Norris, 1983a). These moderate thermophiles are similar to certain strains of *Thiobacillus*. Some of these organisms have substantial activity at 30°C (in chalcopyrite leaching) (Norris, Brierley amd Kelly, 1980). A tolerance of such a wide range of temperature suggests that artificial mixed cultures of moderate thermophiles and mesophiles and also of moderate thermophiles and extreme thermophiles could be of considerable value in heap-leaching where temperatures might vary considerably. It is clear that moderate thermophiles should be very important in the bio-oxidation of minerals (Barrett *et al.*, 1988a) as expensive cooling of the reactor will be unnecessary.

(c) Extreme thermophiles

Several examples of the thermophilic, acidophilic group of organisms *Sulfolobus* have been isolated, usually from hot sulphur springs. These have optimum growth temperatures in the range 65–80°C. The cells are almost spherical in shape, with diameters in the range 0.7–1.0 μm. All *Sulfolobus* spp. oxidize reduced sulphur, but only some oxidize Fe(II) and minerals. They have been reviewed by Brock (1986).

Sulfolobus, like the extreme halophiles and the methanogens, belongs to the classification Archaebacteria, which is fundamentally different from both the true bacteria and the eukaryotes.

Sulfolobus was first isolated from a hot sulphur spring at pH 2.6 and 66–69°C (Brierley and Brierley, 1973). This and other isolates have now been classified into three different species of one genus on the basis of homology studies on their DNA-dependent RNA polymerases (Zillig *et al*, 1980). These are *Sulfolobus brierleyi, Sulfolobus acidocaldarius* (Brock *et al.*, 1972), and *Sulfolobus solfataricus* (de Rosa, Gambacorta and Bu' Lock, 1975). In all cases several different strains of each species have been isolated subsequently. Strains of *Sulfolobus acidocaldarius* grow at temperatures as high as 85°C and pH values as low as 1.0. Growth at high pressure has also been demonstrated for this organism, using a hyperbaric helium reactor (Sturm *et al.*, 1987).

The habitat range of these extreme thermophiles has now been extended to include a coal pile (Birch Coppice) and an Icelandic hot spring (Lake Myvam) (Marsh and Norris, 1983b); these are probably strains of *Sulfolobus acidocaldarius*. Recently an additional species has been isolated, *Sulfolobus ambivalens* (Segerer, Stetter and Klink, 1985), which can grow on sulphur either aerobically or anaerobically. In the latter case the sulphur is reduced to H_2S. The anaerobic mode of growth can be correlated with the amplification of a plasmid (Zillig *et al.*, 1985).

Many isolates of *Sulfolobus* grow autotrophically and heterotrophically, but others only grow heterotrophically. There is also considerable variation in the ability of *Sulfolobus* spp. to oxidize sulphide minerals. The Lake Myvam and Birch Coppice isolates are more effective in solubilizing pyrite than is *Sulfolobus brierleyi* (Marsh, Norris and Le Roux, 1983).

(d) Mixed cultures

The importance of mixed cultures is now recognized (Norris and Kelly, 1982) and is illustrated in Section 7.2.2(a) for *Leptospirillum ferrooxidans* and *Thiobacillus thiooxidans*, which in mixed culture are able to solubilize pyrite effectively. In general, a mixed culture solubilizes pyrite more effectively than any of the pure components present. Mineral dumps will also contain a range of other microbes, such as fungi and yeasts, which grow on the trace organic matter leaked by the autotrophs. The role, if any, of these organisms is uncertain. They could contribute to the overall process by utilization of compounds such as pyruvate which are toxic to iron-oxidizing bacteria, by acting as N-fixers and so supplying nitrogen nutrients, or excreting organic compounds to act as ligands to the metal (Kiel and Schwartz, 1980). Alternatively, heterotrophic organisms could inhibit the leaching process by blocking the flow of the leaching solution, or by competing for sites on the mineral.

Groudev, Genchev and Gaidarjcev (1978) have made a detailed study of the organisms present in a low grade copper ore dump in Bulgaria. In some parts of the dump, *Amoeba* spp. were found in numbers up to $500\,ml^{-1}$. In laboratory experiments this resulted in a major decrease in the number of viable, iron-oxidizing bacteria, and decreases in the rate and extent of leaching. However, the percentage of the total biomass corresponding to the iron oxidizing bacteria increased from 70% to 93% and then to 97% as the pH was decreased from 3.5 to 2.4 and 2.1. This demonstrates the non-viability of other organisms under these conditions of acidity. Populations of organisms will also vary with the temperature of the dump.

The progressive increase in acidity in freshly exposed pyrite ores, which culminates in the dominance of acidophilic bacteria, probably involves growth of a series of autotrophic organisms and other organisms with gradual reduction of pH and sequential domination by organisms with decreasing pH maxima for optimal growth. Iron-oxidizing bacteria of the genus *Metallogenium*, active in the pH range 3.5–4.5, could be involved in the early stages. The need for such a process is decreased if acidic leaching solutions are used initially.

7.3.3 Mechanism of microbial attack upon ores

It is now established that the leaching bacteria attach themselves to the surface of the ore. Detailed microscopic studies have shown the formation of deep pits in crystal surfaces by bacterial attack, leading to the dissolution of the crystal (Bennett and Tributsch, 1978). This has been shown for *Thiobacillus thiooxidans*, *Thiobacillus ferrooxidans* and *Sulfolobus acidocaldarius* (on elemental sulphur, Shivvers and Brock, 1973).

An interesting comparison has been made of the surface pitting of pyrite single crystals brought about by *Thiobacillus ferrooxidans* at 30°C and a thermophilic organism at 55°C with the surface pitting resulting from treatment of crystals with a solution of iron(III) sulphate at pH 2.3 at these two temperatures. Bacterial leaching did not require the addition of Fe(III) and the concentration of Fe(III) present was less than 100 ppm. In contrast, leaching with the acidified iron(III) sulphate solution required nearly 2000 ppm [Fe(III)]. Furthermore, the pattern of surface corrosion was different in the two cases, showing that the two processes are not identical and confirming the operation of the direct mechanism. It is also apparent that crystallographic orientation plays a role in the leaching of pyrite crystals (Keller and Murr, 1982).

Recently, the application of scanning electron microscopy, XPS and X-ray fluorescence techniques to the corrosion pits brought about by *Thiobacillus ferrooxidans* on synthetic pyrite has led to the suggestion that bacterial attack is initiated by secretion of organic material around the contact area between the bacterial cell and the mineral surface (Rodriguez-Leiva and Tributsch, 1988). Further studies on the mechanism of bacterial attack on pyrites in coal are presented in Section 7.3.9(a).

7.3.4 The bacterial oxidation of inorganic compounds of sulphur

The oxidation of sulphide to sulphate is an eight-electron process and must involve the formation of several intermediate species. Compounds of main group elements usually undergo redox reactions in two-electron steps, often by transfer of oxo groups. It is to be expected that sulphur (0) and sulphite will be intermediates and that other reactions will produce intermediates with S–S bonds, in view of the stability of such groupings. Thiosulphate, dithionate and polythionates are all likely intermediates, not necessarily as free species.

It is increasingly being appreciated that several distinct pathways exist for the oxidation of these intermediate species, and that these may differ from organism to organism. These pathways have been reviewed by Kelly (1985, 1988), and are outlined in Equation 7.11 (Silver, 1978; Silver and Dinardo, 1981). Oxidation of sulphide gives sulphur atoms, which may polymerize to give elemental sulphur or react further to give polysulphide species. The mechanism by which sulphur is oxidized is controversial at present. Its properties make sulphur a poor substrate, and oxidation may well occur on the outer surface of the cytoplasmic membrane. A sulphur oxygenase has been purified from *Sulfolobus* and shown to catalyse the oxidation of sulphur to sulphite (Emmel *et al.*, 1986), while for *Thiobacillus ferrooxidans* and other organisms, as noted earlier, it has been claimed that there is an enzyme that uses Fe(III) as an intermediate electron carrier in the oxidation of sulphur to sulphite (Sugio *et al.*, 1985). In the case of *Thiobacillus ferrooxidans* there is contrary evidence to show that oxidation of sulphur involves a cytochrome bc_1 electron transfer system (Corbett and Ingledew, 1987). A further suggestion is that the sulphur produced extracellularly by *Thiobacillus ferrooxidans* is oxidized directly by the sirohaem-containing sulphite reductase. The composition of this sulphur appears to be a hydrophobic S_8 nucleus, covered by a shell of long chain polythionates (Steudel *et al.*, 1987). Clearly, at this stage it is not possible to give a definitive account of the mechanism of oxidation of sulphur. The oxidation of thiosulphate to sulphite and of sulphite to sulphate are better understood. The enzyme responsible for the oxidation of thiosulphate to tetrathionate is present in most thiobacilli, while sulphite oxidizing enzymes have been found in all the thiobacilli studied so far. *Thiobacillus tepidarius* converts thiosulphate stoichiometrically to tetrathionate prior to autotrophic growth accompanying further oxidation of tetrathionate to sulphate. In contrast *Thiobacillus versutus* lacks the thiosulphate-cytochrome *c* oxidoreductase that catalyses the oxidation of thiosulphate to tetrathionate. In this case there is a six-electron oxidation of the –S– fragment of thiosulphate to sulphite, although it is not known whether this occurs before or after the cleavage of the S–S bond. The reverse reaction, the reduction of sulphite to sulphide, is known for the sulphate reducing bacteria.

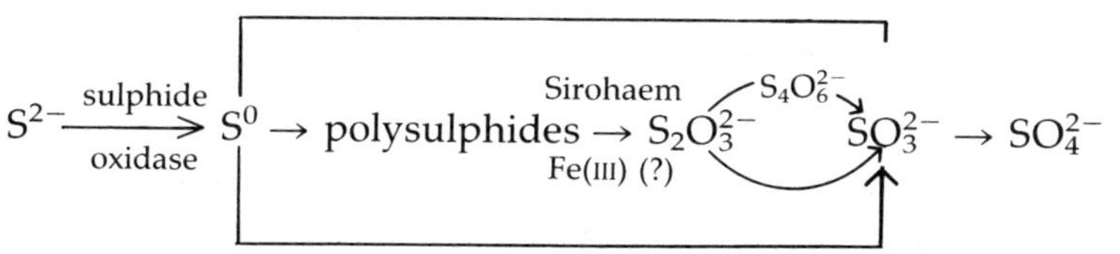

(7.11)

The oxidation of compounds containing sulphur in intermediate oxidation states has also been examined, see for example Silver and Dinardo (1981) and Norris, Marsh and Lindstrom, (1986).

7.3.5 Laboratory investigations of the oxidation of minerals by bacterial cultures

The usual objective of small-scale laboratory experiments is the optimization of conditions for the bio-oxidation of a mineral substrate, as shown by the solubilization of a metal, either the metal to be recovered or iron from pyrites. The rate of solubilization of iron, for example, may then be correlated with particle size, pH, pulp density and temperature. It is also necessary to check that addition of the mineral does not cause a rise in pH of the medium by acting as an ion-exchanger, and that solubilization of the mineral does not lead to the release of toxic materials.

Preliminary experiments may be conveniently carried out at low pulp density in shaken flasks in orbital incubators. The organism is often grown in the 9K medium of Silverman and Lundgren (1959) with pyrites as a substrate, and with an air-flow (0.5% v/v carbon dioxide) at a rate of 50–250 ml min^{-1}. The mineral, appropriately ground, is added to the culture when the pH has reached an appropriate value. It is best added in several portions to allow the organism to become conditioned to the new substrate.

Simulations of heap leaching of materials may be carried out in columns packed with the material, with the culture pumped up from the bottom and returned to a reservoir for recycling. An air/CO_2 stream is admitted from the bottom of the column and helps to prevent consolidation of the material. The progress of the reaction can be monitored by analysis of the recirculating leaching medium.

Small-scale industrial operations may be modelled in the laboratory by the use of stirred pots, air-lift reactors and pachucas. It is advantageous if these experiments can be carried out continuously. Pots need to be fitted with baffles and facilities for following temperature and pH and for withdrawing samples. Stirring rates need to be controlled precisely and defined impellers used. An air-lift reactor consists of two concentric columns. The outer column has a conical bottom section containing a perforated plate as a gas distributor. The inner column is positioned over this plate. The action of the air flow drives the ore up the central column, providing aeration, circulation and ensuring that the ore remains in suspension. The ore returns to the base of the reactor via the outer channel. A pyritic concentrate has been subjected to continuous leaching in an air-lift reactor using a mixed culture of *Thiobacillus ferrooxidans*, *Thiobacillus thiooxidans* and *Leptospirillum*-like bacteria (Helle and Onken, 1988). Interestingly, in continuous culture the presence of the *Leptospirillum* enhanced the leaching rate by a factor of up to 3.9, while in batch culture this effect was not seen. Pachucas operate in a similar fashion, but there is no inner column. Air enters at the conical base of the reactor and keeps the material suspended in the medium.

7.3.6 Large-scale applications of bacterial leaching and oxidation

These are carried out on several different scales with differing degrees of control. These processes involve dumps, heaps, vats and 'in-place' mining, and are shown schematically in Figure 7.1.

(a) Leaching in dumps of low grade copper ore

This process has been used commerically for some time, for example in the USA, where about 15% of the annual production of copper comes from this source. Low grade ores, with less than 0.4% Cu, are usually utilized. The ore is dumped, often in a valley with an impermeable floor to ensure that there is no contamination of ground water by leach solutions. These dumps contain enormous amounts of material: for example, the Kennecott Copper Corporation dump at Bingham Canyon contains about 3.6 × 10^{12} kg of material. Dumps will have built-in aeration facilties and a system for spraying the top surface with acidified

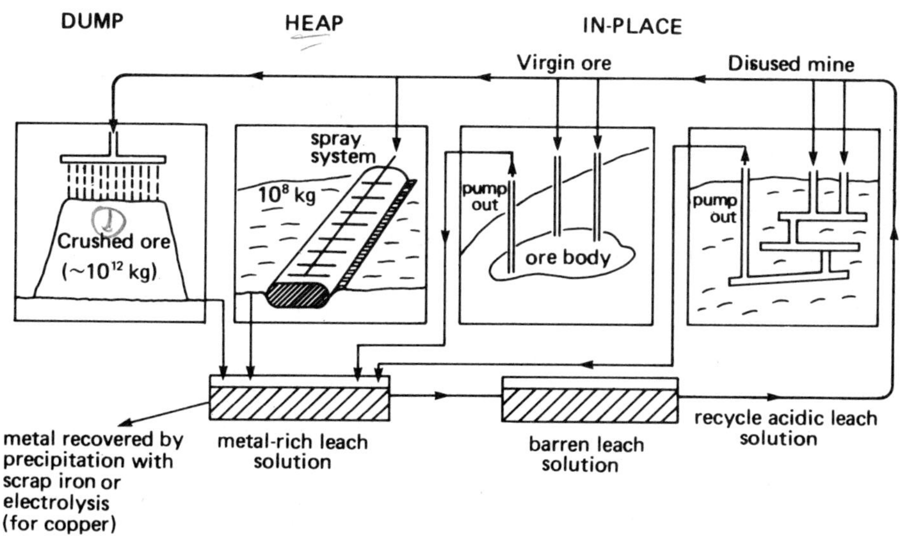

Figure 7.1 Schematic representation of large-scale bacterial leaching processes.

water. The process is started by spraying with water (pH 1.5–3.0). The liquid percolates through the dump, carrying dissolved O_2 and CO_2, and providing a favourable environment for the growth of *Thiobacillus ferrooxidans*, which is normally present in pyrite at about 10^6 cells per gram. The organism breaks down the ore, producing Fe(III) and sulphuric acid, and solubilizing the copper by direct and indirect pathways. Equations for the ores chalcopyrites and chalcocite are shown (7.12 and 7.13 respectively). Nitrogen and phosphate nutrients for the organism will be provided in the ore.

$$4CuFeS_2 + 17O_2 + 2H_2SO_4 \rightarrow 4CuSO_4 + 2Fe_2(SO_4)_3 + 2H_2O \quad (7.12)$$

$$\left.\begin{array}{l} 2Cu_2S + O_2 + 2H_2SO_4 \rightarrow 2CuS + 2CuSO_4 + 2H_2O \\ CuS \rightarrow CuSO_4 \end{array}\right\} \quad (7.13)$$

The process will be retarded if elemental sulphur is produced (by the indirect mechanism), as this will coat the ore and inhibit attack by the micro-organism. The problem is eliminated by *Thiobacillus thiooxidans*, present at about 10^3–10^5 cells per gram. The leach solution, or lixiviant, percolates through the ore body, and is collected at the bottom of the dump, at which stage it contains about 2 g of copper per litre. Treatment of the solution with scrap iron will precipitate the copper, and the resulting barren leach solution is pumped to the top of the dump for recirculation. Alternatively, the copper may be recovered by electrolysis. Figure 7.1 represents dump leaching schematically. The value of the copper leached in the USA amounts to about 350 million dollars each year.

Problems with microbial leaching will be discussed in Section 7.2.7, but large dumps have a number of engineering problems associated with them. These are:

1. The supply of O_2, which is usually the growth limiting factor;
2. The need to ensure even percolation through the dump, without preferential channelling;
3. Overheating in the dump (temperatures may exceed 80°C), which will kill the bacteria;
4. The presence of large boulders, which will be substantially unaffected by the bacterial action;
5. The long time period necessary for the copper to be completely recovered, which may be up to 20 years.

New copper dumps would be much more efficient than existing ones, but the slump in copper prices has discouraged the expansion of dump leaching.

It should be emphasized that dump leaching is much more complicated from a microbiological point of view than previously anticipated. The dump environment is a semi-natural one, having different populations in different parts of the dump, which may well be of different temperatures. The exact role of different organisms is not well understood, nor is the relationship between them. These factors are difficult to select and control.

(b) Heap leaching of ores or concentrates

This is usually carried out on ores of higher quality or which have been concentrated by flotation methods. Heaps are much smaller than dumps (maximum size $\approx 10^8$ kg), and may be controlled much more efficiently. The process is therefore more rapid, taking months rather than years. Heap leaching involves digging out a long narrow depression about a foot deep. This is lined with polythene, and air lines from a compressor positioned. The crushed ore is piled to a height of some 2–3 metres, and the top flooded with acidified water plus culture. The leach solution percolates to the bottom of the heap, and is then continually recycled until recovery of the metal from the heap is complete. Aeration is much more efficient than in dumps. The process is represented schematically in Figure 7.1.

(c) Vat leaching of concentrates

The use of batch or continuous stirred tank reactors offers the possibility of highly efficient leaching under controlled conditions. Several authors have proposed mathematical models to describe the continuous leaching by *Thiobacillus ferrooxidans* of sulphide concentrates, with the ultimate objective of predicting the amount and rate of release of metal as a function of the various operating parameters of the system. In each case experimental leaching studies have been carried out to test the validity of the model.

The mathematical analysis of a heterogeneous reactor system is much more complex than that of a homogeneous one, since mass transfer between the phases must be included. In earlier models (Gormely *et al.*, 1975; Erickson *et al.*, 1970), based upon leaching of zinc sulphide concentrates, it is assumed that attachment of the bacteria to the mineral surface is necessary, and that access to the mineral particle surface is the limiting factor in the rate of growth of the bacteria. The model of Chang and Myerson (1982) for the leaching of iron pyrite allows for the oxidation of Fe(II) in solution by bacteria not attached to the surface. The corresponding experimental study covered dilution rates (0.012–0.22 h^{-1}), pyrite surface areas (18–194 $m^2 l^{-1}$), and inlet soluble substrate Fe^{II} (0–3000 ppm). The bacterial leaching rate increased with increase in each of these variables. The fitting of the results to a Langmuir-type adsorption relationship gave a value for the mean area occupied per bacterium of 86 μm^2, suggesting that bacteria are attached to certain sites on the surface, as the surface area occupied by a cell would be about 1 μm^2.

Recently, the model of Gormely *et al.* (1975) has been tested further with more detailed experimental data on the leaching of a zinc sulphide concentrate (Sanmugasunderam, Branion and Duncan, 1985). The new data fitted the model to a much better extent, due apparently to mixing deficiencies in the reactor used in the earlier work. The area occupied on the sulphide surface by a bacterium was found to be 5.4 μm^2, while the minimum retention time for the reactor was 64.5 h. This retention time is rather high, and implies that commercial scale reactors would have to be large to gave a reasonable overall throughput. The differences in values of the standard constants determined in these studies probably arises as they are dependent on the mineral used.

Continuous leaching tests on the recovery of copper from chalcopyrite concentrates have been carried out using an air-sparged, turbine-agitated 30 l reactor, with *Thiobacillus ferrooxidans* (McElroy and Bruynesteyn, 1978). Leaching for 60 h gave electro-winning grade leach solutions, containing 20–50 $g l^{-1}$. The single stage operation gave approximately 50% recovery, but an overall extraction of greater than 96% was obtained by recovering the chalcopyrite in the residues by oil flotation, regrinding and subjecting to a second leaching operation. Single stage recovery could also be increased to over 80% by adding silver compounds to the ground chalcopyrite at a level of 50 mg (kg concentrate)$^{-1}$. It is suggested that the silver compound stops the oxidation of sulphide at the sulphur stage.

(d) In-place leaching for uranium and copper

In-place leaching is carried out underground, often in disused mines where up to 30% of the metal may remain in the walls and supporting pillars. The mine workings are flooded selectively and, some three to four months later, a metal-rich acidic leachate is pumped out from lower levels for recovery of the metal. Alternatively, access to virgin ore bodies may be obtained by drilling or the use of explosions. The applicability of in-place leaching is very dependent on geological factors. A porous rock structure is necessary to allow the leach solution to percolate through the ore efficiently. Figure 7.1 depicts these processes schematically.

In-place leaching has many advantages in terms of energy costs and environmental impact, and can be used to support conventional mining techniques. It is essential that the flow of the leach liquor is well controlled to ensure maximum recovery without contamination of local water supplies. Leaching bacteria appear to function well at higher pressures.

Several examples of the successful recovery of copper by leaching techniques are known. The acidic leach liquor may be biologically produced elsewhere or it may be generated *in situ* through a growing microbial population. In the latter case, nutrients, O_2 supply, pH and temperature have to be considered. Up to 8000 tonnes of copper per year are produced by leaching techniques from the Miami Mine in Arizona, where conventional mining had ceased. Copper has also been recovered successfully from mine waters at the Britannia Beach Mine in Canada at a rate of 200 tonnes per year. It may well be possible to recover copper from disused mines in Cornwall, UK by these techniques.

The solubilization of uranium by in-place leaching has received much more attention. This involves the oxidation of insoluble UO_2 to the acid-soluble U(VI) compound by bacterially-produced Fe(III) (Equation 7.14). The subsequent reoxidation of Fe(II) by *Thiobacillus ferrooxidans* consumes acid (Equation 7.15), but the effect is limited as the hydrolysis of Fe(III) produces acid.

$$UO_2 + 2Fe^{3+} \rightarrow UO_2^{2+} + 2Fe^{2+} \quad (7.14)$$

$$4Fe^{2+} + O_2 + 4H^+ \rightarrow 4Fe^{3+} + 2H_2O \quad (7.15)$$

Thiobacillus ferrooxidans is able to catalyse the stoichiometric oxidation of UO_2 to UO_2^{2+}, possibly by the same pathway used in the oxidation of Fe(II), with rusticyanin acting as the original electron acceptor, or by the direct reduction by U(IV) of membrane-bound Fe(III). Oxidation of U(IV), like that of Fe(II), is coupled to fixation of carbon dioxide (DiSpirito and Tuovinen, 1982 a,b).

The uranyl ion is toxic to *Thiobacillus* at concentrations greater than 0.5 mM, but the level of resistance can be enhanced by selective culture techniques up to 1 mM UO_2^{2+}. Ultimately, however, accumulation of soluble uranium in the leach solution will limit recycling as the bacteria will be unable to reoxidize the iron(II). Resistance to uranium toxicity is probably linked to a plasmid (Tuovinen *et al.*, 1981). *Thiobacillus ferrooxidans* will take up uranium, largely in the cell wall and membrane fractions. Uptake is about 40% greater in KCN-treated cells, possibly through greater membrane permeability or by the inhibition of an energy-dependent efflux of uranium (DiSpirito, Talnagi and Tuovinen, 1983).

The solubilization of uranium from a range of ores occurs readily, except in the case where there is an excessive amount of alkaline material (Tuovinen, Hiltanen and Vuorinen, 1983).

Underground in-place leaching was carried out at the Agnew Lake uranium mines, Sudbury, Canada over the period 1977–1983. The ore body was fractured with explosives and acidic culture solutions pumped through. The mine effluents were treated to recover the uranium. However, only about 50% of the uranium was recovered, due to inadequate fracturing of the ore, and the operation was stopped in 1984 (Lakshmanan, 1986). A successful venture at the Denison mines in Elliot Lake, Ontario used an in-place bioleaching process to supplement the normal mining operation (Lakshmanan, 1986).

Several in-place operations have experienced decreased production due to heavy accumulation of micro-organisms on equipment and the clogging by micro-organisms of the leachate flow path. This could represent a major problem for in-place leaching. *Bacillus* spp., *Micrococcus* spp., pseudomonads and xanthomonads have been isolated from leach solution and solid samples. Hydrogen peroxide ($0.2\,g\,l^{-1}$) killed the microbes and alleviated the problem (Yates *et al.*, 1983).

There appears no doubt that the selective in-place leaching of uranium and copper is successful, and that current lack of interest in the process is due to the fall in demand for the elements and the continued availability of high grade ores.

(e) Treatment of refractory gold ores

Gold occurs native as the free metal, and is usually extracted by the long-established cyanidation process in which a cyanogold(I) species is formed (Equation 7.16).

$$2Au + 4CN^- + O_2 \rightarrow 2Au(CN)_2^- + O_2^{2-} \qquad (7.16)$$

In some cases, however, the gold occurs in a matrix of pyrites or arsenopyrite, in which the gold is encapsulated by the sulphidic material and so cannot be recovered efficiently on cyanidation. Alternatively, the gold may be present as extremely fine particles trapped within the pyrites lattice. Recovery of gold from these refractory materials is often only in the range 15–60% of the total. The chemical solution is to roast these refractory materials, driving off the sulphide as SO_2. This process is not always effective but in any case is unsatisfactory on environmental grounds, particularly for concentrates containing arsenic, antimony or tellurium.

An attractive, environmentally acceptable solution involves the employment of bacterial leaching techniques to solubilize the pyrites or arsenopyrite (Lawrence and Bruynesteyn, 1983). The liberated gold can then be extracted by the standard cyanidation process. In this case, then, bacterial oxidation of the sulphidic gangue is viewed as a pretreatment of the concentrate prior to the normal cyanidation process. Many reports have shown that gold-bearing pyritic concentrates can be treated at up to 20% pulp density in stirred pot reactors, with enhancement of gold recovery to over 95% of the total (Barrett *et al.*, 1988a). Bio-oxidation of arsenopyrite concentrates is more difficult, as the solubilized arsenic reaches levels that are toxic to the culture when reasonable pulp densities are used. However, it is possible to condition organisms to high levels of arsenic, up to 200 mM, for example (Barrett *et al.*, 1988b). The arsenic appears to inhibit the oxidation of sulphide beyond sulphur, and so less acid is produced. The presence of the sulphur in the bio-oxidized material results in high consumption of cyanide, although it can be removed by mechanical means.

The bio-oxidation of refractory gold ores is attracting much attention, and several pilot plants are in operation (Bruynesteyn, Hackl and Wright, 1986). A major problem of cooling the reactors arises if *Thiobacillus ferrooxidans* is used, and consideration is being given to the use of thermophilic organisms such as *Sulfolobus*. It is anticipated that the extra costs involved in maintaining temperatures around 70°C will be compensated for by much smaller residence times in the reactor, with major savings on capital costs. This appears to be one situation where a moderately thermophilic culture is particularly advantageous (Barrett *et al.*, 1988a). Several estimates of the comparative economics of bacterial oxidation and roasting as a pretreatment step for gold-bearing sulphidic materials suggest that the bacterial process is favoured (Bruynesteyn, Hackl and Wright, 1986; Gilbert, Bounds and Ice, 1988; Livesey-Goldblatt, Norman and Livesey-Goldblatt, 1983). It seems very likely that this process will find commercial application.

(f) Leaching of other metals

A wide range of metals have been leached successfully from various minerals by microbiological techniques, namely arsenic, antimony, bismuth, cadmium, cobalt, gallium, lead, molybdenum, nickel, tin, vanadium and zinc. Candidates for future commerical exploitation are cobalt and nickel, as vast amounts of low grade ores are available.

Thiobacillus ferrooxidans is able to oxidize synthetic pure sulphides of cobalt, nickel and zinc (Torma, 1971; Gormely *et al.*, 1975; Groudev, 1982a,b; Sanmugasunderam, Branion and Duncan, 1985). Other workers

(Sugio *et al.*, 1984) found that a pure strain of *Thiobacillus ferrooxidans* (AP19–3) could not grow on a cobalt sulphide-containing medium unless Fe(II) was present, in which case both Fe(II) and the sulphide group in CoS were utilized for growth, with enhanced solubilization of Co(II). Cell growth on a sulphur medium was strongly inhibited by Co(II), but this was removed by the addition of Fe(II). Cobalt-resistant cells, obtained by subculturing in medium supplemented with both Fe(II) and Co (II), required less Fe(II) for growth in presence of cobalt sulphide. The mechanism(s) of this protective effect of Fe(II) are not yet clarified. Iron as Fe(II) was also found to protect against the toxic effects of Ni(II), Zn(II), Sn(II), Hg(II) and Ag(I) ions.

A number of other sulphides are oxidized to sulphates by *Thiobacillus ferrooxidans*, in some cases with oxidation of the metal (Lundgren and Silver, 1980). It is not always certain if the latter oxidations are used as an energy source. These are shown in Equations 7.17–7.20.

$$Sb_2S_3 + 6O_2 \rightarrow Sb_2(SO_4)_3 \quad (7.17)$$

$$Sb_2(SO_4)_3 + O_2 + 2H_2SO_4 \rightarrow Sb_2(SO_4)_5 + 2H_2O \quad (7.18)$$

$$Ga_2S_3 + 6O_2 \rightarrow Ga_2(SO_4)_3 \quad (7.19)$$

$$2FeAsS + 7O_2 + 2H_2O \rightarrow 2FeAsO_4 + 2H_2SO_4 \quad (7.20)$$

Bacterially-produced iron(III) is also able to oxidize the sulphides of Ga(III) and Bi(III), and arsenopyrite, FeAsS.

Naturally-occurring ores oxidized by *Thiobacillus ferrooxidans* also include sulphides of cadmium (greenockite), cobalt (sycoporite), lead (galena), nickel (millerite and pentlandite, ($(Ni,Co,Fe)_9S_8$) and zinc (sphalerite) (Lundgren and Silver, 1980).

Molybdenum is generally toxic to *Thiobacillus ferrooxidans*, but the organism oxidizes molybdenite (MoS_2) to molybdate (H_2MoO_4). This reaction is also catalysed very strongly by *Sulfolobus* (Brierley, 1974). Scanning electron microscopy shows that the cells were attached to the mineral crystals. *Sulfolobus* seems to have a high tolerance for molybdenum (750 ppm), whereas *Thiobacillus ferrooxidans* has a maximum tolerance of up to 90 ppm.

7.3.7 Optimization of the leaching process

For successful industrial exploitation of microbial leaching it is essential that the process be optimized by careful consideration of factors such as pH, temperature, nutrients, availability of oxygen and carbon dioxide, particle size, pulp density, and flow rate of the leach solution. A particular set of conditions will probably only hold for the one ore body and will have to be re-established even for nominally similar minerals from another location.

(a) Choice of organism

Leaching bacteria may be isolated readily, but different strains of *Thiobacillus ferrooxidans*, for example, may vary considerably in their resistances to adverse factors. Natural mixed cultures offer advantages over pure strains. Other bacteria may also be accidentally introduced as the ore is not sterile. Genetically modified or adapted organisms may be used in the future, but they may be unable to compete with organisms naturally present, particularly in dump or heap situations.

One of the major issues in vat leaching is the possible use of extreme thermophiles. The additional energy costs involved in maintaining temperatures at around 70°C will be compensated for by rapid reaction rates (and hence lower capital costs, as smaller reactors are necessary to give satisfactory throughput of ores), and better process control such as more efficient mass transfer and stirring. There is an urgent need to carry out large-scale pilot plant experiments with extreme thermophiles in order to obtain better estimates of costs.

(b) Control of temperature

Leaching bacteria may be killed if overheating occurs, for example in local 'hot spots' in dumps or by failure to control exothermic reactions in vats. Mixed cultures, especially those with moderate thermophiles, seem best suited to cope with fluctuations in temperatures. It should be noted that remarkably little information is available on temperature variation in dumps.

(c) Control of pH

There are important advantages in keeping the pH of the leaching solution as low as is compatible with the properties of the micro-organism in use. Ores may increase the pH by acting as ion exchangers (M^{n+} for H^+, where M = metal), or by containing acid-consuming groups such as carbonate. Values of pH should be monitored to ensure that unforseen increases in pH value are compensated for by addition of mineral acid.

(d) Prevention of precipitation

Insoluble precipitates may deposit on the ore, preventing attack by the microbe. If this occurs, regrinding of the ore will be necessary to expose fresh surfaces. Such precipitation takes place if the pH value of the medium rises. Thus, ores with high levels of arsenic may precipitate iron arsenate at pH > 2.0. A more serious problem is the precipitation at pH 2.5 of hydroxysulphatocomplexes of iron(III), either in an amorphous form or as crystalline jarosites, of composition $MFe_3(SO_4)_2(OH)_6$, where M is a univalent cation (Equation 7.21) (Ivarson, 1973).

$$3Fe_2(SO_4)_3 + 2NaOH + 10H_2O \rightarrow 2NaFe_3(SO_4)_2(OH)_6 + 5H_2SO_4 \quad (7.21)$$

Lazaroff, Sigal and Wasserman (1982) have interpreted the infrared spectra of these precipitates in terms of hydroxo-bridged and sulphato-bridged polymeric structures for the amorphous form and the jarosites, respectively.

A further precipitation problem involves the deposition of sulphur, possibly by the effect of toxic metals in inhibiting the bacterial oxidation of sulphide. This may be limited by the presence of *Thiobacillus thiooxdans* or by the bacterially mediated oxidation by Fe(III) of sulphur. In some cases, however, regrinding of the ore will be necessary.

(e) Solubilization of toxic elements

Toxic elements may be solubilized from the ore by the action of the leaching bacteria, and may rise to lethal concentrations, especially if the leach solution is recycled. Acidophilic organisms tend to be metal-tolerant, possibly because potential uptake sites for toxic metals on the cell surface are protonated in acidic media. *Thiobacillus ferrooxidans* is very sensitive to low levels of silver and mercuric ions (e.g. about 10 μg l^{-1} for Ag^+; Hoffman and Hendrix, 1976) and also to molybdenum. However, some strains do show enhanced resistance to these toxic effects. A silver-resistant *Thiobacillus thiooxidans* has been obtained by subculturing in autotrophic media supplemented by silver nitrate. This organism grows at 5×10^{-4} M Ag^+ (Sugio, Tano and Imai, 1981). Pooley (1982) has shown that oxidation of several sulphide minerals containing traces of silver by a mixed culture of *Thiobacillus ferrooxidans* and *Thiobacillus thiooxidans* results in the binding of Ag^+ to the cell as Ag_2S, from which it may be recovered by treatment with cyanide. A *Thiobacillus ferrooxidans* strain is adaptively resistant to Hg(II) at concentrations up to 5 μM, due to the synthesis of a mercuric reductase enzyme which reduces Hg(II) ions to volatile elemental mercury (Olson *et al.*, 1981, 1982, Chapter 6). These results suggest that development of strains with enhanced metal resistance may be achieved by genetic manipulation, although little is known at present about the genetics of leaching bacteria.

Other elements that are toxic to various Thiobacilli are uranium, arsenic, antimony, vanadium, chromium (Wong, Silver and Kushner, 1982) and copper (at high concentrations). Little is known about the toxicity of metals to *Sulfolobus*.

Toxic elements can in some cases be recovered from recycling leach solutions. Alternatively, fresh medium may have to be added in place of medium with toxic levels of these elements. It should be noted that the recovery of these toxic elements may be an economically attractive process.

(f) Pulp density in vats

The pulp density should be kept as high as possible, but the maximum will probably be about 20%. The limit on pulp density is probably the mass transfer of nutrients to the organisms, as solids will interfere with this process. High pulp densities will exacerbate problems due to solubilization of toxic elements and pH fluctuation.

(g) Nutrient and oxygen supply in dumps and heaps

Requirements for trace metals and sulphate will be met by solubilization of the ore. It may be necessary in heap leaching to add nitrogen (as ammonia) and phosphate, although the latter may be available. Supply of oxygen may well be the rate-limiting factor.

7.3.8 Future prospects

There are many obvious advantages in the bacterial leaching of sulphidic ores, from economic and environmental viewpoints. However, application of these techniques has not progressed beyond the recovery of copper and uranium by dump and in-place leaching, respectively. This must largely be due to the fall in metal prices and the continuing supply of good quality ores, which has reduced interest in the application of new technology. Markets for metals are unstable, so investors in bacterial leaching processes would need returns over a relatively short time period to ensure that the capital costs of the new technology are recovered. Companies are often locked into their present processes by such factors.

However, there are certain ores which cannot be treated by conventional methods and which must be candidates for bio-oxidation in the long term. It is important that more work be carried out at pilot plant level to allow confirmation of the feasibility of the process on economic and technical grounds. Without doubt, the most promising prospect is the recovery of precious metals, particularly gold, from refractory concentrates. It seems likely that commercial processes will be operating by the early 1990s.

7.3.9 Microbial methods for the desulphurization of coal

The burning of coal in power stations represents a major contribution to the problem of acid rain. Sulphur compounds in the coal are oxidized to sulphur dioxide, which in turn is oxidized to sulphuric acid in the atmosphere. This is eventually precipitated as acid rain. The sulphur dioxide can be removed from waste gases in coal-fired power stations by various catalytic processes, but there is much interest in developing more economic methods. Consideration is now being given to the use of autotrophic bacteria in the controlled oxidation of sulphide in coal before combustion. The method, in principle, is a simple one, with low capital and operating costs (Monticello and Finnerty, 1985).

The amount of sulphur in coal can vary from about 0.05% up to about 7%. The average total sulphur content in US coals has been estimated at 3.0%, although this varies considerably even within the same coal seam. Sulphur is found in coal as organic and inorganic compounds. Organic sulphur is present in a range of compounds, but mainly as thiols, dibenzothiophene and benzothiophene, and is randomly distributed throughout the polymeric coal structure. Inorganic sulphur occurs as iron pyrites, marcasite (a different crystalline form of FeS_2), and sulphate, and tends to be found in discrete pockets. The solubilization of the sulphide by microbial methods involves the same organisms and reactions described earlier for the leaching of metals. The sulphuric acid produced in this process will leach out any metals present in the coal, and ultimately results in a decreased ash content. The removal of organic sulphur compounds is less straightforward. Ideally, this should also be accomplished by the sulphide-oxidizing bacteria, and there is evidence that this may occur with some moderate thermophiles (Gokcay and

Yirteri, 1983). Alternatively, other bacteria may be used to effect this reaction (Kargi and Robinson, 1984).

(a) *Removal of pyritic sulphur from coal by bio-oxidation*

A range of autotrophic organisms have been used to remove pyritic sulphur from coal by catalysing the oxidation of sulphide to sulphate (Andrews and Maczuga, 1984; Kargi and Robinson, 1984; Dugan, 1986). *Thiobacillus ferrooxidans* has been most frequently used, while some workers have employed mixed cultures or extreme thermophiles. It is noteworthy that thermophilic organisms have often been isolated from acidic waters from coal mines and dumps. Temperatures in coal dumps can be high and so will allow the proliferation of thermophilic organisms. Acidophilic heterotrophs have also been found in some cultures. Most studies have used shaken-flask techniques, achieving high removal of sulphide at pulp densities up to 20% provided the coal is crushed to appropriate particle sizes. Other workers have used batch or continuous stirred reactors, and the latter results indicate that large-scale desulphurization of coal might be possible in a coal slurry pipe line (Myerson and Kline, 1984). Heap leaching of coal, although attractive with respect to capital costs and simplicity of technology, seems an unrealistic prospect. Recent work using a packed column filled with coal crushed to −5/8 +3/8 inches suggests that heap leaching would require a residence time of several years (Tillet and Myerson, 1987), although this would be improved on further grinding of the coal. Recent work on coal desulphurization is illustrated by the following examples.

Many studies of the factors that affect the rate of solubilization of pyrites in coal are available. Silverman, Rogoff and Wender (1951, 1963) showed that *Thiobacillus ferrooxidans* leached 60% of the total FeS_2 from coal of particle size less than 43 μm over a period of eight days. Hoffmann *et al.* (1981), working with shaken flasks, and several strains of *Thiobacillus ferrooxidans* and *Thiobacillus thiooxidans*, showed that the major factors affecting the rate were bacterial strain, the availability of CO_2, the N/P molar ratio in the medium, the surface area and the coal source. The most effective nutrient medium contained N/P = 90/1; phosphate inhibited the process, perhaps due to precipitation of iron(III) phosphate on the reactive surface. Coals with high levels of pyritic sulphur were freed from FeS_2 (up to 90–98%) in eight to twelve days, provided the particle size was less than 74 μm and the pulp densities were 20% or less. Increased pulp densities and larger particle size fractions gave slower leaching rates and reduced extents of leaching. This could be overcome by suspending the solids in fresh medium and reinoculating.

Other workers were unable to maintain bacterial growth at pulp densities greater than 20% in stirred reactors or conventional air-lift reactors (Beyer, Ebner and Klein, 1986) but were able to reach 30% in a modified air-lift reactor. This allowed removal of over 90% of the pyritic sulphur at an oxidation rate of 230 (mg S)$l^{-1}d^{-1}$. This compares favourably with the results of Myerson and Kline (1984) who, in continuous culture, achieved rates in the range 260–650 (mg S)$l^{-1}d^{-1}$.

The facultative autotroph *Sulfolobus acidocaldarius* has been used at 75°C, removing about 96% of the inorganic sulphur from coal samples containing 2.1% inorganic sulphur within ten days, with optimal pulp density of about 15%. In this case, the inclusion of organic nutrients in the medium reduced the rate and extent of sulphur removal.

Investigation of the adsorption of *Thiobacillus ferrooxidans* to the surface of coal samples from eight different coal fields shows that this process is extremely selective, occurring predominantly at exposed massive pyrite inclusions and, to a lesser extent, to pyrite crystals and along topographical defects in the coal. A mathematical model has been developed to describe the binding process (Bagdigian and Myerson, 1986). Andrews (1988), commenting on this report, drew attention to earlier experiments on coal desulphurization (Andrews and Maczuga, 1982). These indicated that oxidation of the pyrites did not take place stoichimetrically in the earlier part of the experiment, sulphide being oxidized preferentially over Fe(II). This was attributed to diffusion of sulphur along dislocations in the pyrites where the cells were bound. Kargi and Weissman (1984) have also presented mathematical models for the FeS_2 dissolution process.

The likelihood that microbiological desulphurization of coal will ever become a viable process is difficult to assess. Future, tougher legislation on the level of sulphur emissions from coal burning stations could be important in stimulating the application of this process, at least in types of coal which can be treated effectively. In such a process it seems that additions of minimal nutrients would be necessary, and a crucial factor may be the need to reaerate the system, particularly if a slurry pipeline reactor were used.

7.4 REMOVAL OR RECOVERY OF METAL IONS AND COMPOUNDS FROM SOLUTION BY MICROBIOLOGICAL METHODS

Microbiological methods are being applied increasingly to the recovery or removal of metal ions from aqueous solutions. These applications vary from large-scale processes such as the removal of heavy metals from sewage sludge, industrial effluents and mine water to much smaller operations involving the recovery of precious metals from process streams or waste water from the electroprocessing and jewellery industries. Large-scale operations are often unsophisticated and merely involve the use of large ponds in which the growth of algae and bacteria is encouraged, with resulting precipitation of metallic compounds.

Metal recovery or removal from solution may involve the following pathways, which are represented schematically in Figure 7.2:

1. The binding of metal cations to cell surfaces, or within the cell wall when microprecipitation may enhance uptake;
2. Translocation of the metal into the cell, possibly by active transport;
3. The formation of metal-containing precipitates, by reaction with extracellular polymers or microbially produced anions such as sulphide or phosphate;
4. The volatilization of the metal by biotransformation.

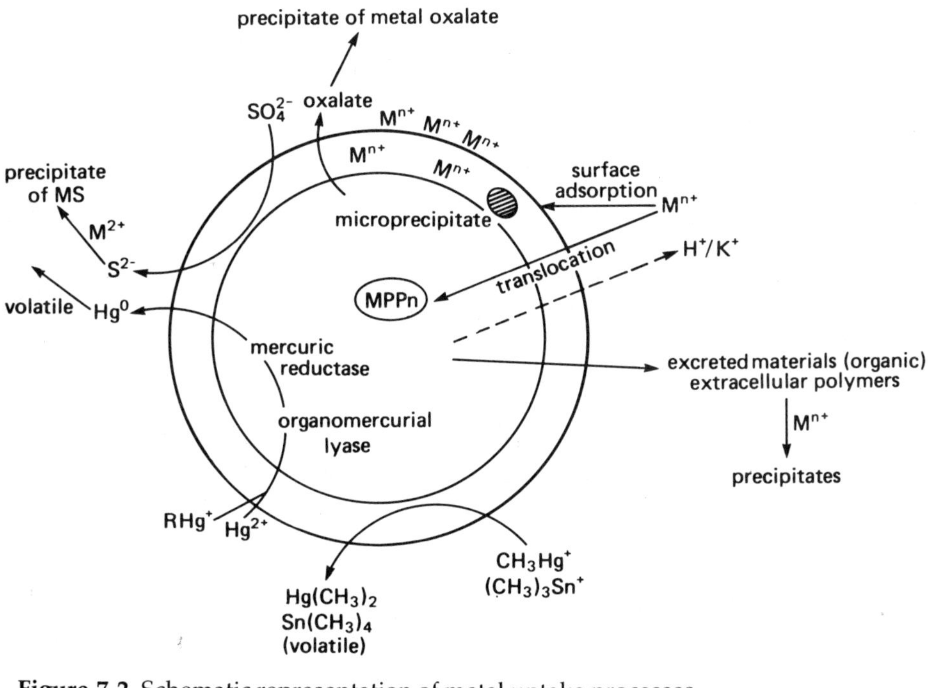

Figure 7.2 Schematic representation of metal uptake processes.
M = metal; R = alkyl group.

In addition, processes have been developed using immobilized extracellular or cellular ligands or more simple chemical models based upon them.

7.4.1 Intracellular and extracellular accumulation of metals by micro-organisms – general considerations

The uptake of metal ions on to cell surfaces and their subsequent translocation into the cell are well known natural processes, central to the accumulation of essential metals by micro-organisms (Chapter 3). Micro-organisms have a range of metal-transport systems that are often highly specific for certain metals, and capable of accumulating metals against large concentration gradients. These adsorption and accumulation processes can be exploited in the accumulation of other metals (Eccles and Hunt, 1986; Thompson, 1987; Brierley and Brierley, 1983; Volesky, 1987).

The terminology used to describe the direct interaction of micro-organisms with metal ions is sometimes ambiguous. The term 'uptake' is usually used to describe the removal of metal cations from solution by microbial cells, but does not convey any mechanistic information. Exposure of microbial cells to metal ions results, first of all, in the rapid binding of cations to negatively charged sites on the cell wall. This process is often complete within seconds. The strength and extent of this cell wall–metal interaction will depend on the chemical constitution of the cell wall, the distribution and number of ligand groups, and the affinity of the particular metal ion for these groups. This process, described as **surface binding** in this chapter, is usually, but not always, a prerequisite for subsequent translocation into the cell.

Translocation of the cations across wall and membrane into the cell will usually be a slower, metabolism-dependent process, and so will cease if the temperature falls or if an inhibitor such as azide or cyanide is added. As discussed in Chapter 3, essential metal ions have specific transport systems associated with their uptake, which may involve gated channels or pores in the membrane. Foreign metal ions may be able to exploit these transport pathways, if they are of similar size and have similar chemistry to a native cation. However, the 'shut-off' mechanisms which control the extent of uptake of the native metal may fail to function for the foreign metal, and so large amounts of a possibly toxic metal may be accumulated by the cell. Occasionally a metal ion (usually a heavy metal) will enter the cell without the requirement for metabolic activity. Thus, *Pseudomonas aeruginosa* accumulates uranium intracellularly very rapidly (<10 s) (Strandberg, Shumate and Parrott, 1981), as does a mixed culture of denitrifying bacteria (Shumate *et al.*, 1981).

Uptake of metal ions by growing cells is sometimes biphasic. It is commonly suggested that the initial very fast phase involves surface binding, while the subsequent slower phase involves intracellular uptake. The latter assumption may be incorrect, as the slow secondary uptake of metal could be due to the simultaneous effects of growth and surface adsorption. Such an effect has been predicted in theoretical studies of a model for metal adsorption with cell growth (Khummangkol, Canterford and Fryer, 1982).

In studies of this type, it is important to assess the speciation of a metal ion in solution from the formation constants for the metal and the ligands present in solution, and to ensure that medium components (notoriously phosphate), do not precipitate the metal from solution, and so appear to hinder uptake.

It is customary to define the extent of uptake of metals by cells in terms of μg(metal) g(dry weight of cells)$^{-1}$. This is misleading and should be presented in terms of moles of metal, particularly for consideration of uptake of high atomic weight elements. It is also helpful if uptake is related to the amount of metal presented to the biomass, that is (for batch systems) as an enrichment factor, the ratio of the concentration of metal in the cell to the concentration in the medium. In this case, concentrations can be expressed in terms of grams; i.e. μg(metal) g(dry weight of cells)$^{-1}$ (μg(metal) ml(solution)$^{-1}$)$^{-1}$.

7.4.2 Binding of metals to cell surfaces

Cell surfaces are anionic due to the presence of ionized groups such as carboxylate, hydroxyl and phosphate in the various cell wall polymers. Uncharged groups such as peptide N atoms may also function as ligands to complete the coordination number requirements of the metal ion. Organisms will have various distributions of charge and geometry for these binding groups, and so may well selectively bind certain metal cations. The extent of uptake of metal may vary with pH, due to the protonation of these anionic groups.

The use of a surface-binding recovery process has a number of advantages over other methods. The process may function equally well or even more effectively with a suspension of dead cells, which can suffer denaturation of the cell wall and allow free access to cell wall binding sites. The cost of nutrient materials is saved, and there is no need to protect against contamination by other organisms. Certain types of waste biomass are easily available, with additional savings in cost. Binding of cations to waste biomass may be less selective but large amounts of cation can be accumulated. The use of a process involving growing cultures of organisms is more complicated. They may, for example, be sensitive to the toxic effects of the metals they accumulate. In this case, resistant strains would have to be used. It should be noted, however, that metal-resistant cells sometimes function by preventing the translocation of toxic cations into the cell by binding them tightly on the surface. Such strains may be excellent for use in small-volume metal recovery work.

(a) Immobilization or aggregation of cells

Packed biomass offers undesirable resistance to liquid flow. It is also important that the biomass can be handled easily for packing into columns or reactors, and that it does not undergo compression during operation. Any aggregation or immobilization treatment to accomplish this should not affect the surface structure of the cells.

Cells may be aggregated into stable particles by cross-linking with a bifunctional reagent, such as glutaraldehyde, by the use of flocculating agents, or by heat or acid treatment. The filamentous fungus *Aspergillus niger* can be readily grown as pellets 4 mm in diameter by adjustment of pH (Galbraith and Smith, 1969). Heat-killed pellets retain their physical structure and absorptive capacity over long periods of time. Again, the bacterium *Zoogloea ramigera* may be flocculated by adjustment of the pH of the medium to 5.5 (Norberg and Persson, 1984).

Cell-immobilization techniques are sometimes based upon adsorption to surfaces such as silica gel, metal oxides, glass helices or porous glass or ceramic material having accurately known pore sizes (Kolot, 1981). However, most studies have been carried out with cells trapped in polymers. Such a process occurs naturally, when cells are trapped in a slime of their own production. The most widely used materials for trapping cells are polyacrylamide, K carrageenan, calcium alginate and cellulose acetate. The use of calcium alginate has many advantages. Cells are suspended in a solution of sodium alginate, and the suspenion run into a solution of a calcium salt (Kierstan and Burke, 1977). Trapped cells have better mechanical strength, and are not compressible. Biomass support particles have been prepared from stainless-steel knitted mesh and polyester foam, which provides a robust, porous support which can retain cells which adhere only weakly (Black *et al.*, 1984).

Techniques for immobilization of algae have been reviewed by Robinson, Mak and Trevan (1986).

(b) Criteria for surface-binding processes

The following criteria need to be satisfied if a metal-recovery process involving surface binding to micro-organisms is to be commercially viable:

1. The organism must be immobilized or granulated readily;
2. It must have a metal removal efficiency over 99%;
3. It must have high extracellular binding capacity;
4. It must be effective over a wide pH range;
5. The metal must be removed readily (by treatment with acid or an appropriate ligand), and the organism must be unaffected by at least 100 recycles;
6. The process must be cheaper than ion exchange or solvent extraction methods.

The metal removed from the organism may be recovered from the eluant if required.

An example of a waste water treatment and metal-recovery product based upon biomass is AMT-BIOCLAIM developed by Advanced Mineral Technologies, Inc. (Hutchins *et al.*, 1986). This is a granulated product with high capacity for accumulation of metal cations, such as Ag^+(0.8 mmol g^{-1}), Cd^{2+}(1.9 mmol g^{-1}), Cu^{2+}(2.39 mmol g^{-1}), Pb^{2+}(2.9 mmol g^{-1}) and Zn^{2+}(2.09 mmol g^{-1}), with $>$ 99% efficiency of removal from dilute solutions. The product is also able to accumulate gold, cadmium and zinc from cyanide solutions. This may be of significance as a replacement for carbon-in-pulp technology. Brierley, Goyak and Brierley (1986) have reviewed the factors relevant to the commerical use of biomass for recovery of metals.

(c) The study of surface-binding processes

The binding of metal ions to the surface of micro-organisms can be described by Freundlich, Scatchard or Langmuir plots, using data for bound and solution metal concentrations. The Langmuir isotherm appears to be of most general application. A plot of bound metal(M_B) against the free metal ion concentration (M_F) gives a rectangular hyperbola, corresponding to saturation of the available sites as metal is bound. The Langmuir isotherm may be expressed as Equation 7.21, where

K and B are constants, namely the binding affinity and the binding capacity, respectively.

$$[M_F]/[M_B] = 1/KB + [M_F]/B \qquad (7.21)$$

A plot of $[M_F]/[M_B]$ against $[M_F]$ should be linear, with slope 1/B and intercept 1/KB. Copper adsorption by certain filamentous fungi obeys the Langmuir isotherm (De Rome and Gadd 1987b). Biphasic curves arising from plots of $[M_F]/[M_B]$ against $[M_F]$ may indicate the availability of more than one site type for the metal.

Cases have been reported in which a plot of $[M_B]$ against $[M_F]$ is sigmoidal, suggesting that the initial binding of the metal has brought about structural changes and an increase in the availability of binding sites (Steiner, McLaren and Forster, 1976). In other cases metal binding is described by the Freundlich isotherm, which involves plotting $\log_e(M_B)$ against $\log_e(M_F)$, for example, uranium binding by *Rhizopus arrhizus* (Tsezos, 1983).

Attention has been focused on the identity of the metal-binding group on the cell surface, both by comparing the binding efficiencies of mutant strains having differences in cell wall structure, and by the use of instrumental techniques such as Fourier Transform Infra Red Spectroscopy (FTIR) (Nichols *et al.*, 1985). Some of these techniques will be illustrated in the following sections.

(d) Binding of metals to gram-positive bacteria

The main structural component of these bacterial walls is peptidoglycan, a polymer made up of carbohydrate strands (a repeated dimer of *N*-acetylglucosamine and *N*-acetylmuramic acid) cross-linked by a small peptide (Figure 7.3). The peptide contains a glutamic acid residue, whose carboxylate side chain is probably important in metal-binding; thus, the capacity of *Bacillus subtilis* walls to bind an osmium(III) complex is reduced dramatically if the carboxylate group is esterified (Beveridge and Jack, 1982). The peptidoglycan is linked to a second polymer, teichoic acid or teichuronic acid. The teichoic acid may be extracted with alkali from isolated cell walls of *Bacillus subtilis*, where it is present as 55% by weight. It is a linear polymer of glycerol phosphate as shown in Figure 7.3, and is well established to function as a metal-binding group (Heptinstall, Archibald and Baddiley, 1970; Doyle, Matthews and Streips, 1980). The phosphodiester group binds monovalent cations on a 1:1 basis, while divalent cations may bridge adjacent phosphate groups or polymer strands. Other organisms may contain more complex teichoic acids, such as polyribitol teichoic acid, which could lead to

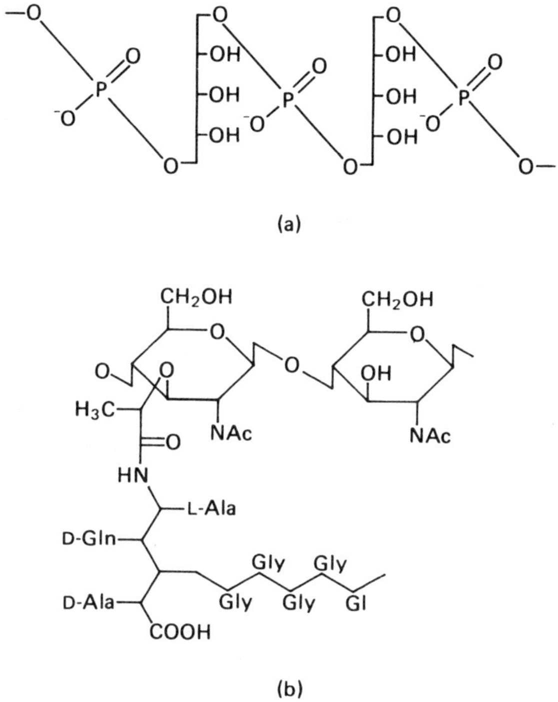

Figure 7.3 Components of bacterial cell walls: (a) ribitol teichoic acid; (b) unit of bacterial peptidoglycan.

different metal-binding properties as the spacing between phosphate groups will vary, as shown in Figure 7.4. Thus, polyribitol teichoic acid binds Mg^{2+} on a 1:1 basis with phosphate.

Cell walls of other gram-positive organisms contain teichuronic acid, a carbohydrate made up of repeating units of a disaccharide. Cell walls of *Bacillus licheniformis* NCTC 6346 contain 52% teichoic acid, 22% peptidoglycan and 26% teichuronic acid.

Gram-positive bacteria thus appear to have several interesting features and some potential in the binding of metals by cell walls. The anionic composition of the wall can be varied, while the nature of the cell wall polymers (about 25 peptidoglycan molecules thick) gives a porous open structure to the cell wall, with enhanced surface area. There is doubt over the identity of the major metal-binding sites. One complication is that the porous structure of the wall allows the build-up of substantial amounts of metal due to nucleation and precipitation of insoluble salts. This accounts for the binding of greater amounts of

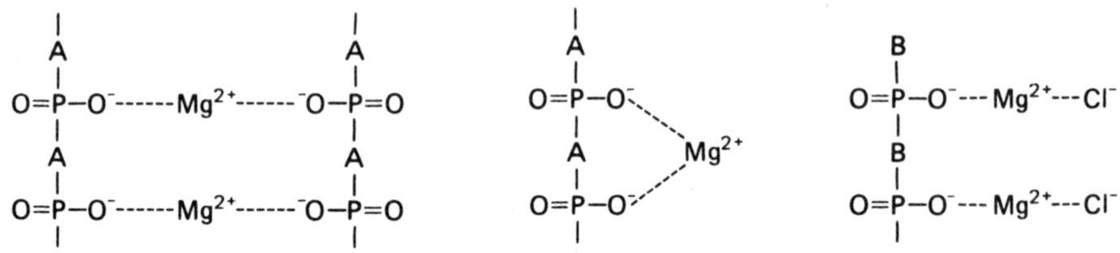

Figure 7.4 Interaction of divalent cations with teichoic acids: A = glycerol; B = ribitol.

metal than allowed by the stoichiometry of binding groups in *Bacillus subtilis* walls (Beveridge and Murray, 1976). The metal deposit will increase in the spaces between the wall polymers until restricted by the wall. This phenomenon is not seen with *Bacillus licheniformis* walls, reflecting their different composition and the possibility of a more open structure.

Data for metal binding by *Bacillus subtilis* 168 and *Bacillus licheniformis* NCTC 6346 walls are included in Table 7.1 (Beveridge and Murray, 1976, Beveridge, Forsberg and Doyle, 1982), and confirm the greater binding capacity of the former organism. The cations Mg^{2+}, Ca^{2+}, Fe^{3+} and Ni^{2+} displace other metals, implying increased affinity and competition for the same site, while partial lysosome digestion of the cells diminishes the retention of Mg^{2+} but not of Ca^{2+}, Fe^{3+} and Ni^{2+}, suggesting the presence of at least two specific sites, each of which will bind a range of metals but with differing affinity.

Bacterial surfaces may well be responsible for concentrating metal ions into marine and river sediments through surface binding, and it has been suggested that they act as templates in the initial mineralization of sediments. There are numerous accounts of possible roles for microbes in the genesis of some ancient sedimentary rock types. An attempt has been made to simulate these processes using metal-loaded *Bacillus subtilis* cells. On mixing with a synthetic sediment, they nucleate the formation of mixed crystalline metal phosphates, metal sulphides and polymeric metal complexes with organic ligands. The events leading to the formation of these mineral phases were followed by transmission electron microscopy and energy dispersive X-ray analysis (Beveridge *et al.*, 1983).

(e) Binding of metals to gram-negative bacteria

Cell walls of gram-negative bacteria are made up of two membrane bilayers. The outer membrane contains lipopolysaccharide and is usually

Table 7.1 Binding of metal ions* by isolated cell walls and membranes

Organism	Component	Na^+	K^+	Mg^{2+}	Ca^{2+}	Mn^{2+}	Fe^{3+}	Ni^{2+}	Cu^{2+}	Au^{3+}
B.subtilis	Wall	2.70	1.94	8.22	0.40	0.80	3.58	0.11	2.99	0.36
B.licheniformis	Wall	0.91	0.56	0.40	0.59	0.66	0.76	0.52	0.49	0.31
E.coli AB264	Peptidoglycan	0.290	0.060	0.035	0.038	0.052	0.010	0.019		
E.coli AB264	Outer membrane	0.081	0.025	0.019	0.020	0.012	0.233	0.019		
E.coli AB264	Cell envelope†	0.042	0.082	2.56	0.035	0.140	0.200	0.002	0.090	0.056

* μmol metal (mg dry weight)$^{-1}$

† Also Sc(0.096), La(0.078), Ce^{4+}(0.10), UO_2^{2+} (0.66), ZrO_2^{2+} (0.212), HfO_2^{2+}(0.940), MoO_2^{2+}(0.225), Zn^{2+}(0.390).

linked chemically to a peptidoglycan monolayer, which lies between the outer and plasma membranes. Gram-negative bacteria bind about one tenth of the amount of metal bound by gram-positive bacteria.

The binding of metal ions to purified cell envelopes of *Escherichia coli* K12 (Beveridge and Koval, 1981), has been investigated by suspending the envelopes in 5 mM metal solutions for 10 minutes at 23°C, and then analysing the envelopes for metal. Details are shown in Table 7.1, where it may be seen that the envelopes bound large amounts of Hf and Os (> 0.9 μmol (mg dry weight)$^{-1}$), intermediate amounts of Pb, Zn, Zr, Fe(III), Mn, Mo, Mg, Co(II) and Ce(IV) (0.1–0.4 μmol (mg dry weight)$^{-1}$) and smaller amounts of other metal cations. Electron microscopy indicates that most metal deposition occurred at the polar head groups of the constituent membranes or on the peptidoglycan. It is noteworthy that the envelopes distinguished between zirconium and hafnium cations, two elements notoriously difficult to distinguish between due to their similar size.

Binding of metal cations to the isolated components of the cell membranes has also been studied. Metal ions bind to one side of the isolated outer membrane of *Escherichia coli* K12 (strain AB264), at the phosphate head group of the lipopolysaccharide as shown by the use of ^{31}P NMR with Eu^{3+} (Hoyle and Beveridge, 1983; Beveridge, 1986). The metal-binding properties of the extracted lipopolysaccharide from mutants of *Escherichia coli*, which synthesize the lipopolysaccharide deficient in the terminal polysaccharide, are associated with the glycosidic diphosphate group, as shown by ^{31}P NMR (Strain, Fesik and Armitage, 1983). Metal ions interact strongly at these sites (Coughlin *et al.*, 1981, Schindler and Osborn, 1979).

The isolated peptidoglycan bound larger amounts of metal (Table 7.1) than did the outer membrane, possibly because of the carboxylate function on the peptide bridge. Precipitation of metal salts on nucleation sites may have occurred, as the amount of bound metal appears to be greater than the availability of binding groups (Hoyle and Beveridge, 1984). This is much less important for gram-negative than gram-positive bacteria in view of the lower levels of peptidoglycan. It is clear that lipopolysaccharide and peptidoglycan form the main cationic binding sites in *Escherichia coli*.

The distribution of naturally occurring metal ions in outer and cytoplasmic membranes and extracted lipopolysaccharides from several *E. coli* K12 strains has been investigated by Coughlin, Tonsager and McGroarty (1983). This shows that the outer and cytoplasmic membranes differ in their associated cations, with a 3-fold enhancement of divalent cations in the outer membrane. This work is discussed more fully in Chapter 4.

(f) The uptake of metals by fungi and yeasts

These cells take up metal ions by binding to the cell surface and intracellularly. Living fungal cells may thus accumulate higher levels of metal than dead biomass, although uptake of uranium and thorium involves surface binding only (Section 7.4.2(h)). The large-scale availability of yeasts means that these organisms have been much used as a source of waste biomass. The main structural component of fungal cell walls is either chitin or chitosan, which are polymers of *N*-acetylated or non-acetylated glucosamine respectively; the latter should be a more effective ligand (Figure 7.5). Other ligand groups may be phosphate, carboxyl, amine and hydroxyl, plus groups in melanin and other pigments (Strandberg, Shumate and Parrott, 1981; Tsezos and Volesky, 1982a, b; Tsezos, 1983; Tobin, Cooper and Neufeld, 1984). Melanin-pigmented chlamydospores of *Aureobasidium pullulans* bind higher amounts of heavy metals than non-pigmented structures (Mowll and Gadd, 1984). A wide range of metals bind to fungal cell walls. Different cell forms of the same organism will have different binding capacities. This topic has been discussed by Gadd (1986).

(g) Binding of metal to waste biomass

Waste microbial biomass is produced in industrial fermentation and waste water treatment processes, and may well be an economic metal-adsorbant material. Examples of biomass that have been studied include *Chlorella vulgaris, Chlamydomonas, Spirulina, Aspergillus niger* (a major waste product of citric acid biosynthesis), *Penicillium chrysogenum, P. digitatum, Pseudomonas fluorescens, Streptomyces nivens, Saccharomyces cerevisiae, Rhizopus arrhizus* and a number of mixed cultures (Eccles and Hunt, 1986; Thompson, 1987; Volesky, 1987; Brierley and Brierley, 1983).

(h) Recovery of uranium and other radionuclides

There is an obvious need to improve the recovery of uranium and other elements from effluents from uranium mines and nuclear waste reprocessing plants.

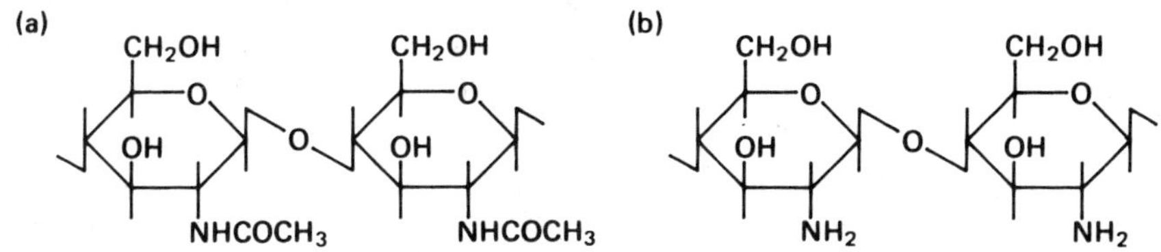

Figure 7.5 The repeat unit in (a) chitin and (b) chitosan.

The solution chemistry of uranium is essentially that of U(IV) and U(VI) species, as U(V) disproportionates and lower oxidation states reduce water. At pH values up to 4.0 the dominant U(VI) species is UO_2^{2+}, but this is present only at 9% at pH 5.0, due to hydrolysis to give $UO_2(OH)^+$. This is the main species at pH 7.0 along with dimers and polymers with bridging hydroxo groups (Baes and Mesmer, 1976; Bulman, 1980). The most important plutonium and thorium species at pH 2.0 will be the M^{4+} cations, and these will still be present at pH 5.0.

Several types of waste microbial biomass have been screened for their ability to remove radium from effluents from Canadian uranium mines, as an alternative to precipitation as a barium-radium sulphate sludge in settling ponds (Tsezos and Keller, 1983). In general, uptake was pH-dependent, with maximum recovery between pH 7.0 and 10.0. Uptake amounts at pH 7.0 from a radium solution of 50 pCi ml^{-1} were (in nCi g^{-1}): mixed culture, 5000; *Penicillium chrysogenum*, 2100; *Rhizopus arrhizus*, 650; *Aspergillus niger*, 60. The most effective examples removed over 99% of the radium in solution.

Greater attention has been paid to the recovery of uranium (Table 7.2). The fungus *Rhizopus arrhizus* is much more effective in recovering uranium(VI) and thorium(IV) than radium, with maximum uptake of over 200 mg g^{-1} occurring at pH 4.0. The uranyl cation UO_2^{2+} is probably the species taken up. The capacity is about 2.5 times greater than that of the ion exchange resins used by uranium mining companies (Tsezos and Volesky, 1981). The use of X-ray techniques shows that the uranium is found only in the cell wall. The non-acetylated amino group of the chitosan is suggested to bind uranium in the *Rhizopus arrhizus* cell wall. The different efficiencies of *Rhizopus arrhizus* and *Aspergillus niger* biomass in removing uranyl ion have been attributed to the absence of chitosan in the cell wall of the latter organism, which results in much lower uptake. In contrast, a chitosan–glucan complex prepared (and deacetylated) by treament of *Aspergillus niger* biomass with alkali binds transition metals effectively (Muzzarelli and Tanfani, 1980). The important role of fungal wall biopolymers, notably chitin, in taking up uranium has also been established for *Penicillium digitatum* (Galun *et al.*, 1983).

Binding of uranium to *Rhizopus arrhizus* chitin occurs at 6 mg g^{-1} (Tsezos, 1983), but this does not account for the high uptake of uranium. It has been suggested that the uranium cation coordinated to the chitosan group acts as a nucleation site for the binding of additional uranium in the chitin network, presumably as polymers. Cations such as Fe(III) and Zn(II) suppress the uptake of uranium, possibly because they bind to these key chitosan sites and so prevent additional uptake. In addition, hydrolysis of the chitin–uranyl complex occurs with precipitation of

Table 7.2* Recovery of U^{VI} by microbial cells or cell products

Organism	Process	[U] ($\mu g\ ml^{-1}$)	Uptake ($mg\ g^{-1}$)	Reference
Rhizopus arrhizus	Waste fungal biomass	150	200	Tsezos and Volesky, 1981
Sacch. cerevisiae	Binding to cell	100	150	Strandberg, Shumate and Parrott, 1981
Chlorella regularis	Adsorption to immobilized cells	20	160	Nakajima, Horikoshi and Sakaguchi, 1982
Streptomyces viridochromogenes	Adsorption to immobilized cells	20	132	Nakajima, Horikoshi and Sakaguchi, 1982
Citrobacter sp.	Precipitation as phosphate	200	9000	Macaskie and Dean, 1985
Penicillium digitatum	Fungal biomass	100	5.7	Galun *et al.*, 1984
Penicillium chrysogenum	Fungal biomass	150	70	Tsezos and Volesky, 1982a
Pseudomonas fluorescens			6	Tsezos and Volesky, 1982a
Municipal sludge			12	Tsezos and Volesky, 1982a
Zoogloea ramigera	Polysaccharide matrix around cell	500	800	Norberg and Persson, 1984

* After Macaskie and Dean, 1985. See also Nakajima and Sakaguchi, 1986.

uranyl compounds within the wall structure, thus allowing additional binding to occur. FTIR studies have shown bands assigned to $\nu_{u=0}$ at 908 cm^{-1}. Uptake of uranium(VI) is rapid, saturation of *Rhizopus arrhizus* sites being complete within one minute in laboratory studies (Tsezos and Volesky, 1982a). Uranium may be eluted from the biomass by treatment with bicarbonate, due to the ready formation of carbonato complexes such as $UO_2(CO_3)_2^{2-}$ and $UO_2(CO_3)_3^{4-}$. Uranium recovery by *Rhizopus arrhizus* biomass has been patented, (Volesky and Tsezos, 1981).

Uptake of thorium by *Rhizopus arrhizus* cells is also rapid. The infrared spectrum of thorium-laden walls shows a new band at 362 cm^{-1}, assigned to $\nu_{Th\text{-}N'}$ suggesting that thorium is also bound by chitosan amino groups. Pure chitin binds thorium with a capacity of 8 mg g^{-1}, but the overall uptake by *Rhizopus arrhizus* is 170 mg g^{-1}. The additional thorium-binding appears to result from the products of hydrolysis of thorium cations, which bind to the cell surface. Uptake of thorium is hardly affected by other cations, as these would only compete for the chitosan amino ligands which in this case do not trigger further binding of metal. Thus, accumulation of uranium and thorium occurs by different mechanisms, reflecting differences in their chemistry (Tsezos and Volesky, 1982b).

Horikoshi, Nakajima and Sakaguchi (1981) compared the ability of a wide range of organisms to accumulate U(VI) from solution, and found actinomycetes > bacteria > yeasts > fungi. Uptake was independent of temperature and metabolic inhibitors and was maximum at pH 6.0 for *Actinomyces levoris* and *Streptomyces viridochromogenes*, the two organisms studied in detail. The latter organism can remove uranium almost quantitatively from a solution at 10 ppm. The metal was surface bound and could be removed readily by treatment with EDTA or bicarbonate. High concentrations of carbonate inhibited uptake. These authors had previously studied the uptake of uranium by *Chlorella regularis* (Nakajima, Horikoshi and Sakaguchi, 1979), for which uptake was a maximum at pH 5.0, suggesting the possible involvement of UO_2^{2+} and UO_2OH^+. Uptake of uranium by immobilized *Chlorella regularis* cells was independent of pH, and increased with temperature, an interesting observation. Uranium could be recovered almost quantitatively from both fresh and sea water containing uranium by use of the immobilized cells in columns (Nakajima, Horikoshi and Sakaguchi, 1982).

Uptake of uranium by *Saccharomyces cerevisiae* is extremely rapid, although dependent upon the medium in which the cells are grown up. Rates were increased at higher temperatures and had maximum values at pH values between 3.0 and 4.0, suggesting UO_2^{2+} is the species taken up. Overall, cells accumulated about 15% of dry weight, but only some 32% of cells showed the presence of uranium so these cells must have had much higher amounts of uranium (Strandberg, Shumate and Parrott, 1981). Long-standing suggestions for the ligand groups on the cell surface are polyphosphate (Rothstein and Meier, 1951) and protein carboxylate (Dounce and Flagg, 1949).

Macaskie and Dean (1985) have used immobilized *Citrobacter* cells to accumulate very large amounts of uranium (Table 7.2). Uranium solutions were pumped upwards through columns containing the cells embedded in shredded polyacrylamide gel. Over 90% of the metal was removed from the flow, and subsequently released with 1 M citrate buffer at pH 4.0. A typical experiment involved pumping 24 l of a solution containing 200 μg ml^{-1} of uranium up a column with about 5 g wet weight of cells (corresponding to about 0.5 g dry weight). The cells accumulated about 4.5 g uranium. A feature of this work is the requirement for an organic phosphate in the flow medium and pregrowth of the organism in a medium with glycerol 2-phosphate as the sole phosphorus source. This induces the presence of a phosphatase enzyme which cleaves the organic phosphate to yield P_i, which precipitates the metal phosphate at the cell surface, and then in the gel until the liquid cannot be pumped through the column. While these recoveries are impressive, the ultimate distinction between this and other systems, for example with waste biomass, must be made on economic grounds. For example, one mole of glycerol 2-phosphate is consumed for each mole of uranium recovered.

(i) Recovery of gold

The recovery of gold from dilute aqueous solution by biomass has been investigated as an alternative to the carbon-in-pulp technolgy used in the gold industry. It is clear that most types of biomass bind gold to some extent when exposed to a solution containing either gold(I), usually $[Au(CN)_2]^-$, or gold(III), usually $[AuCl_4]^-$. Indeed algal mats near the coast of Sri Lanka have been found to accumulate gold up to 1.1 ppm (Dissanayake and Kritsotakis, 1984), and bacteria and algal blooms have been implicated in the laying down of certain gold deposits in South Africa (Zumberge, Sigleo and Nagy, 1978).

High levels of gold recovery have been achieved using waste biomass; thus Kuyucak and Volesky (1987) recovered gold from a dilute solution using an algal biosorbent giving 420 mg Au (g biomass)$^{-1}$, while Darnall *et al.* (1986) recovered 90% of the total gold from a 10^{-9} M solution of Au(III) at pH 2.0 using *Chlorella vulgaris*. The latter work gave gold-laden biomass with 10% dry weight of gold and represents a concentration factor of 5000. The binding of gold does not appear to follow a common pattern. Kuyucak and Volesky (1987) found the gold was present in elemental form, while Watkins *et al.* (1987) established that the nature of *Chlorella*-bound gold depended upon its original oxidation state. Algal-bound gold from sodium gold(I) thiomalate was present as Au(I) with two sulphur ligands 2.30×10^{-10} m away, as shown by EXAFS techniques, although the free sulphydryl content of the biomass is not sufficient to account for the binding of more than a small part of the gold by this set of ligands. It is probable that saturation of this site is followed by the binding of Au(I) to other sites of lower affinity. Algal-bound gold from $[Au(CN)_2]^-$ was present as several species, some of which were bound to S donors, while algal-bound gold from $[AuCl_4]^-$ was partly reduced and was bound to nitrogen plus a heavy atom.

7.4.3 Accumulation of metals by translocation into cells

In theory it is possible for much greater amounts of metal to be accumulated inside the cell than on the cell surface. The intracellular metal could be precipitated as an oxalate or polyphosphate. Active translocation into cells presents a number of problems which make it less attractive as a technological process, as emphasized earlier. The process is likely to be inhibited by low temperatures, uncouplers, toxic metal ions, and various metabolic inhibitors. The requirement for a nutrient medium is not only economically unattractive, but the components may complex the metal ions in solution and hinder uptake. Furthermore, the metal can only be recovered from the cell by destroying it, so recycling is not possible.

Examples of metal ion recovery by translocation into the cell are known. *Klebsiella aerogenes* takes up Ni(II) by active transport, while *Saccharomyces cerevisiae* accumulates Co(II) (Norris and Kelly, 1977). Certain cyanobacteria take up Cd(II), which is precipitated as polyphosphate (Jensen *et al.*, 1982). The more interesting examples of intracellular uptake seem to occur by fast, metabolism-independent transport (Section 7.4.1) as noted for uranium uptake by *Pseudomonas aeruginosa* (Strandberg, Shumate and Parrott, 1981). In this example, the uranium is take up within 10 s and is localized in the cell. Only about 44% of the cells contained visible deposits. It would be instructive to discover if the distinction between the cells was environmental or genetic in origin. If the fraction of the cells able to take in uranium was close to unity, then the usefulness of this organism in recovery of uranium would be significantly enhanced.

7.4.4 Recovery of transition metals, silver, lead and cadmium

There is much information available on the accumulation of these metal ions by bacteria and algae (Norris and Kelly, 1979), although often the site of deposition of the metal is unknown. The observation of biphasic uptake has usually been interpreted in terms of translocation into the cell. In the work described in this section the distinction between surface binding and translocation is not always clear.

Chlorella regularis cells show selectivity towards metals, the amount taken up decreasing in the sequence $UO_2^{2+} >> Cu^{2+} >> Zn^{2+} \geqslant Ba^{2+} \approx Mn^{2+} > Co^{2+} \approx Cd^{2+} \geqslant Ni^{2+} \approx Sr^{2+}$. Similar results were found for live and dead cells, but the total amount adsorbed by dead cells was greater (Nakajima, Horikoshi and Sakaguchi, 1981a,b). Molybdenum was accumulated in a biphasic process that was unaffected by metabolic inhibitors. It is assumed that some of the molybdate(VI) was reduced to Mo(III) or Mo(V) in the algal cells to account for the ESR signals observed after 20 h (Sakaguchi, Nakajima and Horikoshi, 1981).

Copper(II) is taken up on the surface of *Bacillus subtilis*, but the extent depends upon environmental factors. Thus, potassium-limitation of growth leads to the highest uptake of Cu(II). This probably reflects the variation in the structure of the cell wall with environmental conditions (Baldry and Dean, 1981). Copper(II) was also accumulated effectively by surface binding by a mixed culture of activated sludge bacteria (Dunn and Bull, 1983).

Uptake of silver by surface binding and intracellular accumulation occurs rapidly. The Ag^+ cations bind strongly to the cell surface of *Escherichia coli* and to intracellular sites (Ghandour *et al.*, 1988). Uptake of silver by a mixed community of three bacteria was due largely to *Pseudomonas maltophilia* (Charley and Bull, 1979). Silver is also surface-bound by some silver-tolerant micro-organisms isolated from soil materials from a silver mine (Pumpel and Schinner, 1986).

The *Citrobacter* sp. described in Section 7.4.2(h) in the context of recovery of uranium has also been used successfully to accumulate lead (Aickin and Dean, 1979; Aickin *et al.*, 1979) and cadmium (Macaskie and Dean, 1984a,b). In both cases the insoluble metal phosphate was precipitated on the cell surface as a result of the production of phosphate by the phosphatase-catalysed hydrolysis of glycerol-2-phosphate. This has been confirmed directly for $CdHPO_4$ by the use of X-ray micro-analysis and magic angle spinning NMR analysis (Macaskie *et al.*, 1987). It should be pointed out that the insolubility of cadmium phosphate has not always been appreciated by research workers in the area of cadmium accumulation by micro-organisms. Observations that surface-bound cadmium decreases progressively as phosphate in the medium increases may merely reflect the increasing precipitation of cadmium from solution as the phosphate.

Direct binding of Cd^{2+} to bacterial surfaces is well established, although the amount taken up can vary considerably among different strains of related bacteria. Thus *Bacillus megaterium* KM binds 43 mg Cd^{2+} per gram dry weight from a solution of Cd^{2+} containing 112 mg l^{-1}. In contrast, *Bacillus polymyxa* binds only 10 mg Cd^{2+} (g dry weight)$^{-1}$. Uptake of Cd^{2+} by *Saccharomyces cerevisiae* (Norris and Kelly, 1977), unicellular algae (Khummangkol, Canterford and Fryer, 1982) and *Anacystis nidulans* (Singh and Yadava, 1985) is biphasic. In the last case Ca^{2+} and Zn^{2+} showed competitive inhibition, as expected from their chemical properties, but Hg^{2+} enhanced uptake, possibly via the active transport process.

A wide range of organisms accumulate lead from a citrate solution in which the lead will be present as a citrate complex. Some typical data for lead as a percentage of the dry weight are *Aspergillus niger* (13%), *Bacillus subtilis* (10%), *Pseudomonas reptilovora* (8%), *Gluconobacter suboxydans* (7%), *Pseudomonas fluorescens* (1%), *Klebsiella pneumoniae* (0.1%), while natural isolates accumulated up to 36% (Broda, 1972; Tornabene and Edwards, 1972; Aickin and Dean, 1977).

7.4.5 Uptake of metals by micro-organisms in natural environments

Sediments in natural environments may compete well with microbial surfaces for binding metal ions. Typical studies involve: sediment, bacteria and mercury(II) (Ramamoorthy, Springthorpe and Kushner, 1977); sediment, algae, cadmium(II) and copper(II) (Laube, Ramamoorthy and Kushner, 1979); sediment, bacteria and copper(II) (Titus and Pfister, 1982); soil constituents, bacteria and cadmium(II) (Kurek *et al.*, 1982); and marine sediment, marine bacteria and nitrosylruthenium(II) complexes (Gibson *et al.*, 1986). In the work of Titus and Pfister (1982), uptake of Ca^{2+} by a cadmium-resistant *Pseudomonas* sp. was about ten times greater than uptake by sediment.

7.4.6 Immobilization of metals by extracellular compounds and slimes: removal of metals from waste water in the activated sludge process

Many micro-organisms either produce extracellular polymers or have a capsule of extracellular polysaccharide which can become detached from the cell. Micro-organisms in their natural habitat may thus grow as flocs or biofilms or slimes. This acidic polysaccharide material binds metal ions effectively, and has been implicated in the binding and precipitation of heavy metals from sewage sludge (Costerton, Geesey and Cheng, 1978; Rudd, Sterritt and Lester, 1984; Brown and Lester, 1979, 1982a,b). Extracellular polymers from film-forming marine bacteria also bind heavy metals, although added metal does not appear to stimulate the synthesis of larger amounts of polymer (Corpe, 1975).

Bacteria of the genus *Zoogloea* produce a gelatinous or zoogleal matrix in which cells are embedded. Zoogloea strains are important contributors to aerobic waste treatment processes, as they are floc formers in addition to synthesizers of the zoogleal matrix. The organisms settle out of suspension, thus allowing the removal of toxic metals that have complexed to the cell wall or the matrix (Friedman and Dugan, 1968).

Extracellular polymers present in activated sludge include proteins and nucleic acids in addition to polysaccharides. Those obtained from pure bacterial cultures consist of hexose and pentose sugars, hexuronic acid and hexosamines (Brown and Lester, 1979, 1980). An alkali-soluble polysaccharide with uronic acid residues has been extracted from the metal-tolerant alga *Klebshormidium fluitans* (Strong, Madgwick and Ralph, 1982).

Factors which affect the concentrations of extracellular polymers produced in bacterial cultures and activated sludge have been identified as the ratios in the growth medium of C:N, C:P and C:S; amount of dissolved O_2; biological oxygen demand, and biomass loading rates. Capsular materials produced under conditions of acute phosphorus limitation appeared to improve the metal-binding capacity of sludge organisms (Casey and Wu, 1977).

Klebsiella aerogenes produces extracellular polymers and loose slime polymers which can be extracted readily. Comparisons between the binding capacities of intact biomass, stripped biomass and extracted polymer show that the extracellular polymer is responsible largely for the binding and precipitation of added metal.

Inactivation of cells in activated sludge has very little effect on the uptake of a range of metals, confirming the process to be a passive one. Loss of viability does, however, cause a marked decrease in the uptake of nickel, suggesting that some of the nickel is removed by intracellular accumulation (Sterritt and Lester, 1986).

The sludge flocs bind soluble metal species as described previously, but they also help to remove insoluble salts by trapping the precipitates. Stephenson and Lester (1984) found that although nearly all compounds of chromium, copper, lead and zinc in the mixed liquor matrix from a pilot plant for treating sewage were insoluble, less than 10% would settle in the absence of biomass. Between 50 and 90% settled in the presence of activated sludge biomass.

(a) Precipitation of metal sulphides, oxides and oxalates

Metals in aqueous solution may be precipitated as sulphides, either when industrial effluents are depleted of oxygen (Ilyaletdinov, Enker and Loginova, 1977) or when algal blooms, which have accumulated metals, decompose (Jackson, 1978). In both cases this arises from the bacterial reduction of sulphate to give hydrogen sulphide. The sulphate-reducing bacteria are widely distributed in anaerobic environments, and were probably involved in the laying down of metal sulphide minerals. Ponds with algal blooms may be used to remove metals from effluents through this process. In theory, this could be carried out more efficiently. Organisms such as *Desulfovibrio* spp. (Postgate, 1979), grown in continuous culture, could be fed a metal-polluted solution to which sulphate and waste organic nutrient had been added. The H_2S generated by the action of the bacteria would precipitate metal sulphides which would collect in the bottom sludge.

Some species of bacteria may use the formation of insoluble salts as a detoxification mechanism. A strain of *Klebsiella aerogenes* precipitated cadmium sulphide on the cell surface when potentially toxic concentrations of cadmium compounds were added to the medium (Aiking *et al.*, 1982). The plasmid encoding mercury reductase in a strain of *Mycobacterium scrofulaceum* also conferred resistance to copper by sulphate-dependent precipitation as copper sulphide (Erardi, Failla and Falkinham, 1987). Other organisms excrete hydrogen peroxide to precipitate metal cations as oxides on the cell surface, while the production of oxalic acid by fungi leads to the precipitation of insoluble metal oxalate crystals on the cell surface and in the medium (Murphy and Levy, 1983).

7.4.7 Iron and manganese-depositing micro-organisms

A large and varied group of bacteria are able to cause the deposition of $Fe(OH)_3$ and MnO_2 from dissolved Fe(II) and Mn(II). There is strong evidence to suggest that this does not represent an autotrophic growth mode for the bacteria, and that precipitation and oxidation on extracellular polymers in sheaths and capsules are important.

Deposits of manganese and iron oxides are found in many aquatic environments, and may occur as thin deposits on rocks or as crusts of thicker accumulations. The deep sea ferromanganese nodules found in all the world's oceans represent a particularly interesting example. These nodules may contain up to 60% of their mass as MnO_2, with large amounts of iron(III) oxide, with silicate and substantial amounts of elements such as cobalt and nickel. In some cases, the floor of the ocean may take on the appearance of a 'cobbled' street, with packed nodules from 5–10 cm in diameter. The nodules represent a major source of raw materials, and involve the accumulation of these elements from very dilute solutions to abundances far above their normal level (Cronan, 1980). However, new techniques will be required to separate these metals (De Carlo *et al.*, 1982a,b).

Iron- and manganese-oxidizing bacteria have been isolated from various ferromanganese deposits, including nodules (Ehrlich, 1975; Schweisfurth *et al.*, 1978; Ghiorse and Hirsch, 1982), confirming the importance of microbial action in their formation.

Ferromanganese-depositing micro-organisms include bacteria, fungi, algae and protozoa, but bacteria and fungi are the most well-studied. They produce acidic exocellular polymers, sometimes as capsules, sheaths or stalks, on which the metal oxides are deposited. The presence of these polysaccharides was shown by staining experiments with Ruthenium Red (Ghiorse, 1980).

Examples of bacteria that bring about these oxidations are the 'sheathed' bacteria, notably of the genera *Leptothrix* and *Sphaerotilus*. In nature and in cultures these sheaths become encrusted with deposited iron and manganese oxides. The iron-oxidizing and- depositing bacteria include the genera *Gallionella*, which produces a twisted stalk made up of filaments of iron(III) hydroxides, and *Siderocapsa* which deposits iron in mucoid capsules. *Hyphomicrobium* and *Pedomicrobium* bring about the oxidation of Mn^{2+}, as do many isolates from fresh water and salt water sediments. Species of *Metallogenium* are associated with iron and manganese deposition (Fenchal and Blackburn, 1979).

The details of the oxidation pathway for Mn(II) are not known, but it appears to occur after adsorption of the metal on a surface, such as MnO_2 (Arcuri and Ehrlich, 1979), clay particles (Ehrlich, 1982) or bacterial spores, even when dead (Rosson and Nealson, 1982; Kepkay and Nealson, 1982). The biology of the iron- and manganese-depositing bacteria has been reviewed by Ghiorse (1984).

The formation of the ferromanganese nodules appears to be initiated by the bacterial oxidation of Mn^{2+} to MnO_2 on some surface such as a clay or ferric hydroxide precipitates. The MnO_2 absorbs further Mn^{2+}, which is oxidized along with Fe^{2+}. Clay and other cations from the sea water are deposited and the nodule gradually increases in size. A Mn^{2+}-oxidizing protein has been isolated from cell-free spent culture medium used to grow a sheathless strain of *Leptothrix discophora*. This protein could be a component of the sheath normally produced by this organism, explaining why these sheaths usually have ferromanganese encrustations (Ghiorse, 1986). The manganese(II) oxidase from this organism transfers electrons to O_2 via cytochromes.

Ghiorse (1986) has suggested that bacteria such as *Leptothrix, Pedomicrobium* and *Hyphomicrobium* could be used to oxidize Mn(II) to MnO_2, which will coprecipitate other metals for recovery purposes.

It should be noted that MnO_2 can be solubilized in nature by bacterial reduction to Mn(II). It is possible that such processes could be of value in the extraction of metals from manganese nodules.

7.4.8 Biotransformation of metallic species to volatile products

In a few cases river water is detoxified by the microbial transformation of toxic species to volatile products, which escape to the atmosphere.

(a) Biomethylation

A range of metallic and metalloidal species undergoes biomethylation in the environment. The list includes lead, arsenic, tin, selenium, mercury and probably thallium (Section 6.3). The best-studied example involves the conversion of Hg(II) to much more toxic organomercury compounds in river sediment. This reaction was central to the environmental tragedy at Minimata in Japan. Several pathways are available for the production of organometallic compounds in this way. The methylation of Hg(II) involves the intermediacy of the methyl donor methylcobalamin, which is involved in the conversion of homocysteine to methonine. However, the methyl group may be transferred rather to metallic species, with the ultimate formation of volatile compounds, as shown in Equations 7.22 and 7.23 (Summers and Silver, 1978; Robinson and Tuovinen, 1984).

$$Hg^{2+} + CH_3Co \rightarrow Hg^{+}CH_3 \xrightarrow{CH_3Co} Hg(CH_3)_2 \uparrow \qquad (7.22)$$

$$SnR_3^{+} + CH_3Co \rightarrow SnR_4 \uparrow \qquad (7.23)$$

(b) Reduction of mercury(II) to mercury(0)

Mercury compounds may also be volatilized by conversion to elemental mercury. This reaction is carried out by cells resistant to the toxic effects of mercury compounds, and represents an example of a plasmid-determined resistance to toxic elements. The detoxification process is inducible by exposure of the cells to subtoxic levels of mercury compounds, and has been found in a range of organisms such as the enterobacteria, mycobacteria (Meissner and Falkinhan, 1984), pseudomonads and the autotroph *Thiobacillus ferrooxidans*. The plasmid encodes an uptake system for Hg(II) plus the synthesis of the enzyme mercuric reductase. The mercuric reductase is an intracellular enzyme, so the bacterium must first bring Hg(II) into the cell in order to detoxify it (Silver, 1981). Organisms are known which are resistant to a range of organomercury compounds. These organisms also have an inducible enzyme, organomercurial lyase, which converts the organomercury compound to Hg^{2+}, which is then reduced by mercuric reductase (Summers, 1985) (Chpater 6)

$$RHg^{+} \xrightarrow{\text{organomercurial lyase}} Hg^{2+} + RH \qquad (7.24)$$

$$Hg^{2+} \xrightarrow{\text{mercuric reductase}} Hg^{0}\ \text{(vapour)} \qquad (7.25)$$

7.4.9 Removal of metals by complexation to high-affinity ligands of microbial origin or related model compounds

Certain micro-organisms synthesize compounds which bind specific metals with high affinity. Reference has been made in Chapter 3 to the iron-binding siderophores, which are usually phenolate or hydroxamate ligands, and in Chapter 6 to the metallothioneins, which are small cysteine-rich proteins which bind copper, zinc and cadmium strongly. Ligands of this type, or related model compounds, are of interest as metal-recovery agents, although they may have to be bound to solid supports to accomplish this.

Butt and Ecker (1987) have discussed the exploitation of yeast metallothioneins (MT) in the recovery of precious metals. If strains of yeast which accumulate metals intracellularly are modified to overproduce metallothioneins, this should result in substantial accumulation of metals inside the cell. A different strategy must be followed for metals which do not enter the cell. The MT synthesized in the yeast cells can be secreted in the periplasmic space or extracellularly by attaching the appropriate sequences to MT genes. Proteins of molecular weights less than 10 000 are more efficiently secreted than larger ones. These metallothioneins can be engineered to have specific affinity for different metals, perhaps by replacing some cysteine residues with histidines. The metal-loaded yeast can then be washed to release the metal and to regenerate the cells.

Several modified polysaccharides (amino acid-substituted glucans) have been used successfully in the recovery of metal ions from acidic solutions (Muzzarelli *et al.*, 1985). Catecholate ligands have been immobilized on silica gel, with a spacer compound to hold the ligand clear of surface effects from the carrier (Figure 7.6). DeVoe and Holbein have also developed products based upon phenolic siderophores, which are able to complex selectively a wide range of metals, including precious metals and radionuclides. Applications of these materials include recovery of plutonium from nitric acid solutions used to dissolve uranium fuel rods, the recovery of metals from plating effluent, silver from photography waste, mercury from chlor-alkali plant effluents, and more recently the recovery of gold (Chemistry in Britain, 1986).

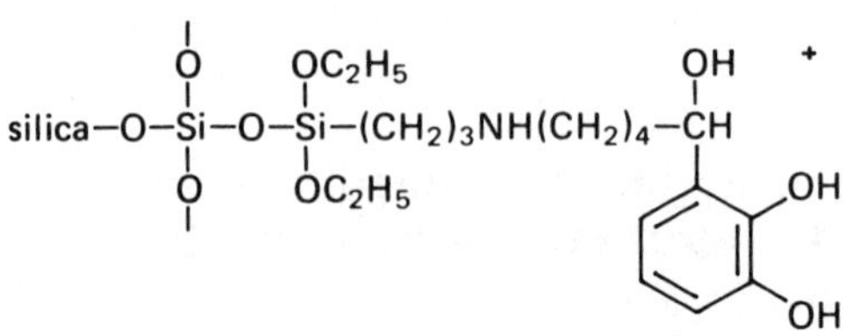

Figure 7.6 A metal-binding model compound based upon catecholate siderophores.

7.5 CORROSION AND OTHER ADVERSE EFFECTS ARISING FROM METAL–MICROBE INTERACTIONS

Several aspects of the interactions between metals and microbes discussed in this chapter are associated with environmental and industrial problems of major economic significance.

7.5.1 Corrosion of metallic structures

Oil and water pipelines and other buried metal structures are subject to corrosion brought about by microbial action: some 50% of all failures in underground pipes result from this. Similar problems occur with aviation fuel tanks and other aluminium alloy structures. Several mechanisms are involved, often utilizing the action of corrosive products of microbes, as discussed in subsequent sections, notably acids and sulphide (Miller, 1981; Booth, 1968).

Growth of organisms on industrial structures usually involves algae, and occurs on irregularities in the metal surface, or (in the case of water coolers) where there is a sudden drop in water velocity. The production of polysaccharide capsules and sheaths leads to slime formation. This may trap non-slime forming organisms, which could be responsible for specific corrosion problems. However, corrosion can also result from the presence of microbial growths on damp or wet metal surfaces, as this results in the absorption of oxygen and nutrients from beneath the growing cells, and the establishment of electrochemical cells. The oxygen-depleted zone beneath the microbial growth is anodic with respect to the exposed metal around the growth, and anodic dissolution will occur.

Micro-organisms may degrade additives or emulsifiers intended to protect against corrosion.

(a) Corrosion by acid production

The ubiquitous thiobacilli produce sulphuric acid, which will corrode underground structures. The likelihood of this taking place will depend upon the local environment of the structures, particularly the presence of reduced sulphur compounds, such as pyrite, and acid conditions. Concrete pipes will also be attacked by *Thiobacillus*.

(b) Corrosion of alloys by carboxylic acids

Non-ferrous metals are often attacked by organisms that produce organic acids. The action may be an indirect one, as the acids damage protective coatings, so exposing the metals to attack by other agents. Integral fuel tanks in aircraft sometimes contain a sludge, which blocks fuel-filters, and causes corrosion-pitting of the aluminium structure. The principal component of the sludge is *Cladosporium resinae*, 'the kerosene fungus', but other organisms are usually present. Corrosion is usually attributed directly to the action of organic acids (McKenzie, Akbar and Miller, 1977), but Iverson (1967) has suggested it results from the attack of sulphate-reducing bacteria with *Cladosporium resinae* providing a matrix, nutrients and an anaerobic environment for the sulphate reducers.

(c) Corrosion by attack of sulphide

The sulphate-reducing bacteria have been implicated in the corrosion of buried and water-immersed iron or steel structures. These anaerobic bacteria use sulphate as an electron acceptor, reducing it to sulphide. Two genera of sulphate-reducing bacteria are recognized, *Desulfovibrio* and *Desulfotomaculum* (Postgate, 1979). Species of the former genus are mainly responsible for anaerobic bacterial corrosion (Bell and Lim, 1981; Booth and Wormwell, 1961).

Corroded material shows the presence of black iron sulphide, which probably contains sulphate-reducing bacteria. Removal of this material reveals a pitted area. Cast iron products are susceptible to this type of attack, undergoing 'graphitization', in which the iron has been leached out leaving a carbon-rich, mechanically weak area. Cast iron pipes of 6 mm thickness have been perforated in under a year because of this problem. The oil industry suffers considerably from this type of corrosion (Farquhar, 1974).

The ability of *Desulfovibrio* spp. to cause corrosion is linked to the activity of an hydrogenase involved in the production of sulphide (Booth and Wormwell, 1961). It appears that corrosion may result from the deposition of iron sulphide on the metallic surface, but the detailed mechanism is uncertain, and a role for H_2S cannot be excluded.

7.5.2 Problems arising from the activity of iron-oxidizing bacteria

A range of iron-oxidizing bacteria are known, in addition to *Thiobacillus* species. Some of these have been described in Section 7.3.6, and are known to deposit ferric compounds and, in some cases, manganese compounds.

(a) Deposition of iron (III) hydroxide

The presence of the iron-oxidizing bacteria *Sphaerotilus* and *Gallionella* in water mains and cooling systems may result in extensive local deposition of iron(III) hydroxide. Eventually, large tubercles are formed, which restrict water flow and allow corrosion to occur underneath. These problems can be treated with biocides.

Iron-oxidizing bacteria are also responsible for unsightly deposits on solids in rivers and streams.

(b) Acidic drainage water

Drainage water from coal and pyritic mines is characterized by low pH values and high concentrations of iron compounds, usually Fe(III), and other metals. If the drainage water contaminates streams or rivers, these too may become acid beyond the tolerance of the freshwater fauna, and excessive precipitation of ferric hydroxide may occur. The drainage water may be stored in settling ponds to allow deposition of the iron(III) compounds to occur. The combination of sodium lauryl sulphate and benzoic acid inhibited iron- and sulphur-oxidizing bacteria in coal refuse at a concentration of 1.1 mg (kg refuse)$^{-1}$ (Dugan, 1987a,b).

Streamers are sometimes found in acidic drainage water or contaminated waters. These are usually found to be aerobic heterotrophic bacteria trapped in extracellular polymeric material (Johnson, Kelso and Jenkins, 1979), but in a recent example the acid streamers were essentially chemolithotrophic in origin, with *Thiobacillus ferrooxidans* embedded in a gelatinous matrix (Wakao *et al.*, 1985).

7.5.3 Mobilization and accumulation of radionuclides

Radioactive waste from the reprocessing of nuclear fuels and other sources is usually disposed of in one of three ways:

1. Highly active waste is concentrated and stored;
2. Weakly active waste is diluted and dispersed, usually by pumping into the sea; and
3. Contaminated items are dumped in land fill sites.

Micro-organisms may interfere with these operations through the processes described in Sections 7.3 and 7.2.

(a) The possibility of uptake of radioactive cations by surface or intracellular binding

The presence of microbes in ground water flowing through landfill sites could result in the adsorption of radionuclides on to cell surfaces or their accumulation into the cell, and their eventual transport away from the landfill site. This possibility could be brought about by a rise in the level of the water table.

Radionuclides in effluent discharged to sea could also be concentrated by micro-organisms, and eventually enter the food chain. Polikarpov (1966) has demonstrated the accumulation of radionuclides by freshwater and marine organisms, particularly algae which often had an uptake several orders of magnitude greater than that of other organisms.

The effluent disposed to sea contains nitro and nitrato complexes of nitrosylruthenium(II), which undergo aquation over some months to give the pentaaquanitrosyl complex. These compounds are taken up by the edible alga *Porphyra umbilicalis*, accounting for the sale in the 1950s and early 1960s of radioactive laver bread. This problem has now been resolved, but there is still concern over the fate of these compounds in the environment. Gibson *et al.* (1986) have confirmed that marine bacteria take up these ruthenium compounds, but have shown that the bacteria are unable to compete with sediment, which binds the compounds more effectively. This is confirmed by the fact that the ruthenium compounds in effluents were not dispersed on being pumped out to sea, but remained in the vicinity bound to sediment.

(b) Possible mobilization of plutonium

Plutonium is present in surface soils in low concentrations (0.07–0.4 pg g^{-1}) due to fallout from nuclear testing. The long half life of plutonium (>20 000 years) means that the effect of soil organisms on the mobilization and concentration of plutonium must be considered carefully. Plutonium is present in soil as the insoluble, stable hydrolysis products of Pu(IV), but solubilization could result from complex formation with metabolites excreted by soil micro-organisms (Wilding and Garland, 1982).

(c) Disposal of highly active nuclear waste

The procedure for disposal of this waste has involved sealing the concentrate in steel drums, which are buried in concrete in deep, disused mines. However, there has been concern over the possibility that the action of chemolithotrophic organisms on the steel drums and concrete may over a long time period cause corrosion and leakage. More recently, these wastes have been sealed in glass, which is much more resistant to bacterial attack.

7.6 SUMMARY

Micro-organisms have important roles in the solubilization and laying down of minerals in the natural environment. Both processes can be exploited in the recovery of metals.

Certain autotrophic, acidophilic bacteria catalyse the oxidation of minerals, converting insoluble metal sulphides to soluble sulphates, with oxidation of the metal and production of acid. The best known of these organisms is *Thiobacillus ferrooxidans*, which functions at around 30–35°C, but other sulphide-oxidizing and iron-oxidizing bacteria are also important. Moderately-thermophilic and thermophilic organisms also effect these oxidations, with temperature optima around 45 and 70°C respectively. *Sulfolobus* is a well-known example of a thermophile. Bacterial leaching of low grade copper ores has been successfully carried out on a large scale, with the copper being solubilized through oxidation of Cu(I) to Cu(II) in some ores and the solubilization of Cu(II) by the slowly percolating acidic liquors. The action of the bacteria involves a direct attack on the surface of the mineral, with pitting and ultimate dissolution, and an indirect effect brought about by the Fe(III) produced by bacterial oxidation. The recovery of uranium from flooded worked-out stopes and tunnels is an example of 'in-place' leaching involving bacterially-produced Fe(III), which oxidizes insoluble UO_2 to soluble UO_2^{2+} which is recovered from the uranium-bearing solutions. These processes are carried out inefficiently as the detailed conditions cannot be controlled.

Most current interest centres on the recovery of high value metals. Gold sometimes occurs in a matrix of pyrites or arsenopyrite, and so cannot be extracted by cyanide-containing solutions. Autotrophic bacteria solubilize the sulphidic material allowing considerable enhancement of gold recovery, typically from 20–40% to over 95%. Cooling of the reactor is a major problem if the mesophilic *Thiobacillus ferrooxidans* is used, and attention is now focused on the use of moderate thermophiles and thermophiles. Commercial exploitation of this process seems very likely.

Removal or recovery of toxic or precious metals from wastes or factory process streams can be accomplished with micro-organisms. Waste, dead biomass binds metals on the cell surface, often with greater capacities than ion-exchange resins and, suitably immobilized or granulated, has been used commercially. Such material is effective in removing a variety of metals from aqueous environments, including gold, radionuclides and typical toxic metals. Cells may also excrete polymers (often polysaccharides) with significant metal-binding properties. Removal of metals from effluents can also involve cellular activity, such as accumulation in the cell, reduction of sulphate to sulphide and precipitation of metal sulphides, and reduction of mercury compounds to volatile Hg^0. There is also the possibility that cells could be engineered to overproduce metal-binding ligands such as the metallothioneins which could be exploited in metal recovery in several ways.

Metal-microbe interactions can cause a range of harmful effects in the environment, including the production of acidic drainage water, the corrosion of underground steel structures, and the oxidation of minerals with the undesired precipitation of metal oxides.

REFERENCES

Abel, T.H. and Barlocher, F. (1984) *Appl. Environ. Microbiol.*, **48**, 245–51.
Adachi, K. and Sells, B.H. (1979) *Biochim. Biophys. Acta*, **563**, 163–70.
Adams, B. and Pain, R.H. (1986) *FEBS Letters*, **196**, 361–4.
Adman, E.T., Sieker, L.C. and Jensen, L.H. (1973) *J. Biol. Chem.*, **248**, 3987–96.
Adman, E.T., Stenkamp, R.E., Sieker, L.C. and Jensen, L.H. (1978) *J. Mol. Biol.*, **123**, 35–47.
Agarwal, N. and Kalra, V.K. (1983) *Biochim. Biophys. Acta*, **727**, 285–92.
Aickin, R.M. and Dean, A.C.R. (1977) *Microbios. Letters*, **5**, 129–33.
Aickin, R.M. and Dean, A.C.R. (1979) *Microbios. Letters*, **9**, 55–56.
Aickin, R.M., Dean, A.C.R., Cheetham, A.K. and Skarnulis, A.J. (1979) *Microbios. Letters*, **9**, 7–15.
Aiking, H. and Tempest, D.W. (1976) *Arch. Microbiol.*, **108**, 117–24.
Aiking, H. and Tempest, D.W. (1977) *Arch. Microbiol.*, **115**, 215–21.
Aiking, H., Sterkenburg, A. and Tempest, D.W. (1977) *Arch. Microbiol.*, **113**, 65–72.
Aiking, H., Van Holst, G.J. and Tempest, D.W. (1977) *Arch. Microbiol.*, **115**, 79–84.
Aiking, H., Hok, K., van Heerikhuizen, H. and van't Riet J. (1982) *Appl. Environ. Microbiol.*, **44**, 938–44.
Aiking, H., Stijnman, A., van Garderen, C., van Heerikhuizen, H. *et al.* (1984) *Appl. Environ. Microbiol.*, **47**, 374–7.
Aiking, H., Govers, H. and van't Riet, J. (1985) *Appl. Environ. Microbiol.*, **50**, 1262–7.
Ainsworth, M.A., Tompsett, C.P. and Dean, A.C.R. (1980) *Microbios.*, **27**, 175–84.
Albert, A. (1979) *Selective Toxicity. The Physico-chemical Basis of Therapy*, (6th Edn), Chapman and Hall, London, pp. 385–442.
Albracht, S.P.J., Graf, E.-G. and Thauer, R.K. (1982) *FEBS Letters*, **140**, 311–13.
Albracht S.P.J., Kalkman, M.L. and Slater, E.C. (1983) *Biochim. Biophys. Acta*, **724**, 309–16.
Aldridge, W.N. (1976) *Am. Chem. Soc., Adv. in Chem. Ser.*, **157**, 186–96.
Ambler, R.P. and Brown, L.H. (1967) *Biochem. J.*, **104**, 784–825.
Ambudkar, S.V., Zlotnick, G.W. and Rosen, B.P. (1984) *J. Biol. Chem.*, **259**, 6142–6.
Amesz, J. (1983) *Biochim Biophys. Acta*, **726**, 1–12.
Amy, M.K. and Rajagopalan, K.V. (1979) *J. Bacteriol.*, **140**, 114–24.
Andersson, K.K., Kent, T.A., Lipscomb, J.D., Hooper, A.B. *et al.* (1984) *J. Biol. Chem.*, **259**, 6833–40.

Andreasen, T.J., Heideman, W., Rosenberg, G.B. and Storm, D.R. (1983) *Biochemistry*, **22**, 2757–62.
Andreesen, J.R. and Ljungdahl, L.G. (1973) *J. Bacteriol.*, **116**, 867–73.
Andrews, G.F. (1988) *Biotech. Bioeng*, **31**, 378–81.
Andrews, G.F. and Maczuga, J. (1982) *Biotech. Bioeng.*, **12**, 337–42.
Andrews, G.F. and Maczuga, J. (1984) *Fuel*, **63**, 297–304.
Appleby, C.A. (1984) *Ann. Rev. Plant Physiol.*, **35**, 443–78.
Arancia, G., Belli, S., Donelli, G. and Trovalusci, P. (1980) *J. Gen. Microbiol.*, **119**, 155–64.
Archibald, F.S. (1983) *FEMS Microbiol. Letters*, **19**, 29–32.
Archibald, F.S. and DeVoe, I.W. (1978) *J. Bacteriol.*, **136**, 35–48.
Archibald, F.S. and Fridovich, I. (1981a) *J. Bacteriol.*, **145**, 442–57.
Archibald, F.S. and Fridovich, I. (1981b) *J. Bacteriol.*, **146**, 928–36.
Archibald, F.S. and Fridovich, I. (1982a) *Arch. Biochem. Biophys.*, **215**, 589–96.
Archibald, F.S. and Fridovich, I. (1982b) *Arch. Biochem. Biophys.*, **214**, 452–63.
Archibald, F.S. and Duong, M.-N. (1984) *J. Bacteriol.*, **158**, 1–8.
Arcuri, E.J. and Ehrlich, H.L. (1979) *Appl. Environ. Microbiol.*, **37**, 915–23.
Arnold, W.N. and Garrison, R.G. (1979) *J. Biol. Chem.*, **254**, 4919–24.
Arnold, W.N., Evans, B.J. and Denniston, M.L. (1983) *J. Gen. Microbiol.*, **129**, 2351–8.
Arst, H.N., Tollervey, D.W. and Sealy-Lewis, H.M. (1982) *J. Gen. Microbiol.*, **128**, 1083–93.
Attramadal, A. and Svatun, B. (1980) *Acta Odontol. Scand.*, **38**, 349–54.
Averill. B.A. (1983) *Structure and Bonding*, **53**, 61–96.
Babich, H. and Stotzky, G. (1980) *CRC Critical Reviews in Microbiol.*, **8**, 99–145.
Babich, H. and Stotzky, G. (1982) *Ecotoxicology and Environ. Safety*, **6**, 577–89.
Babich, H., Schiffenbauer, M. and Stotzky, G. (1983) *Current Microbiology*, **8**, 101–5.
Babich, H., Bewley, R.J.F. and Stotzky, G. (1983) *Arch. Environ. Toxicol.*, **12**, 421–6.
Baes, F.C. and Mesmer, E.R. (1976) *The Hydrolysis of Cations*, Wiley, New York.
Bagdigian, R.M. and Myerson, A.S. (1986) *Biotech. Bioeng.*, **28**, 467–79.
Bagg, A. and Neilands, J.B. (1987) *Biochemistry*, **26**, 5471–7.
Bakker, E.P. (1978) *Biochemistry*, **17**, 2899–904.
Bakker, E.P. and Harold, F.M. (1980) *J. Biol. Chem.*, **255**, 433–40.
Bakker, E.P. and Mangerich, W.E. (1982) *FEBS Letters*, **140**, 177–80.
Bakker, E.P., Booth, R., Dinnbier, U., Epstein, W. and Gajewska, A. (1987). *J. Bacteriol.*, **169**, 3743–49.
Balakrishnan, M.S., Villafranca, J.J. and Brenchley, J.E. (1977) *Arch. Biochem. Biophys.*, **181**, 603–15.
Balakrishnan, M.S. and Villafranca, J.J. (1979) *Biochemistry*, **18**, 1546–51.
Baldry, M.G.C. and Dean, A.C.R. (1980) *Microbios. Letters.*, **15**, 83–7.
Baldry, M.G.C. and Dean, A.C.R. (1981) *Biotechnol. Letters*, **3**, 137–42.
Balschi, J.A., Cirillo, V.P. and Springer, C.S. (1982) *Biophys. J.*, **38**, 323–6.
Banks, E.D., Blake, C.C.F., Evans, P.R., Haser, R., Rice, D.W. *et al.* (1979) *Nature*, **279**, 773–6.
Barber, J. (1984) *Trends in Biochem. Sci.*, **9**, 79–82.
Barclay, R. and Ratledge, C. (1983) *J. Bacteriol.*, **153**, 1138–46.
Barkay, T. and Colwell, R.R. (1983) *J. Gen. Microbiol.*, **129**, 2945–50.
Barnes, R.D., Bull, A.T. and Poller, R.C. (1983) *Pestic. Sci.*, **4**, 305–17.
Barnett, H.L. and Lilly, V.G. (1966) *Mycologia*, **58**, 585–91.

Barratt, D.G. and Cook, R.A. (1978) *Biochemistry*, **17**, 1561–6.
Barrett, J., Ewart, D.K., Hughes, M.N., Nobar, A.M. *et al.* (1988a) *Proceedings of the Perth International Gold and Silver Conference* 148, Randol.
Barrett, J., Ewart, D.K., Hughes, M.N., Nobar, A.M., *et al.* (1988b) *New Technology and Gold Mining*, Western Australian School of Mines, p. 275.
Barug, D. (1981) *Chemosphere*, **10**, 1145–54.
Baryshev, V.A. (1982) *FEMS Microbiol. Letters*, **14**, 139–43.
Baryshev. V.A., Glagolev, A.N. and Skulachev, U.P. (1982) *FEMS Microbiol. Letters*, **13**, 47–50.
Bassilana, M., Damiano, E. and Leblanc, G. (1984) *Biochemistry*, **23**, 1015–22.
Bast, E. (1988) *Arch. Microbiol.*, **150**, 6–10.
Bayley, S.T. and Morton, R.A. (1978) *CRC Critical Rev. Microbiol.*, **6**, 151–205.
Beaman, B.L., Scates, S.M., Moring, S.E., Deem, R. *et al.* (1983) *J. Biol. Chem.*, **258**, 91–6.
Beck, D.J. and Fisch, J.E. (1980) *Mutation Research*, **77**, 45–54.
Beinert, H. (1978) *Methods in Enzymology*, **54**, 435–45.
Beinert, H., Emptage, M.H., Dryer, J.L., Scott, R.A. *et al.* (1983) *Proc. Natl. Acad. Sci. U.S.A.*, **80**, 393–6.
Beinert, H. and Thomson, A.J. (1983) *Arch. Biochem. Biophys.*, **222**, 333–61.
Bell, R.G. and Lim, C.K. (1981) *Can. J. Microbiol.*, **27**, 242–5.
Bell, S.H., Dickson, D.P.E., Johnson, C.E., Cammack, R. *et al.* (1982) *FEBS Letters*, **142**, 143–6.
Belliveau, J.W. and Lanyi, J.K. (1978) *Arch. Biochem. Biophys.*, **186**, 98–105.
Belly, R.T. and Kydd, G.C. (1982) *Dev. Ind. Microbiol.*, **23**, 567–77.
Beltramini, M. and Lerch, K. (1982) *FEBS Letters*, **142**, 219–22.
Benemann, J.R., McKenna, C.E., Lie, R.F., Trayler, T.G. *et al.* (1972) *Biochim. Biophys. Acta*, **264**, 25–38.
Bennett, J.C. and Tributsch, H. (1978) *J. Bacteriol.*, **134**, 310–17.
Benyoucef, M., Rigaud, J.-L. and Leblanc, G. (1982) *Biochem. J.*, **208**, 539–47.
Berk, S.G. and Colwell, R.R. (1981) *J. Exp. Mar. Biol. Ecol.*, **52**, 157–72.
Berman, E. (1980) *Toxic Metals and their Analysis*, Heyden, London.
Beveridge, T.J. (1986) In *Biotechnology and Bioengineering Symposium No.16*, (Eds. H.L. Ehrlich and D.S. Holmes), Wiley, New York. p.126.
Beveridge, T.J. and Murray, R.G.E. (1976) *J. Bacteriol.*, **127**, 1502–18.
Beveridge, T.J. and Koval, S.F. (1981) *Appl. Environ. Microbiol.*, **42**, 325–35.
Beveridge, T.J. and Jack, T. (1982) *J. Bacteriol.*, **149**, 1120–3.
Beveridge, T.J., Forsberg, C.W. and Doyle, R.J. (1982) *J. Bacteriol.*, **150**, 1438–48.
Beveridge, T.J., Meloche, J.D., Fyfe, W.S. and Murray, R.G.E. (1983) *Appl. Environ. Microbiol.*, **45**, 1094–108.
Beyer, M., Ebner, H.G. and Klein, J. (1986) *Appl. Microbiol. Biotechnol.*, **24**, 342–6.
Bezkorovainy, A. and Topouzian, N. (1983) *Int. J. Biochem.*, **15**, 361–6.
Bianchi, M.E., Carbone, M.L., Lucchini, G. and Magni, G.E. (1981) *Current Genetics*, **4**, 215–20.
Bilinski, C.A., Miller, J.J. and Girvitz, S.C. (1983) *J. Bacteriol.*, **155**, 1178–84.
Birchall, J.D. and Chappell, J.S. (1988) *Clinical Chemistry*, **34**, 265–7.
Blair, W.R., Olson, G.J., Brinckman, F.E. and Iverson, W.P. (1982) *Microbial Ecol.*, **8**, 241–51.
Black, G.M., Webb, C., Matthews, T.H. and Atkinson, B. (1984) *Biotech. Bioeng.* **26**, 134–41.
Blakemore, R.P. and Frankel, R.B. (1981) *Sci. Amer.*, **245**, 42–9.

Blakemore, R.P. (1982) *Ann. Rev. Microbiol.*, **36**, 217–38.
Blanchard, J.J. (1984) *Methods in Enzymology*, **104**, 404–14.
Blanco, M. (1982) *Biochim. Biophys. Acta*, **687**, 94–6.
Blond, D.M. and Whittam, R. (1965) *Biochem. J.*, **97**, 523–31.
Blum, H., Poole, R.K. and Ohnishi, T.C. (1980) *Biochem. J.*, **190**, 385–93.
Boggis, W., Kenward, M.A. and Brown, M.R.W. (1979) *J. Appl. Bacteriol.*, **47**, 477–88.
Bohner, H., Sandmann, G. and Böger, P. (1981) *Biochim. Biophys. Acta*, **636**, 65–9.
Bombelka, E., Richter, F-W., Stroh, A. and Kadenbach, B. (1986) *Biochem. Biophys. Res. Commun.*, **140**, 1007–14.
Bonhomme, A., Quintana, C. and Durand, M. (1980) *J. Protozool.*, **27**, 491–7.
Booth, G.H. and Wormwell, F. (1961) *Proc. First Int. Congress of Metallic Corrosion*, London, pp. 341–4.
Booth, G.H. (1968) *Process Biochem.*, **3**, 17–20.
Booth, I.R. (1985) *Microbiol. Rev.*, **49**, 359–78.
Booth, I.R. (1988) In *Bacterial Energy Transduction*, (Ed. C. Anthony), Academic Press, London, pp. 327–428.
Bopp, L.H., Chakrabarty, A.M. and Ehrlich, H.L. (1983) *J. Bacteriol.*, **155**, 1105–9.
Borbolla, M.G. and Rosen, B.P. (1984) *Arch. Biochem. Biophys.*, **229**, 98–103.
Bordas, J., Bray, R.C., Garner, C.D., Gutteridge, S. *et al.* (1980) *Biochem. J.*, **191**, 499–508.
Bordas, J., Koch, M.H.J., Hartmann, H.-J. and Weser, W. (1982) *FEBS Letters*, **140**, 19–21.
Borman, A.H., De Jong, E.W., Huizinga, M. and Westbroek, P. (1983) In *Biomineralization and Biological Metal Accumulation* (Eds Westbroek, P. and De Jong, E.W.) Reidel, pp. 303–5.
Bortels, H. (1936) *Zentr. Bakt Parasit. Kde. Abt.II*, **95**, 193–218.
Bosecker, K. (1984) In *Sulfur; its significance for Chemistry, for the Geo-Bio- and Cosmosphere and Technology* (Eds. A. Müller and B. Krebs) *Studies in Inorg. Chem.*, **5**, 331–48.
Bostian, K.A., Betts, G.F., Man, W.K. and Hughes, M.N. (1982) *Biochem. J.*, **207**, 73–80.
Bourret, J.A. (1982) *Exp. Mycol.*, **6**, 210–15.
Boutry, M., Foury, F. and Goffeau, A. (1977) *Biochim. Biophys. Acta*, **464**, 602–12.
Bowman, B.J., Allen, K.E. and Slayman, C.W. (1983) *J. Bacteriol.*, **153**, 292–6.
Bradley, R., Burt, A.J. and Read, D.J. (1981) *Nature*, **292**, 335–7.
Braun, V. (1985) *Trends Biochem. Sci.*, **10**, 75–8.
Braun, V., Burkhardt, R., Schneider, R. and Zimmermann, L. (1982) *J. Bacteriol.*, **151**, 553–9.
Braun, V., Brazel-Faisst, C. and Schneider, R. (1984) *FEMS Microbiol. Letters*, **21**, 99–103.
Brewer, J.M. (1981) *Crit. Rev. Biochem.*, **11**, 209–54.
Brewer, J.M., Carreira, L.A., Irwin, R.M. and Elliott, J.I. (1981) *J. Inorg. Biochem.*, **14**, 33–44.
Brey, R.N. and Rosen, B.P. (1979) *J. Biol. Chem.*, **254**, 1957–63.
Brey, R.N., Rosen, B.P. and Sorensen, E.N. (1980) *J. Biol. Chem.*, **255**, 39–44.
Bridgen, J., Harris, J.I. and Kolb, E. (1976) *J. Mol. Biol.*, **105**, 333–5.
Brierley, C.L. (1974) *J. Less Common Metals*, **36**, 237–47.

Brierley, C.L. (1978) *CRC Critical Rev. Microbiol.*, **5**, 207–62.
Brierley, C.L. (1982) *Microbiological Mining, Scientific American*. W.H. Freeman & Co., San Francisco, August, **247**, 44–53.
Brierley, C.L. and Brierley, J.A. (1973) *Can. J. Microbiol.*, **19**, 183–8.
Brierley, C.L. and Brierley, J.A. (1983) In *Biomineralisation and Biological Metal Accumulation*, (Eds. P. Westbroek and E.W. de Jong), Reidel, pp. 499–509.
Brierley, J.A., Goyak, G.M. and Brierley, C.L. (1986) Considerations for commercial use of natural products in immobilisation of ions by biosorption. In: *Biosorption* (Eds. H. Eccles and S. Hunt), Ellis Horwood, Chichester, pp. 105–117.
Brierley, J.A. and Lockwood, S.J. (1977) *FEMS Microbiol Letters*, **2**, 163–5.
Brock, T.D. (1986) (Ed.) *Thermophiles*, Wiley, New York.
Brock, T.D., Smith, D.W. and Madigan, M.T. (1984) *Biology of Micro-organisms*, (4th Edn), Prentice-Hall International Inc., London.
Brock, T.D., Brock, K.M., Belly, R.T. and Weiss, R.L. (1972) *Arch. Microbiol.*, **84**, 54–68.
Brock, T.D. and Gustafson, J. (1976) *Appl. Environ. Microbiol.*, **32**, 567–71.
Broda, E. (1972) *Ann. Microbiol. Enzymol.*, **22**, 93–108.
Bromfield, R.J., Dainty, R.H. Gillard, R.D. and Heaton, B.T. (1969) *Nature*, **223**, 735–6.
Brown, A.E. and Swinburne, T.R. (1981) *Trans. Br. Mycol. Soc.*, **77**, 119–24.
Brown, D.A. and Cook, R.A. (1981) *Biochemistry*, **20**, 2503–12.
Brown, H,J. and Gibbons N.E. (1955) *Can. J. Microbiol.*, **1**, 486–94.
Brown, M.J. and Lester, J.N. (1979) *Water Research*, **13**, 817–37.
Brown, M.J. and Lester, J.N. (1980) *Appl. Environ. Microbiol.*, **40**, 179–85.
Brown, M.J. and Lester, J.N. (1982a) *Water Research*, **16**, 1539–48.
Brown, M.J. and Lester, J.N. (1982b) *Water Research*, **16**, 1549–60.
Brown, M.R.W. and Melling, J. (1969) *J. Gen. Microbiol.*, **59**, 263–74.
Brown, S.B. (1980) (Ed.) *An Introduction to Spectroscopy for Biochemists*, Academic Press, London.
Brown, T.A. and Shrift, A. (1982) *Can. J. Microbiol.*, **28**, 307–10.
Bruynesteyn, A., Hackl, R.P. and Wright, F. (1986) *Gold 100. Proceedings of the International Conference on Gold*, Vol. 2, SAIMM, Johannesburg, pp. 353–65.
Bucheder, F. and Broda, E. (1974) *Eur. J. Biochem.*, **45**, 555–9.
Buckel, W. and Semmler, R. (1982) *FEBS Letters*, **148**, 35–8.
Bull, C., Ballou, D.P. and Otsuka, S. (1981) *J. Biol. Chem.*, **256**, 12681–6.
Bulman, R.A. (1980) *Coord. Chem. Revs.*, **31**, 221–50.
Burger, D., Cox, J.A., Comte, M. and Stein, E.A. (1984) *Biochemistry*, **23**, 1966–71.
Burns, R.O., Umbarger, H.E. and Gross, S.R. (1963) *Biochemistry*, **2**, 1053–8.
Butt, T.R. and Ecker, D.J. (1987) *Microbiol. Rev.*, **51**, 351–64.
Butt, T.R., Sternberg, E.V., Gorman, J.A., Clarke, P. *et al.* (1984) *Proc. Natl. Acad. Sci. USA*, **81**, 3332–6.
Bunrman, E.T., Pennock, J., Tempest, D.W., Teixeira de Mattos, M.J. and Neijssel, O.M. (1989) *Archiv. Microbiol.* In press.
Byers, B.R. and Arceneaux, J.E.L. (1977) In *Micro-organisms and Minerals*, (Ed. L.D. Weinberg), Marcel Dekker Inc., New York, pp. 215–49.
Byrd. J.T. and Andreate, M.O. (1982) *Science*, **218**, 565–9.
Byrt, P.N., Irving, H.R. and Grant, B.R. (1982) *J. Gen. Microbiol.*, **128**, 1189–98.
Cain, K. and Griffiths, D.E. (1977) *Biochem. J.*, **162**, 575–80.

Cain, K., Partis, M.D. and Griffiths, D.E. (1977) *Biochem. J.*, **166**, 593–602.
Camacho, M., Ramos, J. and Rodriguez-Navarro, A. (1981) *Current Microbiol.*, **6**, 295–9.
Camadro, J.-M. and Labbe. P. (1982) *Biochim. Biophys. Acta*, **707**, 280–8.
Cammack, R. (1988) *Nature*, **333**, 595–6.
Cammack, R., Rao, K.K. and Hall, D.O. (1981) *Biosystems*, **14**, 57–80.
Caradonna, J.P., Lippard, S.J., Gait, M.J. *et al.* (1982) J. Am. Chem. Soc., **104**, 5793–5.
Carafoli, E. and Crompton, M. (1978) *Current Topics in Membranes and Transport*, **10**, 151–216.
Carter, C.W. (1977) *J. Biol. Chem.*, **252**, 7802–11.
Casey, J.D. and Wu, Y.C. (1977) *Proc. 32nd Ind. Waste Conf. Purdue Univ.*, pp. 141–52.
Cass, A.E.G., Galdes, A., Hill, H.A.O., McClelland, C.E. *et al.* (1978) *FEBS Letters*, **94**, 311–4.
Cavallo, J.J. and Hines, P.L. (1971) *U.S. Patent 3616 249* (Oct. 26 1971; assigned to Atlantic Richfield Co.).
Chan, K., Leung, O.C. and Lee, L.H. (1979) *Microbios*, **24**, 81–91.
Chance, B., Powers, L., Ching, Y., Poulos, T. *et al.* (1984) *Archiv. Biochem. Biophys.*, **235**, 596–611.
Chang, Y.Y. and Myerson, A.S. (1982) *Biotechnol. Bioeng.*, **24**, 889–902.
Charley, R.C. and Bull, A.T. (1979) *Arch. Microbiol.*, **123**, 239–44.
Charlson, A.J., McArdle, N.T. and Watton, E.C. (1981) *Inorg. Chim. Acta*, **56**, L35–L36.
Charp, P.A. and Whitson, G.L. (1982) In: *Genetic Expression in the Cell Cycle* (Eds. G.M. Padilla and K.S. McCarthy), Academic Press, New York, pp. 337–61.
Chemistry in Britain, (1986) *Chelation: DeVoe-Holbein process unveiled*. pp. 893–4.
Chen, C-M., Misra, T.K., Silver, S. and Rosen, B.P. (1986) *J. Biol. Chem.*, **261**, 15030–8.
Cheung, W.Y. (1983) (Ed.) *Calcium and Cell Function*, vols. 1–4. Academic Press, Orlando.
Choudhary, A. and Pirt, S.J. (1966) *J. Gen. Microbiol.*, **43**, 71–81.
Christian, J.H.B. and Waltho, J.A. (1962) *Biochim. Biophys. Acta*, **65**, 506–8.
Christner, J.A., Munck, E., Janick, P.A. and Siegel, L.M. (1983) *J. Biol. Chem.*, **258**, 11147–56.
Claiborne, A. and Fridovich, I. (1979) *J. Biol. Chem.*, **254**, 4245–52.
Claiborne, A., Malinowski, P. and Fridovich, I. (1979) *J. Biol. Chem.*, **254**, 11664–8.
Clare, D.A., Blum, J. and Fridovich, I. (1984) *J. Biol. Chem.*, **259**, 5932–6.
Clark, R.J.H. and Hester, R.E. (1988) (Eds.) *Advances in Infrared and Raman Spectroscopy*, John Wiley, Chichester.
Clarke, M., Bazari, W.L. and Kayman, S.C. (1980) *J. Bacteriol.*, **141**, 397–400.
Cleare, M.J. (1974) *Coord. Chem. Rev.*, **12**, 349–405.
Clegg, R.A. and Garland, P.B. (1971) *Biochem. J.*, **124**, 135–54.
Cleland, W.W. (1982) *Methods Enzymol.*, **87**, 159–63.
Cloud, P. (1969) Mineral resourçes from the sea, in *Resources and Man*, W.H. Freeman & Co., San Francisco.
Coakley, W.T., Bater, A.J. and Lloyd, D. (1977) *Adv. Microb. Physiol.*, **16**, 279–341.
Cobley, J.G. and Haddock, B.A. (1975) *FEBS Letters*, **60**, 29–33.

Cocucci, M.C. and Rossi, G. (1972) *Arch. Microbiol.*, **85**, 267–79.
Cohen, S., Oren, A. and Shilo, M. (1983) *Arch. Microbiol.*, **136**, 184–90.
Cole, S.A. and Smith, D.G. (1981) *FEMS Microbiol. Letters*, **10**, 17–20.
Coleman, J.E. and Gettins, P. (1983) *Adv. Enzymol. Rel. Areas Mol. Biol.*, **55**, 381–403.
Corbett, C.M. and Ingledew, W.J. (1987) *FEMS Microbiol. Letters*, **41**, 1–6.
Corpe, W.A. (1975) *Dev. Ind. Microbiol.*, **16**, 249–55.
Costerton, J.W., Geesey, G.G. and Cheng, K.J. (1978) *Sci. Amer.*, **238** (1), 86–95.
Cotton, F.A., Hazen, E.E. and Legg, M.J. (1979) *Proc. Natl. Acad. Sci. USA*, **76**, 2551–3.
Cotton, F.A. and Wilkinson, G. (1980) *Advanced Inorganic Chemistry* (4th Edn), Wiley-Interscience, Chichester.
Coughlin, R.T., Caldwell, C.R., Haug, A. and McGroarty, E.J. (1981) *Biochem. Biophys. Res. Commun.*, **100**, 1137–42.
Coughlin, R.T., Tonsager, S. and McGroarty, E.J. (1983) *Biochemistry*, **22**, 2002–7.
Cowart, R.E. and Foster, B.G. (1981) *Current Microbiol.*, **6**, 287–90.
Cox, C.D. (1980) *J. Bacteriol.*, **141**, 199–204.
Cox, J.A., Ferraz, C., Demaille, J.G., Perez, R.O. *et al.* (1982) *J. Biol Chem.*, **257**, 10694–700.
Cox, J.C. and Boxer, D.H. (1978) *Biochem. J.*, **174**, 497–502.
Craig, P.A. and Dekker, E.E. (1986) *Biochemistry*, **25**, 1870–6.
Cramer, S.P., Gillum, W.O., Hodgson, K.D., Mortenson, L.E. *et al.* (1978) *J. Am. Chem. Soc.*, **100**, 3814–19.
Cramer, S.P., Wahl. R. and Rajagopalan, R.V. (1981) *J. Amer. Chem, Soc.*, **103**, 7721–7.
Cramer, S.P., Solomonson, L.P., Adams, M.W.W. and Mortensen, L.E. (1984) *J. Amer. Chem. Soc.*, **106**, 1467–71.
Crane, R.K. (1977) *Rev. Physiol. Biochem. Pharmacol.*, **78**, 99–159.
Crawford, B.J., Talburt, D.E. and Johnson, D.A. (1974) *Bioinorg. Chem.*, **3**, 121–33.
Crawford, I.P. and Ito, J. (1964) *Proc. Natl. Acad. Sci. USA*, **51**, 390–4.
Crichton, R.R. and Charloteaux-Wauters, M. (1987) *Eur. J. Biochem.*, **164**, 485–506.
Cronan, D.S. (1980) *Underwater Minerals*. Academic Press, London pp. 61–166.
Cséke, C., Horváth, L.I., Simon, P., Borbely, G. *et al.* (1979) *J. Biochem. (Japan)*, **85**, 1397–404.
Curtin, M.E. (1983) *Biotechnology*, May, 229–31.
Cypionka, H. and Meyer, O. (1983) *Arch. Microbiol.*, **135**, 293–9.
Dailey, H.A. and Lascelles, J. (1977) *J. Bacteriol.*, **129**, 815–20.
Dalgarno, D., Klevitt, R.E., Levine, B.A. and Williams, R.J.P. (1984) *Trends Pharm. Sci.*, **5**, 266–8.
Dalton, H. and Leak, D.J. (1985) In *Microbiol Gas Metabolism: Mechanistic, Metabolic and Biotechnological Aspects*, (Eds. R.K. Poole and C.S. Dow), Academic Press, Orlando, pp. 173–208.
Damper, P.D., Epstein, W., Rosen, B.P. and Sorensen, E.N. (1979) *Biochemistry*, **18**, 4165–9.
Danno, G. (1970) *Agric. Biol. Chem.*, **34**, 1805–14.
Darnall, D.W., Greene, B., Henzl, J,M., McPherson, R.A. *et al.* (1986) *Env. Science. Technol.*, **20**, 206–8.

Date, A.R. and Gray, A.L. (1983) *Spectrochim. Acta*, **38B**, 1256–62.
Davis, T.N., Urdea, M.S., Masiarz, F.R. and Thorner, J. (1986) *Cell*, **47**, 423–31.
Davis, W.B., McCauley, M.J. and Byers, B.R. (1971) *J. Bacteriol.*, **105**, 589–94.
Davison, J. (1988) *Biotechnology*, **6**, 282–6.
De Boer, E., van Kooyk, Y., Tromp, M.G.M., Plat, H. *et al.* (1986) *Biochim. Biophys. Acta*, **869**, 48–53.
De Boer, E., Boon, K. and Wever, R. (1988) *Biochemistry*, **27**, 1629–35.
De Bruyn, J., Weckx, M. and Beumer-Jochmans, M.-P. (1981) *J. Gen. Microbiol.*, **124**, 353–7.
DeCarlo, E.H., Zeitlin, H. and Fernando, Q. (1982a) *Anal. Chem.*, **54**, 898–902.
DeCarlo, E.H., Bleasdell, B.D., Zeitlin, H. and Fernando, Q. (1982b) *Separation Science and Technol.*, **17**, 1205–18.
DeCastro, A.F. and Ehrlich, H.L. (1970) *Antonie van Leeuwenhock*, **36**, 317–27.
Degn, H., Lloyd, D. and Hill, G.C. (1977) (Eds.) *Functions of Alternative Terminal Oxidases*, Pergamon Press, Oxford.
Deitch, E.A., Marino, A.A., Gillespie, T.E. and Albright, J.A. (1983) *Antimicrobial Agents and Chemotherapy*, **23**, 356–9.
Del Campillo-Campbell, A. and Campbell, A. (1982) *J. Bacteriol.*, **149**, 469–78.
Demange, P., Bateman, A., Dell, A. and Abdallah, M.A. (1988) *Biochemistry*, **27**, 2745–52.
Derks, W.J.G. and Borst-Pauwels, G.W.F.H. (1980) *Biochim. Biophys. Acta*, **596**, 381–92.
De Rome, L. and Gadd, G.M. (1987a) *FEMS Microbiol. Letters*, **43**, 283–7.
De Rome, L. and Gadd, G.M. (1987b) *Appl. Microbiol. Biotech.*, **26**, 84–90.
De Rosa, M., Gambacorta, A. and Bu'Lock, J.D. (1975) *J. Gen. Microbiol.*, **86**, 156–64.
Devés, R. and Brodie, A.F. (1981) *Mol. Cell. Biochem.*, **36**, 65–84.
Dibrov, P.A., Lazarova, R.L., Skulachev, V.P. and Verkhovskayo, M.L. (1986) *Biochim. Biophys. Acta*, **850**, 458–65.
Diekert, G. and Ritter, M. (1982) J. Bacteriol., **151**, 1043–5.
Diekert, G., Graf, E.G. and Thauer, R.K. (1979) *Arch. Microbiol.*, **122**, 117–20.
Diekert, G., Klee, B. and Thauer, R.K. (1980) *Arch. Microbiol.*, **124**, 103–6.
Diekert, G., Gilles, H.-H., Jaenchen, R. and Thauer, R.K. (1980) *Arch. Microbiol.*, **128**, 256–62.
Diekert, G., Konheiser, U., Piechulla, K. and Thauer, R.K. (1981) *J. Bacteriol.*, **148**, 459–64.
Dilworth, G.L. (1983) *Arch. Biochem. Biophys.*, **221**, 565–9.
Dimroth, P. (1980) *FEBS Letters*, **122**, 234–6.
Dimroth, P. (1982) *Eur. J. Biochem.*, **121**, 443–9.
Dimroth, P. (1987) *Microbiol. Rev.*, **51**, 320–40.
Dismukes, G.C. (1986) *Photochem. Photobiol.*, **43**, 99–114.
DiSpirito, A.A. and Tuovinen, O.H. (1982a) *Arch. Microbiol.*, **133**, 28–32.
DiSpirito, A.A. and Tuovinen, O.H. (1982b) Arch. Microbiol., **133**, 33–7.
DiSpirito, A.A., Talnagi, J.W. and Tuovinen, O.H. (1983) *Arch. Microbiol.*, **135**, 250–3.
DiSpirito, A.A., Taaffe, L.R., Lipscomb, J.D. and Hooper, A.B. (1985) *Biochim. Biophys. Acta*, **827**, 320–6.
Dissanayake, C.B. and Kristsotakis, K. (1984) *Chem. Geol.*, **42**, 61–76.
Ditlow, C., Johansen, J.T., Martin, B.M. and Svendsen, I.B. (1982) *Carlsberg Res. Commun.*, **47**, 81–91.

Domek, M.J., Lechevallier, M.W., Cameron, S.C. and McFeters, G.A. (1984) *Appl. Environ. Microbiol.*, **48**, 289–93.
Donard, O.F.X. and Weber, J.H. (1988) *Nature*, **332**, 339–41.
Dougherty, H.W., Sadowski, S.J. and Baker, E.E. (1978) *J. Biol. Chem.*, **253**, 5220–3.
Dounce, A.L. and Flagg, J.F. (1949) The chemistry of uranium compounds, in: *Pharmacology and Toxicology of Uranium Compounds* (Eds. C. Voegtlin and H.C. Hodge), McGraw Hill, New York.
Dow, D.S., Walton, K.G. and Fleischer, S. (1970) *J. Bioenerg.*, **1**, 247–71.
Downie, J.A. and Garland, P.B. (1973) *Biochem. J.*, **134**, 1051–61.
Doyle, R.J., Matthews, T.H. and Streips, U.N. (1980) *J. Bacteriol.*, **143**, 471–80.
Drea, W.F. (1956) *Am. Rev. Tuberculosis*, **74**, 145–6.
Druyan, R. and Vallee, B.L. (1962) *Fed. Proc.*, **21**, 247.
Dubinina, G.A. (1978) *Microbiology*, **47**, 471–8.
Dugan, P.R. (1986) In *'Workshop on Biotechnology for the Mining, Metal Refining and Fossil Fuel Processing Industries.'* (Eds. H.L. Ehrlich and D.S. Holmes) *Biotech. Bioeng. Symp.* No. 16, pp. 185–203, John Wiley, London.
Dugan, P.R. (1987a) *Biotech. Bioeng.*, **29**, 41–8.
Dugan, P.R. (1987b) *Biotech. Bioeng.*, **29**, 49–54.
Dunaway-Mariano, D., Benovic, J.L., Cleland, W.W., Gupta, R.K. *et al.* (1979) *Biochemistry*, **18**, 4347–53.
Dunbar, J.C., Holmquist, B. and Johansen, J.T. (1984) *Biochemistry*, **23**, 4324–9.
Dunbar, J.C., Johansen, J.T. and Uchida, T. (1982) *Carlsberg Res. Commun.*, **47**, 163–71.
Dunn, G.M. and Bull, A.T. (1983) *Eur. J. Appl. Microbiol. Biotech.*, **17**, 30–4.
Dupourque, D., Newton, W.A. and Snell, E.E. (1966) *J. Biol. Chem.*, **241**, 1233–8.
Durig, J.R., Danneman J., Behnke, W.D. and Mercer, E.E. (1976) *Chemico-Biol. Interactions*, **13**, 287–94.
Dürre, P. and Andreesen, J.R. (1982) *J. Gen. Microbiol.*, **128**, 1457–66.
Eady, R.R. and Robson, R.L. (1984) *Biochem J.*, **224**, 853–62.
Ebdon, L. (1982) *Introduction to Atomic Absorption Spectroscopy*, Heyden, London.
Eccles, H. and Hunt, S. (1986) *'Immobilisation of Ions by Biosorption'*, Ellis Horwood, Chichester.
Ecker, D.J., Lancaster, J.R. and Emery, T. (1982) *J. Biol. Chem.*, **257**, 8623–6.
Efthymiou, C.J. and Joseph, S.W. (1972) *J. Bacteriol.*, **112**, 627–8.
Ehrlich, H.L. (1975) *Soil Sci.*, **119**, 36–41.
Ehrlich, H.L. (1982) *Can. J. Microbiol.*, **28**, 1389–95.
Ehrlich, H.L. and Holmes, D.S. (1986) *Biotech. Bioeng. Symp.*, **16**, John Wiley.
Eichhorn, G.L. (1980) In *Metal Ions in Biological Systems* **10**, (Ed. H. Siegel), Marcel Dekker Inc., New York, pp. 1–22.
Eichhorn, G.L.(1981) In: *Advances in Inorganic Biochemistry*, (Eds. G.L. Eichhorn and L.G. Marzilli) **3**, Elsevier/North Holland, pp. 1–46.
Eichhorn, G.L. and Marzilli, L.G. (1981) (Eds.) *Metal Ions in Genetic Information Transfer, Adv. Inorg. Biochem.*, **3**, Elsevier, North Holland.
Eilam, Y., Lavi, H. and Grossowicz, N. (1985) *J. Gen. Microbiol.*, **131**, 623–9.
Eisbach, P., Weiss, J. and Kao, L. (1985) *J. Biol. Chem.*, **260**, 1618–22.
Ellefson, W.L., Whitman, W.B. and Wolfe, R.S. (1982) *Proc. Natl. Acad. Sci. U.S.A.*, **79**, 3707–10.
Ellfolk, N. and Soininen, R. (1971) *Acta Chem. Scand.*, **25**, 1535–40.
Ellfolk, N., Rönnberg, M., Aasa, R., Andreasen, L.-E. and Vänngard, T. (1983)

Biochim. Biophys. Acta, **743**, 23–30.
Elliott, B.B. and Mortenson, L.E. (1975) *J. Bacteriol.*, **124**, 1295–301.
Elliott, C.G. (1986) *J. Gen. Microbiol.*, **132**, 2781–5.
Elsbach, P., Weiss, J. and Kao, L. (1985) *J. Biol. Chem.*, **260**, 1618–22.
El Solh, N. and Ehrlich, S.D. (1982) *Plasmid*, **7**, 77–84.
Emery, T. and Hoffer, P.B. (1980) *J. Nucl. Med.*, **21**, 935–9.
Emmel, T., Sand, W., Konig, W.A. and Bock, E. (1986) *J. Gen. Microbiol.*, **132** 3415–20.
England, R.R. and Evans, E.H. (1983) *Biochem. J.*, **210**, 473–6.
Epstein, W., Whitelaw, V. and Hesse, J. (1978) *J. Biol. Chem.*, **253**, 6666–8.
Epstein, W. and Laimins, L. (1980) *Trends in Biochem. Sci.*, **5**, 21–3.
Erardi, F.X., Failla, M.L. and Falkinham, J.O. (1987) *Appl. Environ. Microbiol.*, **53**, 1951–4.
Erecinska, M., Deutsch, C.J. and Davis, J.S. (1981) *J. Biol. Chem.*, **256**, 278–84.
Erickson, L.E., Fan, L.T., Shah, P.S. and Chan, M.S.K. (1970) *Biotech. Bioeng.*, **12**, 713–46.
Espejo, R.T. and Ruiz, P. (1987) *Biotech. Bioeng.*, **30**, 586–92.
Ezra, F.S., Lucas, D.S. and Russell, A.F. (1984) *Biochim. Biophys. Acta*, **803**, 263–76.
Fahnestock, S., Erdman, V. and Nomura, M. (1974) *Methods in Enzymology*, **30**, 544–62.
Failla, M.L. (1977) In *Microorganisms and Minerals* (Ed. E.D. Weinberg) Marcel Dekker Inc., New York, pp. 151–214.
Failla, M.L., Benedict, C.D. and Weinberg, E.D. (1976) *J. Gen. Microbiol.*, **94**, 23–36.
Failla, M.L. and Weinberg, E.D. (1977) *J. Gen. Microbiol.*, **99**, 85–97.
Falchuk, K.H., Fawcett, D.W. and Vallee, B.L. (1975a) *J. Cell Sci.*, **17**, 57–78.
Falchuk, K.H., Krishnan, A. and Vallee, B.L. (1975b) *Biochemistry*, **14**, 3439–44.
Fanchiang, Y.-T., Ridley, W.P. and Wood, J.M. (1979) *J. Am. Chem. Soc.*, **101**, 1442–7.
Farquhar, G.B. (1974) *Corrosion 1974*. National Association of Corrosion Engineers' Annual Conference. Houston, Texas.
Fassel, V.A. (1978) *Science*, **202**, 183–91.
Fekete, F.A. Spence, J.T. and Emery, T. (1983) *Appl. Environ. Microbiol.*, **46**, 1297–300.
Fenchal, T. and Blackburn, T.H. (1979) *Bacteria and Mineral Cycling*, Academic Press, London and New York.
Ferguson, C.A., Murray, R.G.E. and Lancy, P. (1979) *Can. J. Microbiol.*, **25**, 545–59.
Fernandez, E.J. and Clark, D.S. (1987) *Enzyme Microb, Technol.*, **9**, 259–71.
Fernandez, M.P., Correa, J.U. and Cabib, E. (1982) *J. Bacteriol.*, **152**, 1255–64.
Flatau, G.N. and Gauthier, M.J. (1983) *Can. J. Microbiol.*, **29**, 210–17.
Fletcher, J. (1983) *J. Gen. Microbiol.*, **129**, 809–12.
Fogel, S., Welch, J.W., Cathala, G. and Karin, M. (1983) *Current Genetics*, **7**, 347–55.
Foster, P.L. (1977) *Nature*, **269**, 322–3.
Foster, T.J. (1983) *Microbiol. Revs.*, **47**, 361–409.
Foster, T.J. (1987) *Crit. Rev. Microbiol.*, **15**, 117–40.
Foury, F., Boutry, M. and Goffeau, A. (1977) *J. Biol. Chem.*, **252**, 4577–83.
Foye, W.O. (1977) In *Microorganisms and Minerals* (Ed. E.D. Weinberg), Marcel Dekker Inc., New York, pp. 387–419.

Franco, A.A., Peres, J.R.R. and Nery, M. (1978) *Plant and Soil*, **50**, 1–11.
Frankel, R.B. (1984) *Ann. Rev. Biophys. Bioeng.*, **13**, 85–103.
Fridovich, I. (1981) In *Oxygen and Living Processes. An Interdisciplinary Approach* (Ed. D.L. Gilbert), Springer Verlag, New York, pp. 250–72.
Fridovich, I. (1982) *Biochem. Soc. Trans.*, **10**, 67–8.
Friedman, B.A. and Dugan, P.R. (1968) *Dev. Ind. Microbiol.*, **9**, 381–8.
Friedrich, C.C., Schneider, K. and Friedrich, B. (1982) *J. Bacteriol.*, **152**, 42–8.
Fries, L. (1982) *Planta*, **154**, 393–6.
Fukuyama, T. and Ordal, E.J. (1965) *J. Bacteriol.*, **90** 673–80.
Furie, B.C., Borowski, M., Keyt, B. and Furie, B. (1982) In: *Calcium and Cell Function* (Ed. W.Y. Cheung) Academic Press, New York. **2**, 217–232.
Gadd, G.M. and Griffiths, A.J. (1978) *Microbiol. Ecol.*, **4**, 303–17.
Gadd, G.M. and Griffiths, A.J. (1980) *Trans. Br. Mycol. Soc.*, **75**, 91–6.
Gadd, G.M. and Mowll, J.L. (1983) *FEMS Microbiol. Letters*, **16**, 45–8.
Gadd, G.M., Chudek, J.A., Foster, R. and Reed, R.H. (1984) *J. Gen. Microbiol.*, **130**, 1969–75.
Gadd, G.M. (1986) In *Immobilisation of Ions by Biosorption* (Eds. H. Eccles and S. Hunt) Ellis Horwood, Chichester, pp. 135–47.
Gaines, C.G., Lodge, J.S., Arceneaux, J.E.L. and Byers, B.R. (1981) *J. Bacteriol.*, **148**, 527–33.
Galbraith, J.C. and Smith, J.E. (1969) *Trans. Br. Mycol. Soc.*, **52**, 237–46.
Galun, M., Keller, P., Malki, D., Feldstein, H. *et al.* (1983) *Science*, **219**, 285–6.
Galun, M., Keller, P., Malki, D., Feldstein, H. *et al.* (1984) *Water, Air, Soil Pollution*, **21**, 411–14.
Gangola, P. and Rosen, B.P. (1987) *J. Biol. Chem.*, **262**, 12570–4.
Garland, P.B. (1970) *Biochem. J.*, **118**, 329–39.
Garty, H. and Caplan, S.R. (1977) *Biochim. Biophys. Acta*, **459**, 532–45.
Gaunt, D.M., Trinci, A.J.P. and Lynch, J.M. (1984) *Trans. Br. Mycol. Soc.*, **83**, 575–81.
Gaynor, J.J. (1982) *Arch. Biochem. Biophys.*, **218**, 309–19.
Gettins, P. and Coleman, J.E. (1983) *J. Biol. Chem.*, **258**, 396–402.
Ghandour, W., Hubbard, J.A., Deistung, J., Hughes, M.N. *et al.* (1988) *Appl. Microbiol. Biotechnol.*, **28**, 559–64.
Ghiorse, W.C. (1980) in *Biogeochemistry of Ancient and Modern Environments* (Eds. P.A. Trudinger, M.R. Walter and B.J. Ralph). Aust. Acad. Sci., Canberra, and Springer Verlag, New York, pp. 345–54.
Ghiorse, W.C. (1984) *Ann. Rev. Microbiol.*, **38**, 515–50.
Ghiorse, W.C. (1986) *Biotech. Bioeng. Symp.*, **16**, 141–5.
Ghiorse, W.C. and Hirsch, P. (1982) *Appl. Environ. Microbiol.*, **43**, 1464–72.
Ghosh, D., O'Donnell, S., Furey, W., Robbins, A.H. *et al.* (1982) *J. Mol. Biol.*, **158**, 73–84.
Gibson, J.F., Hadfield, S.G., Hughes, M.N. and Poole R.K. (1980) *J. Gen. Microbiol.*, **116**, 99–110.
Gibson, J.F., Poole, R.K., Hughes, M.N. and Rees, J.F. (1982) *J. Gen. Microbiol.*, **128**, 2211–14.
Gibson, J.F., Poole R.K., Hughes, M.N. and Rees, J.F. (1984a) in *Role of Microorganisms on the Behaviour of Radionuclides in Aquatic and Terrestrial Systems and their Transfer to Man.* (Eds. E. Bonnyns-van Gelder and R. Kirchmann), International Union of Radioecologists, Brussels pp. 82–91.
Gibson, J.F., Poole, R.K., Hughes, M.N. and Rees, J.F. (1984b) *Arch. Microbiol.*, **139**, 265–71.

Gibson, J.F., Hughes, M.N., Poole, R.K. and Rees, J.F. (1985) *Chem.-Biol Interactions*, **53**, 351–64.

Gibson, J.F., Poole, R.K., Hughes, M.N. and Rees, J.F. (1986) *Arch. Environ. Cont. & Toxicology*, **15**, 519–28.

Gilbert, S.R., Bounds, C.O. and Ice, R.R. (1988) *CIM Bulletin*, 89–94.

Gilles, H. and Thauer, R.K. (1983) *Eur. J. Biochem.*, **135**, 109–12.

Giri, L. and Ramadoss, C.S. (1979) *J. Biol. Chem.*, **254**, 11703–12.

Giordano, G., Haddock, B.A. and Boxer, D.H. (1980) FEMS Microbiol. Letters, **8**, 229–35.

Giordano, G., Violet, M., Medani, C-L. and Pommier, J. (1984) *Biochim. Biophys. Acta*, **798**, 216–25.

Glazer, A.N. (1982) *Ann. Rev. Microbiol.*, **36**, 173–98.

Glazer, A.N. (1983) *Ann. Rev. Biochem.*, **52**, 125–57.

Gleason, F.K. and Hogenkamp, H.P.C. (1972) *Biochim. Biophys. Acta*, **277**, 466–70.

Gleason, F.K. and Frick, T.D. (1980) *J. Biol. Chem.*, **255**, 7728–33.

Gokcay, C.F. and Yirteri, R.N. (1983) *Fuel*, **62**, 1223–4.

Goldberg, M. and Gilboa, H. (1978) *Biochim. Biophys. Acta*, **538**, 268–83.

Gomez-Puyou, A., Sandoval, F., Tuena De Gomez-Puyou, M., Pena, A. *et al. (1972) Biochemistry*, **11**, 97–102.

Gonzalez, M.A., Webb, M.R., Welsh, K.M. and Cooperman, B.S. (1984) *Biochemistry*, **23**, 797–801.

Görrell, T.E., Yarlett, N. and Müller, M. (1984) *Carlsberg Res. Commun.*, **49**, 259–68.

Gorisch, H., Goss, D.J. and Parkhurst, L.J. (1976) *Biochemistry*, **15**, 5743–53.

Gorman, D.S. and Levine, R.P. (1966) *Plant Physiol.*, **41**, 1637–42.

Gormely, L.S., Duncan, D.W., Branion, R.M.R. and Pinder, K.L. (1975) *Biotechnol. Bioeng.*, **17**, 31–49.

Goss, D.J., Parkhurst, L.J. and Wahba, A.J. (1980) *J. Biol. Chem.*, **255**, 225–9.

Grant, C.L. and Pramer, D. (1962) *J. Bacteriol.*, **84**, 869–70.

Gray, T.A., Garland, P.B., Lowe, D.J. and Bray, R.C. (1975) *Biochem J.*, **146**, 239–46.

Greenwood, C. and Hill, H.A.O. (1982) *Chemistry in Britain*, **18**, 7–9.

Griffin, H.G., Foster, T.J., Silver, S. and Misra, T.K. (1987) *Proc. Natl. Acad. Sci. USA*, **84**, 5106–10.

Gromet-Elhanan, Z. and Khananshvili, D. (1984) *Biochemistry*, **23**, 1022–6.

Grosovsky, Dexter-Dyer, B. (1982) *J. Theor. Biol.*, **97**, 83–6.

Groudev. S.N., Genchev, F.N. and Gaidarjcev, S.S. (1978) in *Metallurgical Applications of Bacterial Leaching and Related Microbiological Phenomena* (Eds. L.E. Murr, A.E. Torma and J.A. Brierley), Academic Press, New York and London, pp. 253–74.

Groudev, S.N. (1982a) *Dokl. Bolg. Akad. Nauk.*, **34**, 217–20.

Groudev, S.N. (1982b) *Dokl. Bolg. Akad. Nauk.*, **35**, 1113–16.

Grunberg-Manago, M., Buckingham, R.H., Cooperman, B.S. and Hershey, J.W.B. (1978) *Symp. Soc. Gen. Microbiol.*, **28**, 27–110.

Guard, H.E., Cobet, A.B. and Coleman, W.M. (1981) *Science*, **213**, 770–1.

Guixe, V. and Babul, J. (1985) *J. Biol. Chem.*, **260**, 11001–5.

Gupta, R.K. (1987) (Ed.) *NMR Spectroscopy of Cells and Organisms*. Vols 1,2. CRC Press.

Gupta, R.K., Gupta, P. and Moore, R.D. (1984) *Ann. Rev. Biophys. Bioeng.*, **13**, 221–46.

Haefeli, C., Franklin, C. and Hardy, K. (1984) *J. Bacteriol.*, **158**, 389–92.
Hallas, L.E. and Cooney, J.J. (1981a) *Appl. Environ. Microbiol.*, **41**, 466–71.
Hallas, L.E. and Cooney, J.J. (1981b) *Dev. Ind. Microbiol.*, **22**, 529–35.
Hallas, L.E., Thayer, J.S. and Cooney, J.J. (1982) *Appl. Environ. Microbiol.*, **44**, 193–7.
Hallas, L.E., Means, J.C. and Cooney, J.J. (1982) *Science*, **215**, 1505–07.
Hallbauer, D.K. (1975) *Mineral Sci. Eng.*, **7**, 111–16.
Hamadi, A.F. and Gallon, J.R. (1981) *J. Gen. Microbiol.*, **125**, 391–8.
Hamann, A., Bossemeyer, D. and Bakker, E.P. (1987) *J. Bacteriol.*, **169**, 3138–45.
Hamer, D.H. (1986) *Ann. Rev. Biochem.*, **55**, 913–51.
Hamilton, F.D. (1974) *J. Biol. Chem*, **249**, 4428–34.
Hancock, R.E.W., Raffle, V.J. and Nicas, T.I. (1981) *Antimicrob. Agents Chemotherap.*, **19**, 777–85.
Hantke, K. (1981) *Mol. Gen. Genetics*, **182**, 288–92.
Hantke, K. (1985) in *Proteins of Iron Storage and Transport* (Eds. G. Spik, J. Montreuil, R.R. Crichton and J. Mazurier), Elsevier, Amsterdam, pp. 231–43.
Hantke, K. (1987) *FEMS Microbiol. Letters*, **44**, 53–7.
Hardie, L.P., Balkwill, D.L. and Stevens, S.E. (1983a) *Appl. Environ. Microbiol.*, **45**, 999–1006.
Hardie, L.P., Balkwill, D.L. and Stevens, S.E. (1983b) *Appl. Environ. Microbiol.*, **45**, 1007–17.
Harold, F.M. and Altendorf, K. (1974) in *Current Topics in Membranes and Transport*. **5**, (Eds. F. Bónner and A. Kleitzeller) Academic Press, New York, pp. 1–50.
Harold, F.M. (1977) *Ann. Rev. Microbiol.*, **31**, 181–203.
Harold, F.M. and Van Brunt, J. (1977) *Science*, **197**, 372–3.
Harold, F.M. (1986) *A Study of Bioenergetics*, W.H. Freeman & Co, New York.
Harrison, R.M. and Laxen, D.P.H. (1978) *Nature*, **275**, 738–40.
Harrison, P.M. and Hoare, R.J. (1980) *Metals in Biochemistry*, Chapman & Hall, London and New York.
Hartmann, A. and Braun, V. (1981) *Arch. Microbiol.*, **130**, 353–6.
Hasan, S.M. and Tsuchiya, T. (1977) *Biochem. Biophys. Res. Commun.*, **78**, 122–6.
Hassan, H.M. and Moody, C.S. (1984) *FEMS Microbiol. Letters*, **25**, 233–6.
Hassett, J.M., Jennett, J.C. and Smith, J.E. (1981) *Appl. Environ. Microbiol.*, **41**, 1097–106.
Hausinger, R.P. (1987) *Microbiol. Rev.*, **51**, 22–42.
Hausinger, R.P., Moura, I., Moura, J.J.G., Xavier, A.V. *et al.* (1982) *J. Biol. Chem.*, **257**, 14192–7.
Hawkes, T.R. and Bray, R.C. (1984) *Biochem. J.*, **219**, 481–93.
Hayashi, F., Tompson, J.G. and Agris, P.F. (1981) *FEBS Letters*, **128**, 79–82.
Haylock, S.J., Thornton, R.J., Buckley, P.D. and Blachwell, L.F. (1982) *Exp. Mycol.*, **6**, 335–44.
Heefner, D.L. and Harold, F.M. (1980) *J. Biol. Chem.*, **255**, 11396.
Heefner, D.L. and Harold, F.M. (1982) *Proc. Natl. Acad. Sci. U.S.A.*, **79**, 2798–802.
Helle, U. and Onken, U. (1988) *Appl. Microbiol. Biotechnol.*, **28**, 553–8.
Henry, M.-F., and Nyns, E.-J. (1977) *Arch. Microbiol.*, **114**, 171–4.
Henry, M.-F., Bonner, W.D. and Nyns, E.-J. (1977) *Biochim. Biophys. Acta*, **460**, 94–100.
Herbert, D. and Pinset, J. (1948) *Biochem J.*, **43**, 193–202.
Hermanns, U. and Wackernagel, W. (1977) *Eur. J. Biochem.*, **76**, 425–32.

Heptinstall, S., Archibald, A.R. and Baddiley, J. (1970) *Nature*, **225**, 519–21.
Hewitt, E.J. (1966) *Sand and Water Culture Methods used in the Study of Plant Nutrition*, Commonwealth Agricultural Bureau, Farnham, Bucks, pp. 139–43 and 455.
Hider, R.C. (1984) *Structure and Bonding*, **58**, 25–87.
Higham, D.P., Sadler, P.J. and Scawen, M.D. (1984) *Science*, **225**, 1043–6.
Hilfiker, M.L. and Higgins, W.J. (1981) *Comp. Biochem. Physiol.*, **68A**, 43–8.
Hill, H.A.O. (1982) in *The Biological Chemistry of Iron*, (Eds. H.B. Dunford, D. Dolphin, K.N. Raymond and L. Seiker) Reidel, Dordrecht, pp. 1–12.
Hilpert, N., Schink, B. and Dimroth, P. (1984) *EMBO J.*, **3**, 1665–70.
Hipkins, M.F. (1983) in *Metals and Micronutrients: Uptake and Utilisation by Plants* (Eds. D.A. Rebb and W.S. Pierpoint), Academic Press, London, New York, pp. 147–68.
Hirota, N., Kotada, M. and Imal, Y. (1981) *FEBS Letters*, **132**, 278–80.
Hmiel, S.P., Snavely, M.D., Miller, C.G. and Maguire, M.E. (1986) *J. Bacteriol.* **168**, 1444–50.
Ho, T.-L., Ho. H.C. and Hamilton, L.D. (1978) *Chem. Biol. Interactions*, **23**, 65–84.
Hoffman, L.E. and Hendrix, J.L. (1976) *Biotech. Bioeng.*, **18**, 1161–5.
Hoffman, M.R., Faust, B.C., Panda, F.A., Koo, H.H. *et al.* (1981) *Appl. Environ. Microbiol.*, **42**, 259–71.
Hogarth, C. and Ellar, D.J. (1978) *Biochem. J.*, **176**, 197–203.
Hogarth, C. and Ellar, D.J. (1979) *Biochem. J.*, **178**, 627–32.
Holmes, R.P. and Stewart, P.R. (1979) *J. Gen. Microbiol.*, **113**, 275–85.
Horikoshi, T., Nakajima, A. and Sakaguchi, T. (1979) *Agric. Biol. Chem.*, **43**, 617–25.
Horikoshi, T., Nakajima, A. and Sakaguchi, T. (1981) *Eur. J. Appl. Microbiol. Biotech.*, **12**, 90–96.
Hoving, H., Koning, J.H. and Robillard, G.T. (1982) *Biochemistry*, **21**, 3128–36.
Hoyle, B.D. and Beveridge, T.J. (1983) *Appl. Environ. Microbiol.*, **46**, 749–52.
Hoyle, B.D. and Beveridge, T.J. (1984) *Can. J. Microbiol.*, **30**, 204–11.
Hryniewiecka, L., Jenek, J. and Michejda, J.W. (1980) *Biochem. Biophys. Res. Commun.*, **93**, 141–8.
Hubbard, J.A.M., Lewandowska, K.B., Hughes, M.N. and Poole, R.K. (1986) *Arch. Microbiol.*, **146**, 80–86.
Hubbard, J.A.M., Poole, R.K. and Hughes, M.N. (1986) *Biochem. Soc. Trans.*, **14**, 1214–15.
Hubbard, M., Bradley, M., Sullivan, P., Shepherd, M. *et al.* (1982) *FEBS Letters*, **137**, 85–8.
Huber, F. and Kirchmann, H. (1978) *Inorg. Chimica Acta*, **29**, L249–50.
Huber, M. and Lerch, K. (1988) *Biochemistry*, **27**, 5610–15.
Hughes, M.N. (1981) *The Inorganic Chemistry of Biological Processes*, (2nd Edn.) John Wiley, Chichester.
Hughes, M.N. (1985) in *Microbial Gas Metabolism* (Eds. R.K. Poole and C. Dow) Academic Press, Orlando, pp. 3–27.
Hughes, M.N. (1987) Coordination Compounds in Biology. in *Comprehensive Coordination Chemistry* (Eds. G. Wilkinson, R.D. Gillard and J. McCleverty), Pergamon Press, Oxford, **6**, 541–754.
Hughes, M.N., O'Reardon, D.J., Poole, R.K., Hursthouse, M.B. *et al.* (1987) *Polyhedron*, **6**, 1711–14.
Hutchins, S.R., Davidson, M.S., Brierley, J.A. and Brierley, C.L. (1986) *Ann.*

Rev. Microbiol., **40**, 311–36.
Hutner, S.H. (1972) *Ann. Rev. Microbiol.*, **26**, 313–46.
Huttermann, J. and Kapple, R. (1987) in *Metal Ions in Biological Systems* (Ed. H. Sigel) **22**, Dekker, New York, pp. 1–80.
Ichihara, K., Kasaora, I., Kusunove, E. and Kusunove, M. (1980). *J. Gen. Appl. Microbiol.*, **26**, 387–93.
Ilyaletdinov, A.N., Enker, P.B. and Loginova, L.V. (1977) *Mikrobiologiya*, **46**, 113–17 (English Trans. 92–95).
Imai, K., Sugio, T., Tsuchida, T. and Tano, T. (1975) *Agric. Biol. Chem.*, **39** 1349–54.
Imperial, J., Ugalde, R.A., Shah, V.K. and Brill, W.J. (1984) *J. Bacteriol.*, **158**, 187–94.
Imura, N., Pan, S.K., Shimizu, M., Ukida, T. *et al.* (1977) *Ecotoxicol. Environ. Safety*, **1**, 255–61.
Ingledew, W.J. (1982) *Biochim. Biophys. Acta*, **688**, 89–117.
Ingledew, W.J. and Poole, R.K. (1984) *Microbiol. Rev.*, **48**, 222–71.
Iorio, R.M. and Plocke, D.J. (1982) *FEMS Microbiol. Letters*, **11**, 77–81.
Irgolic, K.J. and Brinkman, F.E. (1986) in *The Importance of Chemical Speciation in Environmental Processes* (Eds. M. Bernhard, F.E. Brinkman and P.T. Sadler). Springer-Verlag, Berlin, pp. 667–84.
Irving, H.R. and Grant, B.R. (1984) *J. Gen. Microbiol.*, **130**, 1569–76.
Ivarson, K.C. (1973) *Can. J. Soil Sci.*, **53**, 315–23.
Iverson, W.P. (1967) *Electrochem. Technol.*, **5**, 77.
Iwasa, Y., Yonemitsu, K., Matsui, K., Fukunaga, K. *et al.* (1981) *Biochem. Biophys. Res. Commun.*, **98**, 656–60.
Iwasaki, H., Saigo, T. and Matsubara, T. (1980) *Plant Cell Physiol.*, **21**, 1573–84.
Iwasaki, H. and Terai, H. (1982) *J. Gen. Appl. Microbiol.*, **28**, 189–93.
Jackson, T.A. (1978) *Environ. Geol.*, **2**, 173–89.
Jacob, G.S. and Orme-Johnson, W.H. (1979) *Biochemistry*, **18**, 2975–80.
Jacob, G.S., Schaefer, J., Stejskal, E.O. and McKay, R.A. (1985) *J. Biol. Chem.*, **260**, 5899–905.
Jarrell, K.F. and Sprott, G.D. (1982) *J. Bacteriol.*, **151**, 1195–203.
Jarrell, K.F., Bird, S.E. and Sprott, G.D. (1984) *FEBS Letters*, **166**, 357–61.
Jasper, P. and Silver, S. (1977) in *Microorganisms and Minerals* (Ed. E.D. Weinberg) Marcel Dekker Inc., New York, pp. 7–47.
Jasper, P. (1978) *J. Bacteriol.*, **133**, 1314–22.
Jasper, P. and Silver, S. (1978) *J. Bacteriol.*, **133**, 1323–8.
Jeffries, T.W. (1982) *Prog. Indust. Microbiol.*, **16**, 21–75.
Jennings, D.H. (1976) in *Transport in Plants* (Eds. V. Lüttge and M.G. Pitman) **II** Part A. Cells. Springer-Verlag, Berlin, pp. 189–228.
Jensen, T.E., Baxter, M., Rachlin, J.W. and Jani, V. (1982) *Environ. Pollut.*, **27**, 119–27.
Jernelöv, A. and Martin, A.-L. (1975) *Ann. Rev. Microbiol.*, **29**, 61–77.
Johnson, D.B., Kelso, W.I. and Jenkins, D.A. (1979) *Environ. Pollut.*, **18**, 107–15.
Johnson, D.B. (1985) *Indust. Biotech.*, **5**, 60–62.
Johnson, D.B., MacVicar, J.H.M. and Rolfe, S. (1987) *J. Microbiol. Methods*, **7**, 9–18.
Johnson, E.A. and Demain, A.L. (1984) *Arch. Microbiol.*, **137**, 135–8.
Johnson, G.W., Taylor, H.E. and Skogerboe, R.K. (1979) *Spectrochim. Acta*, **34B** 197–212.
Johnson, J.L. (1980) in *Molybdenum and Molybdenum-containing Enzymes* (Ed.

M.P. Coughlin) Pergamon Press, Oxford, pp. 345–83.
Johnson, M.K., Czernusz-Ewicz, R.S., Spiro, T.G., Ramsay, R.R. *et al.* (1983) *J. Biol. Chem.*, **258**, 12771–4.
Johnstone, K., Ellar, D.J. and Appleton, T.C. (1980) *FEMS Microbiol. Letters*, **7**, 97–101.
Johnstone, K., Stewart, G.S.A.B., Barratt, M.D. and Ellar, D.J. (1982) *Biochim. Biophys. Acta*, **714**, 379–81.
Jones, C.W. (1982) *Bacterial Respiration and Photosynthesis*, Nelson, Walton-on-Thames, Surrey.
Jones, C.W. and Poole, R.K. (1984) in *Methods in Microbiology*, **19** (Ed. G. Gottschalk), Academic Press, London.
Jones, J.B. and Stadtman, T.C. (1977) *J. Bacteriol.*, **130**, 1404–6.
Jones, J.G., Gardener, S. and Simon, B.M. (1983) *J. Gen. Microbiol.*, **129**, 131–9.
Jones, J.G., Davison, W. and Gardener, S. (1984) *FEMS Microbiol. Letters*, **21**, 133–6.
Jones, O.T.G (1983) in *Metals and Micronutrients: Uptake and Utilisation by Plants* (Eds. D.A. Robb and W.S. Pierpoint) Academic Press, London, pp. 125–44.
Kakinuma, Y. and Harold, F.M. (1985) *J. Biol. Chem.*, **260**, 2086–91.
Kakutani, T., Watanabe, H., Arima, K. and Beppu, T. (1981) *J. Biochem. (Japan)*, **89**, 453–61.
Kalitzer, H.R. (1987) in: *Metal Ions in Biological Systems* (Ed. H. Sigel) **22**, Dekker, New York, pp. 81–103.
Kamekura, M. and Onishi, H. (1982) *Can. J. Microbiol.*, **28**, 155–61.
Kanematsu, S. and Asada, K. (1978) *Arch. Biochem. Biophys.*, **185**, 473–82.
Kanematsu, S. and Asada, K. (1979) *Arch. Biochem. Biophys.*, **195**, 535–45.
Kargi, F. and Robinson, J.M. (1982) *Appl. Environ. Microbiol.*, **44**, 878–83.
Kargi, F. and Robinson, J.M. (1984) *Biotech, Bioeng.*, **26**, 687–90.
Kargi, F. and Weissman, J.G. (1984) *Biotech. Bioeng.*, **26**, 604–12.
Kashket, E.R. (1979) *J. Biol. Chem.*, **254**, 8129–31.
Kasumi, T., Hayashi, K. and Tsumura, N. (1982a) *Agric. Biol. Chem.*, **46**, 21–30.
Kasumi, T., Hayashi, K. and Tsumura, N. (1982b) *Agric. Biol. Chem.*, **46**, 31–40.
Kay, W.W. and Ghei, O.K. (1981) *Can. J. Microbiol.*, **27**, 1194–201.
Keller, L. and Murr, L.E. (1982) *Biotech. Bioeng.*, **24**, 83–96.
Kelly, D.P. (1985) *Microbiol. Science*, **2**, 105–9
Kelly, D.P. (1988) in *The Nitrogen and Sulphur Cycles* (Eds. J.A. Cole and S. Ferguson), SGM Symposium, **42**, 65–98.
Kelly, D.P., Norris, P.R. and Brierley, C.L. (1979) in *Microbial Technology; Current State, Future Prospect* (Eds. A.T. Bull, D.C. Ellwood and C. Ratledge) SGM Symposium **29**, 267–308, Academic Press, London.
Kempner, E.S. and Miller, J.H. (1972) *J. Protozool.*, **19**, 343–6.
Kent, T.A., Moura, I., Moura, J.J.G., Lipscomb, J.D. *et al.* (1982) *FEBS Letters*, **138**, 55–8.
Kenward, M.A., Brown, M.R.W. and Flyer, J.J. (1979) *J. Appl. Bacteriol.*, **47**, 489–503.
Kepkay, P.E. and Nealson, K.H. (1982) *J. Bacteriol.*, **151**, 1022–6.
Ketchum, P.A. and Somerville-Owens, M. (1975) *J. Bacteriol.*, **122**, 412–17.
Keyhani, E. and Chance, B. (1971) *FEBS Letters*, **17**, 127–32.
Keyhani, E. and Keyhani, J. (1975) *Arch. Biochem. Biophys.*, **167**, 596–602.
Keyhani, E. and Keyhani, J. (1980) *Biochim. Biophys. Acta*, **633**, 211–27.
Khanna, G., DeVoe, L., Brown, L., Niven, D.F. *et al.* (1984) *J. Bacteriol.*, **157**, 59–63.

Khazadi, M.B. and Mitra, R.S. (1981) *Appl. Environ. Microbiol.*, **41**, 46–50.
Khummongkol, D., Canterford, G.S. and Fryer, C. (1982) *Biotech. Bioeng.*, **24**, 2643–60.
Kiel, H. and Schwartz, W. (1980) *Z. Allg. Mikrobiologie*, **20**, 627–36.
Kierstan, M. and Bucke, C. (1977) *Biotech, Bioeng.*, **19**, 387–97.
Kino, K. and Usami, S. (1982) *Agric. Biol. Chem.*, **46**, 803–5.
Kirby, T.W., Lancaster, J.R. and Fridovich, I. (1981) *Arch. Biochem. Biophys.*, **210**, 140–8.
Kirkbright, G.F. and Sargent, M. (1974) *Atomic Absorption and Fluorescence Spectroscopy*, Academic Press, London.
Kisser, M., Kubicek, C.P. and Röhr, M. (1980) *Arch. Microbiol.*, **128**, 26–33.
Kita, K., Konishi, K. and Anraku, Y. (1984) *J. Biol. Chem.*, **259**, 3368–74.
Kleiner, D. (1978) *FEBS Letters*, **96**, 364–6.
Kloepper, J.W., Leong, J., Teintze, M. and Schroth, M.N. (1980) *Nature*, **386**, 885–6.
Klucas, R.V., Hanus, F.V., Russell, S.A. and Evans, H.J. (1983) *Proc. Natl. Acad. Sci. U.S.A.*, **80**, 2253–7.
Klug, A. and Rhodes, D. (1987) *Trends Biochem. Sci.*, **12**, 464–9.
Knight, W.B., Dunaway-Mariano, D., Ransom, S.C. and Villafranca, J.J. (1984) *J. Biol. Chem.*, **259**, 2886–95.
Knowles, C.J. (1980) in *Diversity of Bacterial Respiratory Systems* **II**, (Ed. C.J. Knowles), CRC Press, Boca Raton, Florida, pp. 139–58.
Knowles, F.C. and Benson, A.A. (1983) *Trends Biochem. Sci.*, **8**, 178–80.
Knowles, M. and Frary, B.D. (1988) *International Laboratory*, **4**, 52–64.
Kobayashi, H., Van Brunt, J. and Harold, F.M. (1978) *J. Biol. Chem.*, **253**, 2085–92.
Kobayashi, H. (1982) *J. Bacteriol.*, **150**, 506–11.
Kojima, N., Fox, J.A., Hausinger, R.P., Daniels, L. *et al.* (1983) *Proc. Natl. Acad. Sci. U.S.A.*, **80**, 378–82.
Kojima, Y. and Kagi, J.H.R. (1978) *Trends Biochem. Sci.*, **3**, 90–3.
Kolot, B. (1981) *Process. Biochem.*, **16**, 2–9, 30–3.
Konetzka, W.A. (1977) in *Microorganisms and Minerals* (Ed. E.D. Weinberg) Marcel Dekker Inc., New York, pp. 317–42.
Kono, Y. and Fridovich, I. (1983a) *J. Biol. Chem.*, **258**, 6015–19.
Kono, Y. and Fridovich, I. (1983b) *J. Bacteriol.*, **155**, 742–6.
Koshland, D.E. (1979) in *The Bacteria*, **7**, (Ed. J.R. Sokatch), Academic Press, New York, pp. 111–66.
Kretsinger, R.H. (1980) *CRC Crit. Rev. Biochem.*, **8**, 119–135.
Krone, W.J.A., Stegehuis, F., Konigstein, G., deGraf, F.K. *et al.* (1985) *FEMS Microbiol. Letters*, **26**, 153–61.
Krulwich, T.A. (1983) *Biochim. Biophys. Acta*, **726**, 245–64.
Krulwich, T.A., Guffanti, A.A., Borstein, R.F. and Hoffstein, J. (1982) *J. Biol. Chem.*, **257**, 1885–9.
Krulwich, T.A. (1983) *Biochim. Biophys. Acta*, **726**, 245–64.
Krulwich, T.A., Federbush, J.G. and Guffanti, A.A. (1985) *J. Biol. Chem.*, **260**, 4055–8.
Kubicek, C.P. and Röhr, M. (1977) *Eur. J. Appl. Microbiol. Biotech.*, **4**, 167–75.
Kubicek, C.P., Hampel, W. and Röhr, M. (1979) *Arch. Microbiol.*, **123**, 73–9.
Kung, F.-C., Raymond, J. and Glaser, D.A. (1976) *J. Bacteriol.*, **26**, 1089–95.
Kuo, L.C., Lipscomb, W.N. and Kantrowitz, E.R. (1982) *Proc. Natl. Acad. Sci. USA*, **79**, 2250–4.

Kurek, E., Czaban, J. and Bollag, J.M. (1982) *Appl. Environ. Microbiol.*, **43**, 1011–15.

Kuroda, R. and Kuroda, H. (1980) *J. Cell Science*, **44**, 75–85.

Kurowski, W.M., Fensom, A.H. and Pirt, S.J. (1973) *J. Gen Microbiol. Proc*, **75**, 15–16.

Kurtz, D.M., McMillan, R.S., Burgess, B.K., Mortenson, L.E. *et al.* (1979) *Proc. Natl. Acad. Sci. USA*, **76**, 4986–9.

Kusunose, E., Ichihara, K., Noda, Y. and Kusunose, M. (1976) *J. Biochem.*, **80**, 1343–52.

Kuyucak, N. and Volesky, B. (1987) in *Precious Metals 1987* (Eds. Vermeyten and Verbeek).

Kyrtopoulos, S.A. and Satchell, D.P.N. (1972) *Biochem J.*, **129**, 1163–6.

Labuda, D., Nicoghosian, K. and Cedergren, R. (1985) *J. Biol. Chem.*, **260**, 1103–7.

Laddaga, R.A., Bessen R. and Silver, S. (1985) *J. Bacteriol.*, **162**, 1106–10.

Laddaga, R.A. and Silver, S. (1985) *J. Bacteriol.*, **162**, 1100–5.

Laddaga, R.A., Chu, L., Misra, T.K. and Silver, S. (1987) *Proc. Natl. Acad. Sci. USA*, **84**, 5106–10.

Laimins, L.A., Rhoads, D.B. and Epstein, W. (1981) *Proc. Natl. Acad. Sci. USA*, **78**, 464–8.

Lakey, J.H., Lea, E.J.A., Rudd, B.A.M., Wright, N.M. *et al.* (1983) *J. Gen. Microbiol.*, **129**, 3565–73.

Lakshmanan, V.I. (1986) in *Workshop on Biotechnology for the Mining, Metal-refining and Fossil Fuel Processing Industries* (Eds. H.L. Ehrlich and D.S. Holmes). *Biotech. Bioeng. Symp.*, No.16, John Wiley, Chichester.

Lammers, M. and Follmann, H. (1983) *Structure & Bonding*, **54**, 27–91.

Lammers, P.J. and Sanders-Loehr, J. (1982) *J. Bacteriol.*, **151**, 288–94.

Lancaster, J.R. (1980) *FEBS Letters*, **115**, 285–8.

Lancaster, J.R. (1982) *Science*, **216**, 1324–5.

Lancaster, J.R. (1988) (Ed.) *The Bioinorganic Chemistry of Nickel* VCH, Weinheim.

Landner, L. (1971) *Nature*, **230**, 452–4.

Langman, L., Young, I.G., Frost, G.E., Rosenberg, H. *et al.* (1972) *J. Bacteriol.*, **112**, 1142–9.

Lankford, C.E. (1973) *CRC Crit. Rev. Microbiol.*, **2**, 273–331.

Lanyi, J.K. (1979) *Biochim. Biophys. Acta*, **559**, 377–97.

Lascelles, J. and Burke, K.A. (1978) *J. Bacteriol.*, **134**, 585–9.

Laube, V.S., Ramamoorthy, S. and Kushner, D.J. (1979) *Bull. Environ. Contam. Toxicol.*, **21**, 763–9.

Lawford, H.G., Pik, J.R., Lawford, G.R., Williams, T. and Kligerman, A. (1980a) *Can. J. Microbiol.*, **26**, 64–70.

Lawford, H.G., Pik, J.R., Lawford, G.R., Williams, T. and Kligerman, A. (1980b) *Can. J. Microbiol.*, **26**, 71–76.

Lawrence, R.W. and Bruynesteyn, A. (1983) *CIM Bull.*, **76**, 107–10.

Lazaroff, N., Sigal, W. and Wasserman, A. (1982) *Appl. Environ. Microbiol.*, **43**, 924–38.

Lee, S.-H., Kalra, V.K. and Brodie, A.F. (1979) *J. Biol. Chem.*, **254**, 6861–4.

LeGall, J., Prickril, B.C., Moura, I., Xavier, A.V. *et al.* (1988) *Biochemistry*, **27**, 1636–42.

Legerton, T.L., Kanamori, K., Weiss, R.L. and Roberts, J.D. (1983) *Biochemistry*, **22**, 899–903.

Lehn, J.M. (1978a) *Acc. Chem. Res.*, **11**, 49–60.
Lehn, J.M. (1978b) *Pure Appl. Chem.*, **50**, 871–99.
Lehninger, A.L., Reynafarje, B., Vercesi, A. and Tew, W.P. (1978) *Ann. N.Y. Acad. Sci.*, **307**, 160–76.
Leonards, K.S. and Haug, A. (1981) *Biochemistry*, **20**, 1784–9.
Leong, J. and Neilands, J.B. (1976) *J. Bacteriol.*, **126**, 823–30.
Leong, S.A. and Neilands, J.B. (1982) *Arch. Biochem. Biophys.*, **218**, 351–9.
Leontis, N.B., Ghosh, P. and Moore, P.B. (1986) *Biochemistry.*, **25**, 7386–92.
Leppla, S. (1982) *Proc. Natl. Acad. Sci. U.S.A.*, **79**, 3162–6.
Lerch, K. (1980) *Nature*, **284**, 368–70.
Le Trant, N., Meshnick, S.R., Kitchener, K., Eaton, J.W. *et al.* (1983) *J. Biol. Chem.*, **258**, 125–30.
Levine, B.A. and Williams, R.J.P. (1982) in *Calcium and Cell Functions* (Ed. W.Y. Cheung) Academic Press, New York. **II**, p.2.
Levinson, H.S. and Hyatt, M.T. (1955) *J. Bacteriol.*, **70**, 368–74.
Liberman, E., Reddy, P., Gazdar, C. and Peterkofsky, A. (1985) *J. Biol. Chem.*, **260**, 4075–81.
Lichko, L.P., Okarokov. L.A. and Kulaev, I.S. (1980) *J. Bacteriol.*, **144**, 666–71.
Liddell, P.R. and Brodie, K.G. (1980) *Anal. Chem.*, **52**, 1256–62.
Lidstrom, M.E., Engebrecht, J. and Nealson, K.H. (1983) *FEMS Microbiol. Letters*, **19**, 1–6.
Light, P.A. (1972) *FEBS Letters*, **19**, 319–22.
Light, P.A. and Garland, P.B. (1971) *Biochem. J.*, **124**, 123–34.
Lindblow-Kull, C., Shrift, A. and Gherna, R.L. (1982) *Appl. Environ. Microbiol.*, **44**, 737–43.
Lindskog, S. (1982) *Adv. Inorg. Biochem.* **4**, 115–70.
Lindström, K. (1980) *Arch. Hydrobiol.*, **89**, 110–17.
Lipscomb, J.D. and Hooper, A.B. (1982) *Biochemistry*, **21**, 3965–72.
Lipscomb, J.D., Andersson, K.K., Münck, E., Kent, T.A. *et al.* (1982a) *Biochemistry*, **21**, 3973–6.
Lipscomb, J.D., Whittaker, J.W. and Arciero, D.M. (1982b) in *Oxygenases and Oxygen Metabolism* (Eds. M. Nozaki, S. Yamamoto, Y. Ishimura, M.J. Coon *et al.* Academic Press, New York. pp. 27–85.
Little, C., Aakre, S.-E., Rumsby, M.G. and Gwarsha, K. (1982) *Biochem. J.*, **207**, 117–21.
Little, J.W. and Mount, D.W. (1982) *Cell*, **29**, 11–22.
Livesey-Goldblatt, E., Norman, P. and Livesey-Goldblatt, D.R. (1983) *Progress in Biohydrometallurgy*, Cagliari, pp. 627–41.
Lloyd, D. (1974) *The Mitochondria of Microorganisms*, Academic Press, London.
Lloyd, D., Poole, R.K. and Edwards, S.W. (1982) *The Cell Division Cycle – Temporal Organisation and Control of Cellular Growth and Reproduction*, Academic Press, New York and London.
Lodge, J.S. and Emery, T. (1984) *J. Bacteriol.*, **160**, 801–4.
Loeb, L.A. and Mildvan, A.S. (1981) *Adv. Inorg. Biochem*, **3**, 125–42.
Lohmeyer, M. and Friedrich, C.G. (1987) *Arch. Microbiol.*, **149**, 130–35.
Londesborough, J. and Suoranta, K. (1983) *J. Biol. Chem.*, **258**, 2966–72.
Long, T.V., Loehr, T.M., Allkins, J.R. and Lovenberg, W.N. (1971) *J. Amer. Chem. Soc.*, **93**, 1909–15.
Lovenberg, W. (1973) (Ed.) *Iron-Sulfur Proteins*, Vols. 1–3. Academic Press, New York.

Lovenberg, W. (1974) in *Microbial Iron Metabolism* (Ed. J.B. Neilands), Academic Press, New York, pp. 161–85.
Lovley, D.R. and Phillips, E.J.P. (1987) *Appl. Environ. Microbiol.*, **53**, 1536–40.
Lovley, D.R., Stolz, J.F., Nord, G.L. and Phillips, E.J.P. (1987) *Nature*, **330**, 252–4.
Lowe, D.J. and Thorneley, R.N.F. (1984) *Biochem. J.*, **224**, 877–86, 895–902.
Lumsden, J. and Hall, D.O. (1975) *Nature*, **257**, 670–2.
Lumsden, J., Cammack, R. and Hall, D.O. (1976) *Biochim. Biophys. Acta*, **438**, 380–92.
Lundell, D.J. and Howard, J.B. (1981) *J. Biol. Chem.*, **256**, 6385–91.
Lundgren, D.G. and Silver, M. (1980) Ann. Rev. Microbiol., **34**, 263–83.
Lusk, J.E. and Kennedy, E.P. (1972) *J. Bacteriol.*, **109**, 1034–46.
Lyalikova, N.N., Vedenina, I.Y. and Romonova, A.K. (1976) *Mikrobiologiya*, **45**, 552–4.
Macaskie, L.E. and Dean, A.C.R. (1984a) *J. Gen. Microbiol.*, **130**, 53–62.
Macaskie, L.E. and Dean, A.C.R. (1984b) *Environ. Technol. Letters*, **5**, 177–86.
Macaskie, L.E. and Dean, A.C.R. (1984c) *Biotechnol. Letters*, **6**, 71–6.
Macaskie, L.E. and Dean, A.C.R. (1985) *Biotechnol. Letters*, **7**, 457–62.
Macaskie, L.E., Dean, A.C.R., Cheetham, A.K., Jakeman, R.J.B. *et al.* (1987) *J. Gen. Microbiol.*, **133**, 539–44.
MacDonald, T.L. and Martin, R.B. (1988) *Trends Biochem. Sci.*, **13**, 15–19.
MacDonald, R.E., Greene, R.V., Clark, R.D. and Lindley, E.V. (1979) *J. Biol. Chem.*, **254**, 11831–35.
Macham, L. (1976) *Process. Biochem.*, **11**, 12–17.
Macham, L., Stephenson, M.C. and Ratledge, C. (1977) *J. Gen. Microbiol.*, **101**, 41–9.
Macleod, R.A. and Matula, T.I. (1982) *Can. J. Microbiol.*, **8**, 883–96.
Makemson, J.C. and Hastings, J.W. (1982) *Current Microbiol.*, **7**, 181–6.
Makinen, M.W., Wells, G.B. and Kang, S.-O. (1984) *Adv. Inorg. Biochem.*, **6**, 1–70.
Malikayil, J.A., Sweeney, W.V., McCracken, J. and Peisach, J. (1985) *Biochem. Biophys. Res. Commun.*, **133**, 1119–24.
Malmstrom, B.G. (1982) *Ann. Rev. Biochem.*, **51**, 21–59.
Malpartida, F. and Serrano, R. (1981) *J. Biol. Chem.*, **256**, 4175–7.
Manchee, R. (1979) *Trends in Biochem. Sci.*, **4**, 77–80.
Marsh, R.M. and Norris, P.R. (1983a) *Biotechnol. Letters*, **5**, 585–90.
Marsh, R.M. and Norris, P.R. (1983b) *FEMS Microbiol. Letters*, **17**, 311–15.
Marsh, R.M., Norris, P.R. and Le Roux, N.W. (1983) in *Recent Progress in Biohydrometallurgy* (Eds. G. Rossi and A.E. Torma) Associazona Mineraria Sarda, Italy, pp. 71–81.
Martin, R.B. (1986) *Clin. Chem.*, **32**, 1797–806.
Martin, J.P. and Fridovich, I. (1981) *J. Biol. Chem.*, **256**, 6080–9.
Martin, M.E., Strachan, R.C., Aranha, H., Evans, S.L. *et al.* (1984) *J. Bacteriol.*, **159**, 745–9.
Martinkus, K., Kennelly, P.J., Rea, T. and Timkovich, R. (1980) *Arch. Biochem. Biophys.*, **199**, 465–72.
Mas, J. Glender, W. and Pina, E. (1980) *J. Gen. Microbiol.*, **121**, 433–40.
Mas, M.T., Resplandor, Z.E. and Riggs, A.H. (1987) *Biochemistry*, **26**, 5369–77.
Masuko, M., Iwasaki, H., Sakurai, T., Suzuki, S. *et al.* (1984) *J. Biochem.*, **96**, 447–54.

Matsubara, T. and Zumft, W.G. (1982) *Arch. Microbiol.*, **132**, 322–8.
Matsubara, T., Frunzke, K. and Zumft, W.G. (1982) *J. Bacteriol.*, **149**, 816–23.
Matsumoto, K., Maeda, H. and Okamura, R. (1984) *J. Biochem.*, **96**, 1165–73.
Matthews, B.W. and Weaver, L.H. (1974a) *Biochemistry*, **13**, 1719–25.
Matthews, B.W., Weaver, L.H. and Kester, W.R. (1974b) *J. Biol. Chem.*, **249**, 8030.
Matthews, B.W., Fenna, R.E., Bolognesi, M.C., Schmid *et al.* (1979) *J. Mol. Biol.*, **131**, 259–85.
Maurer, P.J. and Miller, M.J. (1982) *J. Am. Chem. Soc.*, **104**, 3096–101.
Mazus, B., Falchuk, K.H. and Vallee, B.L. (1984) *Biochemistry*, **23**, 42–7.
McBride, B.C. and Wolfe, R.S. (1971) Biochemistry, **10**, 4312–7.
McCandliss, R.J. and Herrmann, K.M. (1978) *Proc. Natl. Acad. Sci. USA*, **75**, 4810–13.
McCormick, D. (1988) *Biotechnology*, **6**, 383–5.
McCracken, J., Desai, P.R., Papadopoulos, N.J., Villafranca, J.J. *et al.* (1988) *Biochemistry*, **27**, 4133–7.
McCray, J.W. and Herrmann, K.M. (1976) *J. Bacteriol.*, **125**, 608–15.
McElroy, R.O. and Bruynesteyn, A. (1978) in *Metallurgical Applications of Bacterial Leaching and Related Microbiological Phenomena* (Eds. L.E. Murr, A.E. Torma and J.A. Brierley) Academic Press, New York, pp. 441–62.
McIntosh, M.A. and Earhart, C.F. (1977) *J. Bacteriol.*, **131**, 331–9.
McKenzie, P., Akbar, S.A. and Miller, J.D.A. (1977) Microbiology Group Symposium: *Microbial corrosion affecting the oil industry*, Institute of Petroleum, London, p.37.
Meers, J.L. (1971) *J. Gen. Microbiol.*, **67**, 359–61.
Meier, B., Barra, D., Bossa, F., Calabrese, L. *et al.* (1982) *J. Biol. Chem.*, **257**, 13977–80.
Meijer, E.M., Schuitenmaker, M.G., Boogerd, F.C., Wever, R. *et al.* (1978) *Arch. Microbiol.*, **119**, 119–27.
Meisch, H.U., Benzschawel, H. and Bielig, H.J. (1977) *Arch. Microbiol.*, **114**, 67–70.
Meisch, H.U. and Bauer, J. (1978) *Arch. Microbiol.*, **117**, 49–52.
Meisch, H.U. and Becker, L.J.M. (1981) *Biochim. Biophys. Acta*, **636**, 119–25.
Meissner, P.S. and Falkinham, J.O. (1984) *J. Bacteriol.*, **157**, 669–72.
Mercer, A.A. and Loutit, J.S. (1979) *J. Bacteriol.*, **140**, 37–42.
Mergeay, M., Gerits, J. and Houba, C. (1978) *Comptes rendus des seance de la Soc. de Biologie*, **172**, 575–9.
Mertz, W. (1969) *Physiol. Rev.*, **49**, 163–239.
Meury, J., Lebail, S. and Kepes, A. (1980) *Eur. J. Biochem.*, **113**, 33–8.
Meury, J. and Kepes, A. (1982) *The EMBO Journal*, **1**, 339–43.
Meyer, O. (1985) In *Microbial Gas Metabolism: Mechanistic, Metabolic and Biotechnological Aspects* (Eds. R.K. Poole and C.S. Dow), Academic Press, Orlando, pp. 131–51.
Michel, H. (1982) *J. Mol. Biol.*, **158**, 567–72.
Mildvan, A.S., Sloan, D.L., Fung, C.H., Gupta, R.K. *et al.* (1976) *J. Biol. Chem.*, **251**, 2431–4.
Mildvan, A.S. and Loeb, C.A. (1981) *Adv. Inorg. Biochem.*, **3**, 103–23.
Miledi, R., Parker, I. and Schalow, G. (1977) *Proc. Roy. Soc.*, **B198**, 201–10.
Miles, A.A., Khimji, P.L. and Maskell, J. (1979) *J. Med. Microbiol.*, **12**, 17–28.
Miller, J.D.A. (1981) In *Microbial Biodeterioration* (Ed. A.H. Rose), Academic

Press, London, pp. 150–203.
Miller, J.H. (1972) *Experiments in Molecular Genetics*, Cold Spring Harbour Laboratory, New York.
Milne, J.A. and Cook, R.A. (1979) *Biochemistry*, **18**, 3604–10.
Misra, H.P. and Keele, B.B. (1975) *Biochim. Biophys. Acta*, **379**, 418–25.
Misra, H.P. and Fridovich, I. (1977) *J. Biol. Chem.*, **252**, 6421–3.
Misra, H.P. and Fridovich, I. (1978) *Arch. Biochem. Biophys.*, **189**, 317–22.
Misset, O. and Robillard, G.T. (1982) *Biochemistry*, **21**, 3136–42.
Miyakawa, T., Tachikawa, T., Akada, R.A., Tsuchiya, E. *et al.* (1986) *J. Gen. Microbiol.*, **132**, 1453–7.
Mitra, R.S., Gray, R.H., Chin, B. and Bernstein, A. (1975) *J. Bacteriol.*, **121**, 1180–8.
Mobley, H.L.T. and Rosen, B.P. (1982) *Proc. Natl. Acad. Sci. U.S.A.*, **79**, 6119–22.
Mobley, H.L.T. and Summers, A.O. (1987) in *Ion Transport in Prokaryotes* (Eds. B.P. Rosen and S. Silver). Academic Press, San Diego, pp. 305–26.
Mollinedo, F., Lopez-Moratalla, N., Pivel, J.P., Larraga, V. *et al.* (1981) *Eur. J. Biochemistry*, **119**, 183–8.
Mokhele, K., Tang, Y.J., Clark, M.A. and Ingraham, J.L. (1987) *J. Bacteriol.*, **169**, 5721–4.
Moncany, M.L.J. and Kellenberger, E. (1981) *Experientia*, **37**, 846–7.
Monti-Bragadin, C., Giraldi, T., Cantini, M., Zassinovich, G. *et al.* (1974) *FEBS Letters*, **43**, 13–16.
Monticello, D.J. and Finnerty, W.R. (1985) *Ann. Rev. Microbiol.*, **39**, 371–89.
Moody, C.S. and Hassan, H.M. (1984) *J. Biol. Chem.*, **259**, 12821–5.
Moody, M.D. and Dailey, H.A. (1984) *Arch. Biochem. Biophys.*, **234**, 178–86.
Moore, A.L., Linnett, P.E. and Beechey, R.B. (1980) *J. Bioenerg. Biomembranes*, **12**, 309–23.
Moore, G.R. and Williams, R.J.P. (1977) *FEBS Letters*, **79**, 229–32.
Moore, G.R., Pettigrew, G.W. and Rogers, N.K. (1986) *Proc. Natl. Acad. Sci. U.S.A.*, **83**, 4998–9.
Mortenson, L.E. and Thorneley, R.N.F. (1979) *Ann. Rev. Biochem.*, **48**, 387–418.
Morton, S.D. and Lee, T.H. (1974) *Environ. Sci. Tech.*, **8**, 673–4.
Moulis, J.-M. and Meyer, J. (1982) *Biochemistry*, **21**, 4762–71.
Moura, I., Moura, J.J.G., Santos, H., Xavier, A.V. *et al.* (1983) *Biochim. Biophys. Acta*, **742**, 84–90.
Moura, J.J.G., Moura, I., Huynh, B.H., Krüger, H.-J. *et al.* (1982) *Biochem. Biophys. Res. Commun.*, **108**, 1388–93.
Moura, J.J.G., Teixeira, M., Moura, I., Xavier, A.V. *et al.* (1984) *J. Mol. Catal.*, **23**, 303–14.
Mowll, J.L. and Gadd, G.M. (1983) *J. Gen. Microbiol.*, **129**, 3421–5.
Mowll, J.L. and Gadd, G.M. (1984) *J. Gen. Microbiol.*, **130**, 279–84.
Mulder, M.M., van der Gulden, H.M.L., Postma, P.W. and van Dam, K. (1988) *J. Gen. Microbiol.*, **134**, 777–83.
Müller, G. and Raymond, K.N. (1984) *J. Bacteriol.*, **160**, 304–12.
Müller, R., Schmitt, S. and Lingens, F. (1982) *Eur. J. Biochem.*, **125**, 579–84.
Munro, G.F. and Sauerbier, W. (1973) *J. Bacteriol.*, **116**, 488–90.
Murata, N. and Miyao, M. (1985) *Trends Biochem. Sci.*, **10**, 122–4.
Murao, S., Kasai, N., Kimura, Y., Oda, K. *et al.* (1982a). *Agric. Biol. Chem.*, **46**, 855–7.
Murao, S., Kasai, N., Kimura, Y. and Oda, K. (1982b) *Agric. Biol. Chem.*, **46**, 855–7.

Murasugi, A., Wada-Nakagawa, C. and Hayashi, Y. (1985) *J. Biochem.*, **96**, 1375–9.

Murphy, R.J. and Levy, J.F. (1983) *Trans. Brit. Mycol. Soc.*, **81**, 165–8.

Murphy, T.P., Lean, D.R.S. and Nalewajko, C. (1976) *Science*, **192**, 900–2.

Murray, A.D. and Kidby, D.K. (1975) *J. Gen. Microbiol.*, **86**, 66–74.

Murvanidze, G.V. and Glagolev, A.N. (1981) *FEMS Microbiol. Letters*, **12**, 3–6.

Muyzer, G., de Bruyn, A.C., Schmedding, D.J.M., Bos, P. *et al.* (1987) *Appl. Env. Microbiol.*, **53**, 660–4.

Muzzarelli, R.A.A. and Tanfani, F. (1980) *Biotechnol. Bioeng.*, **23**, 885–96.

Muzzarelli, R.A.A., Tanfani, F., Emanuelli, M. and Bolognini, L. (1985) *Biotechnol. Bioeng.*, **27**, 1115–21.

Myerson, A.S. and Kline, P.C. (1984) *Biotech. Bioeng.*, **26**, 92–9.

Nagano, T., Miwa, M., Suketa, Y. and Okada, S. (1984) *J. Inorg. Biochem.*, **21**, 61–71.

Nakajima, A., Horikoshi, T. and Sakaguchi, T. (1979) *Agric. Biol. Chem.*, **43**, 625–9.

Nakajima, A., Horikoshi, T. and Sakaguchi, T. (1981a) *Eur. J. Microbiol. Technol.*, **12**, 76–83.

Nakajima, A., Horikoshi, T. and Sakaguchi, T. (1981b) *Agric. Biol. Chem.*, **45**, 903–8.

Nakajima, A., Horikoshi, T. and Sakaguchi, T. (1982) *Eur. J. Appl. Microbiol. Biotechnol.*, **16**, 88–91.

Nakajima, A. and Sakaguchi, T. (1986) *Appl. Microbiol. Biotechnol.*, **24**, 59–64.

Nakajima, H., Yamato, I. and Anraku, Y. (1979) *J. Biochem. (Japan)*, **85**, 303–10.

Nakamura, Y., Katayama, S., Okada, Y., Suzuki, F. *et al.* (1981) *Agric. Biol. Chem.*, **45**, 1167–72.

Nakamura, T., Fujita, K., Eguchi, Y. and Yazawa, M. (1984) *J. Biochem. (Japan)*, **95**, 1551–7.

Nealson, K.H. (1983a) in *Microbial Geochemistry* (Ed. W.E. Krumbein) Blackwell, Oxford, pp. 191–221.

Nealson, K.H. (1983b) in *Microbial Geochemistry* (Ed. W.E. Krumbein) Blackwell, Oxford, pp. 159–90.

Neilands, J.B. (1980) in *Iron in Biochemistry and Medicine, II* (Eds. A. Jacobs and M. Worwood), Academic Press, London, pp. 529–72.

Neilands, J.B. (1981) *Ann. Rev. Biochem.*, **50**, 715–31.

Neilands, J.B. (1982) *Ann. Rev. Microbiol.*, **36**, 285–309.

Neilands, J.B. (1983) in *Adv. Inorg. Biochem.* 5 (Eds. E.C. Theil, G.L. Eichhorn and L.G. Marzilli). Elsevier, New York, 137–66.

Neilands, J.B. (1984) *Microbiol. Sci.*, **1**, 9–14.

Nelson, D.L. and Kennedy, E.P. (1972) *Proc. Natl. Acad. Sci. U.S.A.*, **69**, 1091–3.

Nelson, M.J., Levy, M.A. and Orme-Johnson, W.H. (1983) *Proc. Natl. Acad. Sci. U.S.A.-Biol. Sci.*, **80**, 147–50.

Nettleton, C.J., Bull, C., Baldwin, T.O. and Fee, J.A. (1984) *Proc. Natl. Acad. Sci. USA*, **81**, 4970–3.

Newsome, A.L. and Wilhelm, W.E. (1983) *Appl. Environ. Microbiol.*, **45**, 665–8.

Newton, W.A. and Snell, E.E. (1964) *Proc. Natl. Acad. Sci. USA.*, **251**, 382–6.

Nicas, T.I. and Hancock, R.E.W. (1983) *J. Gen. Microbiol.*, **129**, 509–17.

Nicholas, D.J.D. (1952) *Analyst*, **77**, 629–42.

Nicholls, D.G. (1982) *Bioenergetics. An Introduction to Chemiosmotic Theory*, Academic Press, London.

Nichols, P.D., Henson, J.M., Guckert, J.B., Nivens, D.E. *et al.* (1985) *J. Microbiol.*

Methods, **4**, 79–94.
Nikaido, H. and Nakae, T. (1979) *Adv. Microbial Physiol.*, **20**, 163–250.
Nies, D. and Schlegel, H.G. (1982) *J. Gen. Appl. Microbiol.*, **28**, 311–19.
Nilsson, J.R. and Coleman, J.R. (1977) *J. Cell. Sci.*, **24**, 311–25.
Nilsson, J.R. (1979) *J. Cell. Sci.*, **39**, 383–96.
Nilsson, J.R. (1981) *Protoplasma*, **109**, 359–70.
Norberg, A. and Persson, H. (1984) *Biotechnol. Bioeng.*, **26**, 239–46.
Norris, P.R., Man, W.K., Hughes, M.N. and Kelly, D.P. (1976) *Arch. Microbiol.*, **110**, 279–86.
Norris, P.R. and Kelly, D.P. (1977) *J. Gen. Microbiol.*, **99**, 317–24.
Norris, P.R. and Kelly, D.P. (1979) *Dev. Ind. Microbiol.*, **20**, 299–308.
Norris, P.R., Brierley, J.A. and Kelly, D.P. (1980) *FEMS Microbiol. Letters*, **7**, 119–22.
Norris, P.R. and Kelly, D.P. (1982) In *Microbial Interactions and Communities* (Eds. A.T. Bull and J.H. Slater), Academic Press, London, **1**, pp. 443–74.
Norris, P.R., Marsh, R.M. and Lindstrom, E.B. (1986) *Biotech. Appl. Biochem.*, **8**, 318–29.
Nossal, P.M. (1951) *Biochem. J.*, **49**, 407–13.
Nozaki, M., Yamamoto, S., Ishimura, Y., Coon, M.J. *et al.* (1982) (Eds) *Oxygenases and Oxygen Metabolism*, Academic Press, New York.
Nriagu, J.O. and Pacyna, J.M. (1988) *Nature*, **333**, 134–9.
O'Brien, R.W. and Stern, J.R. (1969) *J. Bacteriol.*, **98**, 388–93.
Obuekwe, C.O., Westlake, D.W.S. and Cook, F.D. (1981) *Can. J. Microbiol.*, **27**, 692–7.
Oertling, W.A. and Babcock, G.T. (1988) *Biochemistry*, **27**, 3331–8.
Ogawa, S., Shulman, R.G., Glynn, P., Yamane, T. *et al.* (1978) *Biochim. Biophys. Acta*, **502**, 45–53.
Ogino, T., den Hollander, J.A. and Shulman, R.G. (1983) *Proc. Natl. Acad. Sci. USA*, **80**, 5185–9.
Ogram, A., Sayler, G.S. and Barkay, T. (1983) *J. Microbiol. Methods*, **7**, 57–66.
Ohsumi, Y. and Anraku, Y. (1985) *J. Biol. Chem.*, **260**, 10482–5.
Ohya, Y., Ohsumi, Y. and Anraku, Y. (1984) *Mol. Gen. Genetics*, **193**, 389–94.
Ohya, Y., Ohsumi, Y. and Anraku, Y. (1986) *J. Gen. Microbiol.*, **132**, 979–88.
Oka, T., Udagawa, K. and Kinoshita, S. (1968) *J. Bacteriol.*, **96**, 1760–7.
Okorokov, L.A., Lichko, L.P., Kadomtseva, V.M., Kholodenko, V.P. *et al.* (1974) *Microbiology*, **43**, 347–51.
Okorokov, L.A., Lichko, L.P. and Kulaev, I.S. (1980) *J. Bacteriol.*, **144**, 661–5.
Olafson, R.W., Abel, K. and Sim, R.G. (1979) *Biochem. Biophys. Res. Commun.*, **89**, 36–43.
Olson, G.J., Iverson, W.P. and Brinckman, F.E. (1981) *Curr. Microbiol.*, **5**, 115–18.
Olson, G.J., Porter, F.D., Rubinstein, J. and Silver, S. (1982) *J. Bacteriol.*, **151**, 1230–6.
Oltmann, L.F., Claassen, V.P., Kastellin, P., Reijinders, W.N.M. *et al.* (1979) *FEBS Letters*, **106**, 43–6.
Ondrias, M.R., Findsen, E.W., Gaul, D.F. and Knaff, D.B. (1984) *Biochim. Biophys. Acta*, **788**, 87–98.
Onishi, H., Kobayashi, T., Morita, N. and Baba, M. (1984) *Agric. Biol. Chem.*, **48**, 2441–8.
Öquist, G. (1974) *Physiol. Plant*, **31**, 55–8.

Ordal, G.W. (1977) *Nature*, **270**, 66–7.
Ordal, G.W. (1980) *Bioscience*, **30**, 408–11.
Orthofer, R., Kubicek, C.P. and Röhr, M. (1979) *FEMS Letters*, **5**, 403–6.
Ose, D.E. and Fridovich, I. (1979) *Arch. Biochem. Biophys.*, **194**, 360–4.
Oshino, R., Oshino, N., Chance, B. and Hagihara, B. (1973) *Eur. J. Biochem.*, **35**, 23–33.
Ottow, J.C.G. and Von Klopotek, A. (1969) *Appl. Microbiol.*, **18**, 41–3.
Ottow, J.C.G. (1970) *Z. Allg. Mikrobiol.*, **10**, 55–62.
Padmanabhan, S., Richey, B., Anderson, C.F. and Record, M.T. (1988) *Biochemistry*, **27**, 4367–76.
Page, W.J. and Doran, J.L. (1981) *J. Bacteriol.*, **146**, 33–40.
Page, W.J. and Von Tigerstrom, M. (1982) *J. Bacteriol.*, **151**, 237–42.
Painter, T.J. (1983) in *The Polysaccharides*, (Ed. G.O. Aspinall) Academic Press, New York, **2**, p.195.
Painter, G.R. and Pressman, B.C. (1985) *Metal Ions in Biological Systems*, **19**, 230–94.
Pan-Hou, H.S.K. and Imura, N. (1981) *Arch. Microbiol.*, **129**, 49–52.
Pan-Hou, H.S.K. and Imura, N. (1982a) *Arch. Microbiol.*, **131**, 176–7.
Pan-Hou, H.S.K. and Imura, N. (1982b) *Bull. Environ. Contam. Toxicol.*, **29**, 290–7.
Pan-Hou, H.S.K., Nishimoto, M. and Imura, N. (1981) *Arch. Microbiol.*, **130**, 93–5.
Park, M.H., Wong, B.B. and Lusk, J.E. (1976) *J. Bacteriol.*, **126**, 1096–103.
Parkin, M.J. and Ross, I.S. (1985) *Microbics. Letters*, **29**, 115–20.
Partridge, C.D.P. and Yates, M.G. (1982; *Biochem. J.*, **204**), 339–44.
Paton, W.H.N. and Budd, K. (1972) *J. Gen. Microbiol.*, **72**, 173–84.
Paulus, H. and Gray, E. (1964) *J. Biol. Chem.*, **239**, 4008–9.
Payer, H.D. and Trultzsch, U. (1972) *Arch. Microbiol.*, **84**, 43–53.
Payne, S.M., Niesel, D.W., Peixotto, S.S. and Lawlor, K.M. (1983) *J. Bacteriol.*, **155**, 949–55.
Pearson, R.G. (1965) *J. Am. Chem. Soc.*, **85**, 3533–9.
Pederson, R.L. and Gupta, R.K. (1979) *Biophys. J.*, **27**, 1–14.
Pedrosa, F.O. and Yates, M.G. (1983) *FEMS Letters*, **17**, 101–6.
Peisach, J., Levine, W.G. and Blumberg, W.E. (1967) *J. Biol. Chem.*, **242** 2847–58.
Peisach, J., Beinert, H., Emptage, M.H., Mims, W.B. *et al.* (1983) *J. Biol. Chem.*, **258**, 131014–16.
Penã, A., Mora, M.A. and Carrasco, N. (1979) *J. Membrane Biol.*, **47**, 261–84.
Penman, C.S. and Duffus, J.H. (1975) *J. Gen. Microbiol.*, **90**, 76–80.
Perry, R.D. and Silver, S. (1982) *J. Bacteriol.*, **150**, 973–6.
Perski, H.-J., Moll, J. and Thauer, R.K. (1981) *Arch. Microbiol.*, **130**, 319–21.
Perski, H.-J. Schöneit, P. and Thauer, R.K. (1982) *FEBS Letters*, **143**, 323–6.
Peschek, G.A. (1979) *FEMS Microbiol. Letters*, **6**, 371–4.
Peschek, G.A., Schmetterer, G., Lauritsch, G., Nitschmann, W.H. *et al.* (1982) *Arch. Microbiol.*, **131**, 261–5.
Pestka, S. (1972) *Proc. Natl. Acad. Sci. USA*, **69**, 624–8.
Pettigrew, G.W. and Moore, G.R. (1987) *Cytochromes c: Biological Aspects*, Springer-Verlag, Berlin, Heidelberg.
Phipps, D.A. (1976) *Metals and Metabolism*. Oxford University Press, Oxford.
Piccinni, E., Irato, R., Coppellotti, O. and Guidolan, L. (1987) *J. Cell Science*, **88**, 283–93.

Pichuantes, S., Cofre, G., Venegas, A. and Rodriguez, M. (1986) *Biotech. Appl. Biochem.*, **8**, 276–83.
Pienkos, P.T., Shah, V.K. and Brill, W.J.C. (1977) *Proc. Natl. Acad. Sci. USA*, **74**, 5468–71.
Pienkos, P.T. and Brill, W.J. (1981) *J. Bacteriol.*, **145**, 743–51.
Pirt, S.J. (1975) *Principles of Microbe and Cell Cultivation.* Blackwell, Oxford.
Pitt, D. and Ugalde, U.O. (1984) *Plant, Cell and Environ*, **7**, 467–75.
Pivovarova, T.A., Piskunov, V.P., Dorofeev, G. and Karavaiko, G.L. (1987) *Mikrobiologia*, **56**, 514–16. (Chem Abs **107**, 180668h).
Plack, R.H. and Rosen, B.P. (1980) *J. Biol. Chem.*, **255**, 3824–5.
Plateau, P., Mayaux, J.F. and Blanquet, S. (1981) *Biochemistry.*, **20**, 4654–62.
Polikarpov, G.G. (1966) *Radioecology of aquatic organisms.* (Translated from Russian by Scripta Technica) Reinhold, New York.
Poole, R.K. and Haddock, B.A. (1975) *Biochem. J.*, **152**, 537–46.
Poole, R.K., Lloyd, D. and Chance, B. (1974) *Biochem. J.*, **138**, 201–10.
Poole, R.K., Lloyd, D. and Chance, B. (1979) *Biochem. J.*, **184**, 555–63.
Poole, R.K. (1983) *Biochim. Biophys. Acta*, **726**, 205–43.
Poole, R.K., Baines, B.S., Curtis, S.J., Williams, H.D. *et al.* (1984) *J. Gen. Microbiol.*, **130**, 3055–8.
Poole, R.K., Baines, B.S., Hubbard, J.A.M. and Williams. H.D. (1985) in *Microbial Gas Metabolism: Mechanistic, Metabolic and Biotechnological Aspects* (Eds. R.K. Poole and C.S. Dow). Academic Press, Orlando, pp. 31–62.
Poole, R.K. and Dow, C.S. (1985) (Eds) *Microbial Gas Metabolism: Metabolic, Mechanistic and Biotechnogical Aspects.* Academic Press, Orlando.
Poole, R.K., Baines, B.S. and Appleby, C.A. (1986) *J. Gen. Microbiol.*, **132**, 1525–39.
Poole, R.K. (1988) in *Energy Transduction in Bacteria* (Ed. C. Anthony), Academic Press, London, pp. 231–91.
Pooley, F.D. (1982) *Nature*, **296**, 642–3.
Poora, R.J. and Ross, B.D. (1965) *Biochem. J.*, **94**, 557–62.
Postgate, J.R. (1978) *Nitrogen Fixation*, Edward Arnold, London.
Postgate, J.R. (1979) *The Sulphate-reducing Bacteria.* Cambridge University Press.
Poulos, T.L., Freer, S.T., Alden, R.A., Edwards, S.L. *et al.* (1980) *J. Biol. Chem.*, **255**, 575–80.
Pretlow, T.P. and Sherman, F. (1967) *Biochim. Biophys. Acta*, **148**, 629–44.
Prince, R.H. (1979) *Adv. Inorg. Chem. Radiochem.*, **22**, 349–426.
Prinz, R. and Weser, U. (1975) *FEBS Letters*, **54**, 224–9.
Proniewicz, L.M., Bajdor, K. and Nakomoto, K. (1980). *J. Phys. Chem.*, **90**, 1760–6.
Pugh, S.Y.R., Diguiseppi, J.C. and Fridovich, I. (1984) *J. Bacteriol.*, **160**, 137–42.
Pumpel, T. and Schinner, F. (1986) *Appl. Microbiol. Biotechnol.*, **24**, 244–7.
Putrament, A., Baranowska, H., Ejchart, A. and Jachymczyx, W. (1977) *Mol. Gen. Genetics*, **151**, 69–76.
Que, L. (1983) *Adv. Inorg. Biochem.*, **5**, 167–98.
Quinn, J.P., Barker, T.W. and Marchant, R. (1981) *J. Appl. Bacteriol.*, **51**, 149–57.
Ragsdale, S.W. and Ljungdahl, L.G. (1984) *J. Bacteriol.*, **157**, 1–6.
Ragsdale, S.W., Ljungdahl, L.G. and Dervartanian, D.V. (1982) *Biochem. Biophys. Res. Commun.*, **108**, 658–63.
Ragsdale, S.W., Clark, J.E., Ljungdahl, L.G., Lundie, L.L. *et al.* (1983) *J. Biol. Chem.*, **258**, 2364–9.

Rainnie, D.J. and Bragg, P.D. (1973) *J. Gen. Microbiol.*, **77**, 339–49.
Raitio, M., Jalli, T. and Saraste, M. (1987) *EMBO J.*, **6**, 2825–33.
Ramamoorthy, S. and Kushner, D.J. (1975) *Microbial Ecol.*, **2**, 162–76.
Ramamoorthy, S., Springthorpe, S. and Kushner, D.J. (1977) *Bull. Environ. Contam. Toxicol.*, **17**, 505–11.
Ramasarma, T. and Crane, F.L. (1981) *Current Topics In Cellular Regulation.* **20**, (Eds. B.L. Hotecker and E.R. Stadtman) Academic Press, New York, pp. 247–301.
Ratledge, C. and Winder, F.G. (1964) *J. Bacteriol.*, **87**, 823–7.
Ratledge, C. and Chaudhary, M.A. (1971) *J. Gen. Microbiol.* **66**, 71–78.
Ratledge, C., Macham, L.P., Brown, K.A. and Marshall, B.J. (1974) *Biochim. Biophys. Acta*, **372**, 39–51.
Ratledge, C., Patel, P.V. and Mundy, J. (1982) *J. Gen. Microbiol.*, **128**, 1559–65.
Ratledge, C. (1987) in *Iron Transport in Microbes, Plants and Animals* (Eds. G. Winkelmann, D. van der Helm and J.B. Neilands) VCH Weinheim, pp. 207–33.
Rauh, J.J. and Nelson, D.L. (1981) *J. Cell. Biol.*, **91**, 860–5.
Raven, J.A. (1976) in *Transport in Plants* **II, Part A**, *Cells*. (Eds. U.Lütlge and M.G. Pitman). Springer Verlag, Berlin, pp. 129–88.
Ravindranath, S.D. and Fridovich, I. (1975) *J. Biol. Chem.*, **250**, 6107–12.
Ray, B.D. and Rao, B.D.N. (1988a) *Biochemistry*, **27**, 5574–8.
Ray, B.D. and Rao, B.D.N. (1988b) *Biochemistry*, **27**, 5579–85.
Raymond, K.N. and Carrano, C.J. (1979) *Acc. Chem. Res.*, **12**, 183–90.
Reamer, D.C. and Zoller, W.H. (1980) *Science*, **208**, 500–2.
Reed, R.H., Rowell, P. and Stewart, W.D.P. (1981) *FEMS Microbiol. Letters*, **11**, 233–6.
Rees, T.A.V. and Bekheet, I.A. (1982) *Planta*, **156**, 385–7.
Rees, D.A., Morris, E.R., Thom, D. and Madden, J.K. (1982) in *The Polysaccharides* **1** (Ed. G.O. Aspinall). Academic Press, New York, p. 195.
Reeves, M.W., Pine, L., Hutner, S.H., George, J.R. *et al.* (1981) *J. Clin. Microbiol*, **13**, 688–95.
Reeves, M.W., Pine, L., Neilands, J.B. and Balows, A. (1983) *J. Bacteriol.*, **154**, 324–9.
Reinards, R., Kubicki, J. and Ohlenbusch, H.D. (1981) *Eur. J. Biochem.*, **120**, 329–37.
Reisinger, K., Stoeppler, M. and Nürnberg, H.W. (1981) *Nature*, **291**, 228–30.
Reissig, J.L. and Kinney, S.G. (1983) *J. Bacteriol.*, **154**, 1397–402.
Rhoads, D.B. and Epstein, W. (1977) *J. Biol. Chem*, **252**, 1394–1401.
Rhoads, D.B., Woo. A. and Epstein, W. (1977) *Biochim. Biophys. Acta*, **469**, 45–51.
Riddle, L.M., Rauh, J.J. and Nelson, D.L. (1982) *Biochim. Biophys. Acta*, **688**, 525–40.
Ridley, W.P., Dizikes, L.J. and Wood, J.M. (1977) *Science*, **197**, 329–32.
Rinderle, S.J., Booth, J.E. and Williams, J.W. (1983) *Biochemistry*, **22**, 869–76.
Ringbom, A. (1963) *Complexation in Analytical Chemistry*. John Wiley, New York.
Ringe, D., Petsko, G.A., Yamakura, F., Suzuki, K. and Ohmori, D. (1983) *Proc. Natl. Acad. Sci. U.S.A.*, **80**, 3879–83.
Roberts, J.E., Cline, J.F., Lum, V., Freeman, H. *et al.* (1984) *J. Am. Chem. Soc.*, **106**, 5324–30.
Robinson, F.A. (1976) Chemistry and the New Industrial Revolution. *Chem. Soc.*

Rev., **5**, 317–35.
Robinson, J.B. and Tuovinen, O.H. (1984) *Microbiol. Rev.*, **48**, 95–124.
Robinson, P.D. and Levy, H.R. (1979) *Arch. Biochem. Biophys.*, **193**, 242–51.
Robinson, P.D., Nowak, T. and Levy, H.R. (1979) *Arch. Biochem. Biophys.*, **193**, 252–9.
Robinson, P.K., Mak, A.L. and Trevan, M.D. (1986) *Process Biochemistry*, 122–7.
Robson, R.L. and Postgate, J.R. (1980). *Ann. Rev. Microbiol.*, **34**, 183–207.
Robson, R.L., Eady, R.R., Richardson, T.H., Miller, R.W. *et al.* (1986) *Nature*, **322**, 388–90.
Rode, H. and Giffhorn, F. (1982) *J. Bacteriol.*, **151**, 1602–4.
Röderer, G. and Schnepf, E. (1977) *Naturwissenschaften*, **64**, 588–9.
Rodriguez-Leiva, M. and Tributsch, H. (1988) *Arch. Microbiol.*, **149**, 401–5.
Rodriguez-Navarro, A. and Ortega, M.D. (1982) *FEBS Letters*, **138**, 205–8.
Rodriguez-Navarro, A. and Ramos, J. (1984) *J. Bacteriol.*, **159**, 940–5.
Roessler, P.G. and Nadler, K.D. (1982) *J. Bacteriol.*, **149**, 1021–6.
Rogers, H.J., Perkins, H.R. and Ward, J.B. (1980) *Microbial Cell Walls and Membranes*, Chapman and Hall, London, New York.
Rogers, H.J., Synge, C. and Woods, U.E. (1980) *Antimicrob. Agents. Chemotherapy*, **18**, 63–8.
Rogers, H.J., Woods, V.E. and Synge, C. (1982) *J. Gen. Microbiol.*, **128**, 2389–94.
Rogers, H.J., Plaha, D.S. and Woods, V.E. (1984) *FEMS Microbiol. Letters*, **24**, 5–8.
Roomans, G.M. and Sevews, L.A. (1976) *J. Cell. Sci.*, **21**, 119–27.
Rose, S.L., Dickinson, L.C. and Westhead, E.W. (1984) *J. Biol. Chem.*, **259**, 4405–13.
Rosen, B.P. and Brey, R.M. (1979) in *Microbiology 1979*, American Society of Microbiology, pp. 62–6.
Rosen, B.P. and Borbolla, M.G. (1984) *Biochem. Biophys. Res. Commun.*, **124**, 760–5.
Rosen, B.P. (1978) *Biochim. Biophys. Acta*, **906**, 101–10.
Rosen, B.P. and Silver, S. (1987) (Eds) *Ion Transport in Prokaryotes*, Academic Press.
Rosenberg, B. (1978) *Interdiscipl. Sci. Rev.*, **3**, 134–47.
Rosenberg, B., Van Camp, L. and Krigas, T. (1965) *Nature*, **205**, 698–9.
Rosenberg, B., Renshaw, E., Van Camp, L., Hartwick, J. *et al.* (1967) *J. Bacteriol.*, **93**, 716–21.
Rosenberg, B., Van Camp, L., Trosko, J.E. and Mansour, V.H. (1969) *Nature*, **222**, 385–6.
Rosner, A. and Paulus, H. (1971) *J. Biol. Chem.*, **246**, 2965–71.
Ross, I.S. (1975) *Trans. Br. Mycol. Soc.*, **64**, 175–93.
Ross. I.S. (1977) *Trans. Brit. Mycol. Soc.*, **69**, 77–81.
Rosson, R.A. and Nealson, K.H. (1982) *J. Bacteriol.*, **151**, 1027–34.
Rothstein, A. and Meier, R. (1951) *J. Cell. Comp. Physiol.*, **38**, 245–70.
Rothstein, A., Hayes, A., Jennings, D. and Hooper, D. (1958) *J. Gen. Physiol.*, **41**, 585–94.
Rothstein, A. (1960) In Ciba Foundation Study Group 6, *Regulation of the Inorganic Ion Content of Cells*. J. & A. Churchill Ltd., London, pp. 53–68.
Rothstein, A. (1974) *J. Gen Physiol.*, **64**, 608–21.
Rudd, T., Sterritt, R.W. and Lester, J.N. (1984) *J. Water Poll. Control. Fed.*, **56**, 1260–8.

Rüegg, C. and Lerch, K. (1981) *Biochemistry*, **20**, 1256–62.

Russell, L.M. and Rosenberg, H. (1979) *Biochem. J.*, **184**, 13–21.

Russell, L.M. and Rosenberg, H. (1980) *Biochem. J.*, **188**, 715–23.

Sachan, D.S. and Stern, J.R. (1971) *Biochem. Biophys. Res. Commun.*, **45**, 402–8.

Sadler, P., Bakka, A. and Beynon, P.J. (1978) *FEBS Letters*, **94**, 315–19.

Sagers, R.D. (1974) in *Microbial Iron Metabolism* (Ed. J.B. Neilands), Academic Press, London, New York, pp. 445–53.

Sakaguchi, T., Nakajima, A. and Horikoshi, T. (1981) *Eur. J. Appl. Microbiol. Biotechnol.*, **12**, 84–9.

Salhany, J.M., Yamane, T., Shulman, R.G. and Ogawa, S. (1975) *Proc. Natl. Acad. Sci. USA*, **72**, 4966–9.

Sanmugasunderam, V., Branion, R.M.R. and Duncan, D.W. (1985) *Biotechnol. Bioeng.*, **27**, 1173–84.

Sarathchandra, S.U. and Watkinson, J.H. (1981) *Science*, **211**, 600–1.

Sato, S. and Harris, J.I. (1977) *Eur. J. Biochem.*, **73**, 373–81.

Sato, S. and Nakazawa, K. (1978) *J. Biochem. (Tokyo)*, **83**, 1165–71.

Saunders, B.C., Holmes-Siedle, A.G. and Stark, B.P. (1964) *Peroxidases*. Butterworths, London.

Savvaidis, I., Nobar, A.M., Hughes, M.N. and Poole, R.K. (1988) *Science and Technol. Letters*, pp. 563–4.

Scherer, P., Lippert, H. and Wolff, G. (1983) *Biol. Tr. Elem. Res.*, **5**, 149–63.

Scheffler, J.E. and Cohn, M. (1986) *Biochemistry*, **25**, 3785–96.

Schimpff-Weiland, G., Follman, H. and Auling, G. (1981) *Biochem. Biophys. Res. Commun.*, **102**, 1276–82.

Schimz, A. (1982) *FEBS Letters*, **139**, 283–6.

Schindler, M. and Osborn, M.J. (1979) *Biochemistry.*, **18**, 4425–30.

Schmid, J. and Auling, G. (1987) *J. Bacteriol.*, **169**, 3385–7.

Schneider, J. and Kaltwasser, H. (1984) *Arch. Microbiol.*, **139**, 355–60.

Schneider, K., Cammack, R. and Schlegel, H.G. (1984) *Eur. J. Biochem.*, **142**, 75–84.

Schottel, J.L. (1978) *J. Biol. Chem.*, **253**, 4341–9.

Schuldiner, S. and Fishkes, H. (1978) *Biochemistry*, **17**, 706–11.

Schultz, J.E. and Klumpp, S. (1982) *FEMS Microbiol. Letters*, **13**, 303–6.

Schulze, H. and Brand, J.J. (1978) *Plant Physiol.*, **62**, 727–30.

Schwab, A.J. (1973) *FEBS Letters*, **35**, 63–6.

Schwartz, A.K. and Bonner, D.M. (1964) *Biochim. Biophys. Acta*, **89**, 337–47.

Schwartz, E. and Oesterhelt, D. (1985) *EMBO J.*, **4**, 1599–603.

Schwarzenbach, G. (1957) *Complexometric Titrations*, Methuen, London.

Schweisfurth, R., Eleftheriadis, D., Gundlach, H., Jacobs, M. *et al.* (1978) in *Environmental Biogeochemistry and Geomicrobiology* **3** (Ed. W.E. Krumbein) Ann Arbor Science, pp. 923–8.

Scott, R.E., Jones, A. and Gaucher, G.M. (1984) *Biotech. Letters*, **6**, 231–6.

Scott, R.H., Sperl, G.T. and DeMoss, J.R. (1979) *J. Bacteriol.*, **137**, 719–26.

Seah, T.C.M. and Kaplan, J.G. (1973) *J. Biol. Chem.*, **248**, 2889–93.

Seah, T.C.M., Bhatti, A.R. and Kaplan, J.G. (1973) *Can. J. Biochem.*, **51**, 1551–5.

Seah, T.C.M., Bhatti, A.R. and Kaplan, J.G. (1978) *FEBS Letters*, **85**, 305–9.

Sedgwick, E.G. and Macleod, R.A. (1980) *Can. J. Biochem.*, **58**, 1206–14.

Seelig, A., Ludwig, B., Seelig, J., and Schatz, G. (1981) *Biochim. Biophys. Acta*, **636**, 162–7.

Seeman, S.L. and Jack, T.R. (1980) *Can J. Microbiol.*, **26**, 636–9.

Segerer, A., Stetter, K.O. and Klink, F. (1985) *Nature*, **313**, 787–8.
Selby, M. and Hieftje, G.M. (1987) *International Laboratory*, **10**, 23–38.
Selwyn, M.J. (1976) *Am. Chem. Soc. Adv. in Chem. Ser.*, **157**, 204–26.
Senda, M., Kanazawa, H., Tsuchiya, T. and Futai, M. (1983) *Arch. Biochem. Biophys.*, **220**, 398–404.
Senior, A.E., Richardson, L.V., Backer, K. and Wise, J.G. (1980) *J. Biol. Chem.*, **255**, 7211–17.
Serpersu, E.H., Shortle, D. and Mildvan, A.S. (1986) *Biochemistry*, **25**, 68–77.
Seto-Young, D., Garcia, M.L. and Krulwich, T.A. (1985) *J. Biol. Chem.*, **260**, 11393–5.
Seto-Young, D. and Ellar, D.J. (1981) *Microbios*, **30**, 191–208.
Shah, V.K., Ugalde, R.A., Imperial, J. and Brill, W.J. (1984) *Ann. Rev. Biochem.*, **53**, 231–57.
Shattuck, R.L., Oldenburg, D.J. and Storm, D.R. (1985) *Biochemistry*, **24**, 6356–62.
Sheridan, R.P., Allen, L.C. and Carter, C.W. (1981) *J. Biol. Chem.*, **256**, 5052–7.
Sherman, D.M. and Sherman, L.A. (1983) *J. Bacteriol.*, **156**, 393–401.
Shimizu, T., Hatano, M., Nagao, S. and Nozawa, Y. (1982) *Biochem. Biophys. Res. Commun.*, **106**, 1112–18.
Shimomura, O. and Johnson, F.H. (1976) in *Calcium in Biological Systems* (Ed. C.J. Duncan), SEB Symposium 30, pp. 41–54.
Shivvers, D.W. and Brock, T.D. (1973) *J. Bacteriol.*, **114**, 706–10.
Shulman, R.G., Brown, T.R., Ugurbil, K., Ogawa, S. *et al.* (1979) *Science*, **205**, 160–6.
Shumate, S.E., Strandberg, G.W., McWhirter, D.A., Parott, J.R. *et al.* (1981) *Biotechnol. Bioeng. Symp.* **No. 10**, 27–34.
Siedel, J. and Shelton, E. (1979) *Arch. Biochem. Biophys.*, **192**, 214–24.
Siegel, F.L., Carafoli, E., Kretsinger, R.H., MacLennon, D.H. *et al.* (1980) *Calcium Binding Proteins: Structure and Function*. Elsevier, North Holland, New York.
Sigel, H. (1983) (Ed.) *Metal Ions in Biological Systems* **15**: *Zinc and its Role in Biology and Nutrition*, Marcel Dekker Inc., New York.
Sigel, H. (1985) (Ed.) *Metal Ions in Biological Systems* **19** *Antibiotics and their Complexes*, Marcel Dekker Inc., New York.
Sigel, H. (1987a) *Metal Ions in Biological Systems* **21**: *Applications of NMR to Paramagnetic System*, Marcel Dekker Inc., New York.
Sigel, H. (1987b) *Metal Ions in Biological Systems*, **22**: *Endor, EPR and Electron Spin Echo for Probing Coordination Spheres*, Marcel Dekker Inc. New York.
Sigel, S.P. and Payne, S.H. (1982) *J. Bacteriol.*, **150**, 148–55.
Sillen, L. and Martell, A.E. (1971) *Stability Constants*. Special publication No. 25, Royal Society of Chemistry, London.
Silver, M. (1978) in *Metallurgical Applications of Bacterial Leaching and Related Microbiological Phenomena* (Eds. L.E. Murr, A.E. Torma and J.A. Brierley), Academic Press, New York, pp. 3–17.
Silver, M. and Dinardo, O. (1981) *Appl. Environ, Microbiol.*, **41**, 1301–9.
Silver, S., Johnseine, P., Whitney, E. and Clark, D. (1972) *J. Bacteriol.*, **110** 186–95.
Silver, S. (1977) in *Micro-organisms and Minerals* (Ed. E.D. Weinberg), Marcel Dekker Inc., New York, pp. 49–103.
Silver, S. and Jasper, P. (1977) in *Micro-organisms and Minerals* (Ed. E.D. Weinberg) Marcel Dekker Inc., New York, pp. 105–49.

Silver, S. (1978) in *Bacterial Transport* (Ed. B.P. Rosen) Marcel Dekker Inc., New York, pp. 221–324.
Silver, S. (1981) in *Molecular Biology, Pathogenicity and Ecology of Bacterial Plasmids* (Eds. S.B. Levy, R.C. Clowes and E.L. Koenig), Plenum Press, New York, pp. 179–89.
Silver, S. and Keach, D. (1982) *Proc. Natl. Acad. Sci. U.S.A.*, **79**, 6114–18.
Silver, S. (1983) in *Biomineralisation and Biological Metal Accumulation. Biological and Geological Perspectives* (Eds. P. Westbroek and E.W. de Jong) D. Reidel Publishing Co., Dordrecht, pp. 439–57.
Silver, S. and Perry, R.D. (1983) in *Membranes and Transport*. (Ed. A.N. Martonosi), **2**, Plenum Press, New York, pp. 115–21.
Silver, S., Rosen, B.P. and Misra, T.K. (1986) in *Fifth International Symposium on the Genetics of Industrial Micro-organisms* (Eds. M. Alačevič, D. Hranueli and Z. Toman) pp. 357–76.
Silver, S. and Lusk, J.E. (1987) in *Ion Transport in Prokaryotes* (Eds. B.P. Rosen and S. Silver), Academic Press, San Diego, pp. 165–80.
Silver, S., Laddaga, R.A. and Misra, T.K. (1989) in *Metal-Microbe Interactions*, (Eds R.K. Poole and G.M. Gadd) IRL Press, London, pp. 49–63.
Silverberg, B.A., Wong, P.T.S. and Chau, Y.K. (1977) *Arch. Environ. Contam. Toxicol.*, **5**, 305–13.
Silverman, M.P., Rogoff, M.H. and Wender, I. (1951) *Appl. Microbiol.*, **9**, 491–6.
Silverman, M.P., Rogoff, M.H. and Wender, I. (1963) *Fuel*, **42**, 113–24.
Silverman, M.P. and Lundgren, D.G. (1959) *J. Bacteriol.*, **77**, 642–7.
Silverman, M.P. (1967) *J. Bacteriol.*, **94**, 1046–51.
Singh, A.P. and Bragg, P.D. (1974) *FEBS Letters*, **40**, 200–2.
Singh, A.P. and Bragg, P.D. (1978) *Biochem Biophys. Res. Commun.*, **81**, 161–7.
Singh, A.P. and Bragg, P.D. (1979a) *Can. J. Biochem.*, **57**, 1376–83.
Singh, A.P. and Bragg, P.D. (1979b) *Can. J. Biochem*, **57**, 1384–91.
Singh, D.P. (1985) *J. Gen. Appl. Microbiol.*, **31**, 277–84.
Singh, S.P. and Pandey, A.K. (1982) *Mol. Gen. Genet.*, **187**, 240–3.
Singh, S.P. and Yadava, V. (1985) *J. Gen. Appl. Microbiol.*, **31**, 39–48.
Sinha, A.P. (1976) *Structure and Bonding*, **25**, 69–149.
Sisken, J.E. (1980) in *Nuclear Cytoplasmic Interactions in the Cell Cycle*. (Ed. G.L. Whitson), Academic Press, New York, pp. 271–92.
Sivaraman, H. and Sivaraman, C. (1979) *FEBS Lett.*, **105** 267–71.
Skulachev, V.P. (1984) *Trends Biochem. Sci.*, **9**, 483–5.
Skulachev, V.P. (1987) in *Ion Transport in Prokaryotes* (Eds. B.P. Rosen and S. Silver), Academic Press, pp. 131–64.
Slaby, I., Lind, B. and Holmgren, A. (1984) *Biochem. Biophys. Res. Commun.*, **122**, 1410–17.
Slavin, W. (1986) *Anal. Chem.*, **56**, 589A.
Slayman, C.C. (1977) In *Water Relations in Membrane Transport in Plants and Animals*, (Eds. A.M. Jungreis, T. Hodges, A.M. Kleinziller and S.G. Schultz), Academic Press, New York, pp. 69–86.
Slonczewski, J.L., Rosen, B.P., Alger, J.R. and McNab, R.M. (1981) *Proc. Natl. Acad. Sci. USA*, **78**, 6271–5.
Slykhouse, T.O. and Fee. J.A. (1976) *J. Biol. Chem.*, **251**, 5472–7.
Smith, A.K. and Emery, T. (1982) *Biochim. Biophys. Acta*, **719**, 606–11.
Smith, B.E., O'Donnell, M.J., Lang, G. and Spartalian, K. (1980) *Biochem. J.*, **191**, 449–55.
Smith, D.G. (1974) *J. Gen. Microbiol.*, **83**, 389–92.

Smith, R.A., Latchnay, L.R. and Senior, A.E. (1985) *Biochemistry*, **24**, 4490–4.
Smulevich, G., Mauro, J.M., Fishel, L.A., English, A.M. *et al.* (1988) *Biochemistry*, **27**, 5477–85; 5486–92.
Snell, W.J., Buchanan, M. and Clausell, A. (1982) *J. Cell. Biol.*, **94**, 607–12.
Snyder, S.W. and Hollocher, T.C. (1984) *Biochem. Biophys. Res. Commun.*, **119**, 588–92.
Solaiman, D. and Wu. F. Y.-H. (1984) *Biochemistry*, **23**, 6369–77.
Song, P.-S. (1981) *Biochim. Biophys. Acta*, **639**, 1–29.
Sorensen, J. (1982) *Appl. Environ. Microbiol.*, **43**, 319–24.
Spartalian, L., Oosterhuis, W.T. and Smarra, N. (1975) *Biochim. Biophys. Acta*, **399**, 203–12.
Spiro, T.G. (1982) (Ed.) *Iron-sulfur Proteins*, Wiley, New York.
Spiro, T.G. (1984) (Ed.) *Calcium in Biology*, Wiley, New York.
Sprott, G.D., Shaw, K.M. and Jarrell, K.I. (1984) *J. Biol. Chem.*, **259**, 12602–8.
Stadtman, T.C. (1974) *Science*, **183**, 915–22.
Stadtman, T.C. (1980) *Trends Biochem. Sci.*, **5**, 203–6.
Stallings, W.C., Powers, T.B., Partridge, K.A., Fee. J.A. *et al.* (1983) *Proc. Natl. Acad. Sci. U.S.A.*, **80**, 3884–8.
Stallings, W.C., Partridge, K.A., Strong, R.K. and Ludwig, M.L. (1984) *J. Biol. Chem.*, **259**, 10695–9.
Stanley, S.H., Prior, S.D., Leak, D.J. and Dalton, H. (1983) *Biotechnol. Letters*, **5**, 487–92.
Steers, E., Barnett, A. and Lee, C.E. (1981) *Comp. Biochem. Physiol.*, **70B**, 185–91.
Steffens, G.C.M., Biewald, R. and Buse, G. (1987) *Eur. J. Biochem.*, **164**, 295–300.
Steiner, A.E., McLaren, D.A. and Forster, C.F. (1976) *Water Research*, **10**, 25–30.
Steinman, H.M. (1982) *J. Biol. Chem.*, **257**, 10283–93.
Stellwagen, E. (1978) *Nature*, **275**, 73–4.
Stephens, P.J., Morgan, T.V., Stout, C.D. and Burgess, B.K. (1985) *Front. Bioinorg. Chem.*, (Ed. A.V. Xavier), VCH, Weinheim, pp. 637–46.
Stephenson, M.C. and Ratledge, C. (1979) *J. Gen. Microbiol.*, **110**, 193–202.
Stephenson, T. and Lester, J.N. (1984) *Proceedings of the International Conference on Environmental Contamination*, UN Environment Programme, London, pp. 800–5.
Sterritt, R.M. and Lester, J.N. (1980a) *Science of Total Environ.*, **14**, 5–17.
Sterritt, R.M. and Lester, J.N. (1980b) *Bull. Environ, Contam. Toxicol.*, **24**, 196–203.
Sterritt, R.M. and Lester, J.N. (1986) in *Immobilisation of Metals by Biosorption*, (Eds. H. Eccles and S. Hunt), Ellis Horwood, Chichester, pp. 121–34.
Stetter, K.O. and Kandler, O. (1973) *FEBS Letters*, **36**, 5–8.
Steudel, R., Holdt, G., Gobel, T. and Hazeu, W. (1987) *Ang. Chem. Int. Ed.*, **26**, 151–3.
Stewart, G.S.A.B., Eaton, M.W., Johnstone, K., Barrett, M.D. *et al.* (1980a) *Biochim. Biophys. Acta*, **600**, 270–90.
Stiefel, E.I. and Watt, G.D. (1979) *Nature*, **279**, 81–3.
St. John, A.C. and Goldberg, A.L. (1980) *J. Bacteriol.*, **143**, 1223–33.
Stoffyn. M. (1979) *Science*, **203**, 651–3.
Stouthamer, A.H. (1973) *Antonie van Leeuwenhoek*, **39**, 545–65.
Stouthamer, A.H. and Bettenhausen, C.W. (1980) *Arch. Microbiol.*, **125**, 239–44.
Strachan, R.C., Aranha, H., Lodge, J.S., Arceneaux, J.E.L. *et al.* (1982) *Appl. Environ. Microbiol.*, **43**, 257–60.

Strain, S.M., Fesik, S.W. and Armitage, I.M. (1983) *J. Biol. Chem.*, **258**, 13466–77.
Strandberg, G.W., Shumate, S.E. and Parrott, J.R. (1981) *Appl. Environ. Microbiol.*, **41**, 237–45.
Stratton, G.W. and Corke, C.T. (1979) *Chem. Sphere*, **8**, 731–40.
Strong, J.R.P., Madgwick, J.C. and Ralph, B.J. (1982) *Biotechnol. Letters*, **4**, 239–42.
Stumm, W. and Bilinski, H. (1972) *Advanced Water Pollution Research* (Ed. S.H. Jenkins), Pergamon Press, Oxford, pp. 39–52.
Sturm, F.J., Hurwitz, S.A., Deming, J.W. and Kelly, R.M. (1987) *Biotech. Bioeng*, **29**, 1066–74.
Subramanyam, C., Venkateswerlu, G. and Rao, S.L.N. (1983) *Appl. Environ, Microbiol.*, **46**, 585–90.
Suelter, C.H. (1970) *Science*, **68**, 789–95.
Sugio, T., Tano, T. and Imai, K. (1981) *Agric. Biol. Chem.*, **45**, 2037–51.
Sugio, T., Domatsu, C., Tano, T. and Imai, K. (1984) *Appl. Environ. Microbiol.*, **48**, 461–7.
Sugio, T., Domatsu, C., Munakata, O., Tano, T. *et al.* (1985) *Appl. Environ. Microbiol*, **49**, 1401–6.
Sugiura, Y., Kuwahara, J. and Suzuki, T. (1983) *Biochem. Biophys. Res. Commun.*, **115**, 878–81.
Sugiyama, Y., Kitano, K. and Kanzaki, T. (1973) *Agric. Biol. Chem.*, **37**, 1837–47.
Summers, A.O. and Silver, S. (1978) *Ann. Rev., Microbiol.*, **32**, 637–72.
Summers, A.O., Jacoby, G.A. Swartz, M.N., McHugh, G. *et al.* (1978) *Microbiology*, (Ed. D. Schlessinger), American Society for Microbiology, pp. 128–31.
Summers, A.O. (1984) in *Antimicrobial Drug Resistance* (Ed. L.E. Bryan) Academic Press, Orlando, pp. 345–67.
Summers, A.O. (1985) *Trends in Biotechnol.*, **3**, 122–5.
Takai, K., Müller, R., Zavala, F., Inoue, S. *et al.* (1982) in *Oxygenases and Oxygen Metabolism* (Eds. M. Nozaki, S. Yamamoto, Y. Ishimura, M.J. Cook, L. Ernster and R.W. Estabrook), Academic Press, New York, pp. 159–72.
Takeda, K., Tezuka, C., Fukuoka, S. and Takahara, Y. (1976) *J. Ferm. Technol.*, **54**, 557–62.
Takeda, T. and Yamamoto, M. (1987) *Proc. Natl. Acad. Sci. USA*, **84**, 3580–4.
Tanaka, Y., Fukumori, Y. and Yamanaka, T. (1983) *Arch. Microbiol.*, **135**, 265–71.
Teeter, M.M., Quigley, G.J. and Rich, A. (1981) *Adv. Inorg. Biochem.* (Eds. G. Eichhorn and L.G. Marzilli) **3**, Elsevier, New York, pp. 233–72.
Teixeira, M., Moura, I., Xavier, A.V., Dervartanian, D.V. *et al.* (1983) *Eur. J. Biochem.*, **130**, 481–4.
Tempest, D.W., Hunter, J.R. and Sykes, J. (1965) *J. Gen. Microbiol.*, **39**, 355–66.
Tempest, D.W., Dicks, J.W. and Meers, J.L. (1967) *J. Gen. Microbiol.*, **49**, 139–47.
Tempest, D.W. (1969) *Symp. Soc. Gen. Microbiol.*, **19**, 87–111.
Terner, J., Sitter, A.J. and Reczek, C.M. (1985) *Biochim. Biophys. Acta*, **828**, 73–80.
Tetaz, T.J. and Luke, R.K.J. (1983) *J. Bacteriol.*, **154**, 1263–8.
Thauer, R.K., Diekert, G. and Schönheit, P. (1980) *Trends Biochem. Sci.*, **5**, 304–6.
Thayer, J.S. (1984) *Organometallic Compounds and Living Organisms*. Academic Press, Orlando.
Thayer, J.S. and Brinckman, F.E. (1982) *Adv. Organometallic Chem.*, **20**, 313–57.
Theil, E.C. (1987) *Ann. Rev. Biochem.*, **56**, 289–315.
Theil, E., Eichhorn, G.L. and Marzilli, L.G. (1983) (Eds.) *Adv. Inorg, Biochem.*, **5**.

Thelander, L. and Reichard, P. (1979) *Ann. Rev. Biochem.*, **48**, 133–58.

Thiers, R.E. (1957) *Methods Biochem. Anal.*, **5**, 273–335.

Thomas, K.C. and Dawson, P.S.S. (1977) *J. Bacteriol.*, **132**, 36–43.

Thomas, K.C. and Dawson, P.S.S. (1978) *Can. J. Microbiol.*, **24**, 440–7.

Thompson, G.W. and Medve, R.J. (1984) *Appl. Environ. Microbiol.*, **48**, 556–60.

Thompson, R. (1987) (Ed.) *Trace Metal Removal from Aqueous Solution*. Special Publication No. 61, Royal Society of Chemistry, London.

Thorneley, R.N.F. and Lowe. D.J. (1983) *Biochem. J.*, **215**, 393–401.

Thorneley, R.N.F. and Lowe. D.J. (1984) *Biochem. J.*, **224**, 887–94, 903–10.

Tillet, D.M. and Myerson, A.S. (1987) *Biotech. Bioeng.*, **29**, 146–50.

Timkovich, R., Cork, M.S., Gennis, R.B. and Johnson, P.Y. (1985) *J. Am. Chem. Soc.*, **107**, 6069–75.

Tinker, P.B. and Gildon, A. (1983) in *Metals and Micronutrients: Uptake and Utilisation by Plants* (Eds. D.A. Robb and W.S. Pierpoint), Academic Press, London, pp. 21–32.

Titus, J.A. and Pfister, R.M. (1982) *Bull. Environ. Contam. Toxicol.*, **28**, 697–704.

Tobin, J.J., Cooper, D.G. and Neufeld, R.J. (1984) *Appl. Environ. Microbiol.*, **41**, 237–45.

Toghrol, F. and Southerland, W.M. (1983) *J. Biol. Chem.*, **258**, 6762–6.

Tokuda, H. and Unemoto, T. (1982) *J. Biol. Chem.*, **257**, 10007–14.

Tokuda, H. and Unemoto, T. (1984) *J. Biol. Chem.*, **259**, 7785–90.

Torma, A.E. (1971) *Rev. Can. Biol.*, **30**, 209–16.

Torma. A.E. (1977) *Adv. Biochem. Eng.*, **6**, 1–37.

Torma, A.E. and Bosecker, K. (1982) *Progress Ind. Microbiol.*, **16**, 77–118.

Tornabene, T.G. and Edwards, H.W. (1972) *Science*, **176**, 1334–5.

Trevors, J.T. (1987a) *Enzyme Microb. Technol.*, **9**, 331–3.

Trevors, J.T. (1987b) *Microbiol. Sci.*, **4**, 29–31.

Tributsch, H. (1976) *Naturwissenschaften*, **63**, 88.

Trick, C.G., Andersen, R.J., Gillam, A. and Harrison, P.J. (1983) *Science*, **219**, 306–8.

Tripathy, B.C. and Mohanty, P. (1980) *Plant Physiol.*, **66**, 1174–8.

Tromalla, H.W. (1978) *Planta*, **138**, 243–8.

Tromballa, H.W. (1980) *Z. Pflanzenphysiol.*, **96**, 123–33.

Trutko, S.M., Medentsev, A.G. and Akimenko, V.K. (1980) *Microbiology.*, **49**, 543–8.

Tsezos, M. and Volesky, B. (1981) *Biotechnol. Bioeng.*, **23**, 583–604.

Tsezos, M. and Volesky, B. (1982a) *Biotechnol. Bioeng.*, **24**, 385–401.

Tsezos, M. and Volesky, B. (1982b) *Biotechnol. Bioeng.*, **24**, 955–69.

Tsezos, M. (1983) *Biotechnol. Bioeng.*, **25**, 2025–40.

Tsezos, M. and Keller, D.M. (1983) *Biotechnol. Bioeng.*, **25**, 201–15.

Tsuchiya, T. and Takeda, K. (1979a) *J. Biochem.*, **85**, 943–51.

Tsuchiya, T. and Takeda, K. (1979b) *J. Biochem..*, **86**, 225–30.

Tsuchiya, T., Oho, M. and Shiota-Niiya, S. (1983) *J. Biol. Chem.*, **258**, 12675–7.

Tsukihara, T., Fukuyama, K., Nakamura, M., Katsube, Y. *et al.* (1981) *J. Biochem.*, **90**, 1763–73.

Tsujibo, H. and Rosen, B.P. (1983) *J. Bacteriol.*, **154**, 854–8.

Tsukuda, K., Kido, T., Shimasue, Y. and Soda, K. (1983) *Agric. Biol. Chem.*, **47**, 2865–70.

Tsverkov, Y.D. and Dikanov, S.A. (1987) in *Metal Ions in Biological Systems* (Ed. H. Sigel) **22**, Dekker, New York, pp. 207–68.

Tuovinen, O.H. and Kelly, D.P. (1972) *Z. Allg. Mikrobiologie*, **12**, 311–46.
Tuovinen, O.H. and Kelly, D.P. (1974) *Int. Metallurgical Rev.*, **19**, 21–9.
Tuovinen, O.H., Silver, M., Martin, P.A.W. and Dugan, P.R. (1981) in *Proceedings of the International Conference on Use of Micro-organisms in Hydrometallurgy*, Hungarian Academy of Sciences, Pecs, pp. 59–69.
Tuovinen, O.H., Hiltanen, P. and Vuorinen, A. (1983) *Eur. J. Appl. Microbiol. Biotechnol*, **17**, 327–33.
Tuovinen, O.H., Puhakka, J., Hiltunen, P. and Dolan, K.M. (1985) *Biotechnol. Letters*, **7**, 389–94.
Tweedie, J.W. and Segel, I.H. (1970) *Biochim. Biophys. Acta*, **196**, 95–106.
Tynecka, Z., Gos, Z. and Zajac, J. (1981a) *J. Bacteriol.*, **147**, 305–12.
Tynecka, Z., Gos, Z. and Zajac, J. (1981b) *J. Bacteriol.*, **147**, 313–19.
Uchida, K., Shimizu, T., Makino, R., Sakaguchi, K. *et al.* (1983) *J. Biol. Chem.*, **258**, 2526–33.
Ugalde, R.A., Imperial, J., Shah, V.K. and Brill, W.J. (1984) *J. Bacteriol.*, **159**, 888–93.
Utkilen, H.C. (1982) *J. Gen. Microbiol.*, **128**, 1849–62.
Utkilen, H.C. (1983) *J. Gen. Microbiol.*, **129**, 1713–15.
Uwajima, T., Shimizu, Y. and Terada, O. (1984) *J. Biol. Chem.*, **259**, 2748–53.
Vallee, B.L. and Williams, R.J.P. (1968) *Proc. Natl. Acad. Sci. U.S.A.*, **59**, 498–505.
Vallee, B.L. and Ulmer, D.D. (1972) *Ann. Rev. Biochem.*, **41**, 91–128.
Van Baalen, C. and O'Donnell, R. (1978) *J. Gen. Microbiol.*, **105**, 351–3.
Van. Brunt, J., Caldwell, J.H. and Harold, F.M. (1982) *J. Bacteriol.*, **150**, 1449–61.
Van Buskirk, J.J., Low, M. and Kirsh, W.M. (1979) in *Vitamin K Metabolism and Vitamin K-dependent Proteins* (Ed. J.W. Suttie) University Park Press, Baltimore, pp. 274–8.
Vandenbergh, P.A., Gonzalez, C.F., Wright, A.M. and Kunka, B.S. (1983) *Appl. Environ. Microbiol.*, **46**, 128–32.
Van Divender, J.M. and Grisham, C.M. (1985) *J. Biol. Chem.*, **260**, 14060–9.
Veillon, C. and Vallee, B.L. (1978) *Methods Enzymol.*, **54**, 446–84.
Verstreate, D.R., Storch, T.A. and Dunham, V.L. (1980) *Physiologia Plantarum*, **50**, 47–51.
Villafranca, J.J. and Balakrishnan, M.S. (1979) *Int. J. Biochem.*, **10**, 565.
Villafranca, J.J. and Raushel, F.M. (1982) in *Adv. Inorg. Biochem.*, (Eds. G.L. Eichhorn and L.G. Marzilli), **4**, 289–320.
Vojak, P.W.L., Edwards, C. and Jones, M.V. (1984) *Microbios.*, **41**, 39–47.
Volesky, B. and Tsezos, M. (1981) *Uranium separation by biosorption*. U.S. Patent No. 4 320 093.
Volesky, B. (1987) *Trends in Biotechnol.*, **5**, 96–101.
Vrdoljak, M., Maric, V. and Johanides, V. (1984) *Biotechnol. Letters*, **6**, 501–6.
Wagner, G., Hartmann, R. and Oesterhelt, D. (1978) *Eur. J. Biochem.*, **89**, 169–79.
Wagner, G., Cammack, R. and Andreesen, J.R. (1984) *Biochim. Biophys. Acta*, **791**, 63–74.
Wakabayashi, S., Matsubara, H. and Webster, D.A. (1986) *Nature*, **322**, 481–3.
Wakao, N., Tachibana, H., Tanaka, Y., Sakurai, Y. *et al.* (1985) *J. Gen. Appl. Microbiol.*, **31**, 17–28.
Walderhaug, M.O., Dosch, D.C. and Epstein, W. (1987). in *Ion Transport in Prokaryotes* (Eds. B.P. Rosen and S. Silver) Academic Press, pp. 85–130.
Walker, G.M. and Duffus, J.H. (1980). *J. Cell Sci.*, **42**, 329–356.

Walker, G.M. and Zeuthen, E. (1980) *Exp. Cell. Res.*, **127**, 487–90.
Walker, G.M., Birch-Andersen, A., Hamburger, K. and Kramhoft, B. (1982) *Carlsberg Rev. Commun.*, **47**, 205–14.
Walker, G.M. and Duffus, J.H. (1983) *Magnesium*, **2**, 1–16.
Walsh, A. (1955) *Spectrochim. Acta*, **7**, 108–17.
Wampler, D.E. and Westhead, E.W. (1968) *Biochemistry*, **7**, 1661–71.
Ware, G., Thompson, A.S. and Light, P.A. (1970) *Lab. Prac.*, **19**, 181–2.
Waring, W.S. and Werkman, C.H. (1942) *Arch. Biochem.*, **1**, 303–10.
Waring, W.S. and Werkman, C.H. (1944) *Arch. Biochem.*, **4**, 75–87.
Watabe, K. and Freese, K. (1979) *J. Bacteriol.*, **137**, 773–8.
Watkins, J.W., Elder, R.C., Greene, B. and Darnall, D.W. (1987) *Inorg Chem.*, **26**, 1147–51.
Watson, H.C., Walker, N.P.C., Shaw, P.J., Bryant, T.N. *et al.* (1982) *EMBO. J.*, **12**, 1635–40.
Webb, M. (1968) *J. Gen. Microbiol.*, **51**, 325–35.
Webley, D.M., Duff, R.B. and Anderson, G. (1962) *J. Gen. Microbiol.*, **29**, 179–87.
Weinberg, E.D. (1977) In *Micro-organisms and Minerals* (Ed. E.D. Weinberg). Marcel Dekker Inc., New York, pp. 289–316.
Weinberg, E.D. (1982) in *Overproduction of Microbial Products* (Eds. V. B. Sikyra and Z. Vanit), Academic Press, London, pp. 181–94.
Weininger, M.S. and Mortenson, L.E. (1982) *Proc. Natl. Acad. Sci. U.S.A.*, **79**, 378–80.
Weintraub, S.T., Muhoberac, B.B. and Wharton, D.C. (1982) *J. Biol. Chem*, **257**, 4940–6.
Weisiger, R.A. and Fridovich, I. (1973) *J. Biol. Chem.*, **248**, 3582–92.
Weiss, R.L. (1977) *J. Cell Biol.*, **73**, 505–19.
Welsh, K.M., Jacobyansky, A., Springs, B. and Cooperman, B.S. (1983) *Biochemistry*, **22**, 2243–8.
Weser, U., Hartmann, H.-J., Fretzdorff, A. and Strobel, G.J. (1977) *Biochim. Biophys. Acta*, **493**, 465–77.
West, I.C. and Mitchell, P. (1974) *Biochem. J.*, **144**, 87–90.
Westbroek, P. and De Jong, E.W. (1983) (Eds.) *Biomineralisation and biological metal accumulation.* (Proceedings of the Fourth International Symposium on Biomineralisation), D. Reidel Publishing Co,
Wever, R., de Boer, E., Plat, H. and Kren, B.E. (1987) *FEBS Letters*, **216**, 1–3.
White, R.E. and Coon, M.J. (1980) *Ann Rev. Biochem.*, **49**, 315–56.
Wieczorek, L. and Altendorf, K. (1979) *FEBS Letters*, **98**, 233–6.
Wiegel, J. (1978) *Biochem. Biophys. Res. Commun.*, **82**, 907–12.
Wihlidal, H. and Broda, E. (1977). *Z. Allg. Mikrobiol.*, **17**, 635–8.
Wikstrom. M., Krab, K. and Saraste, M. (1981) *Cytochrome Oxidase. A Synthesis.* Academic Press, London.
Wilding, R.E. and Garland, T.R. (1982) *Appl. Environ. Microbiol.*, **43**, 418–23.
Willard, H.H., Merritt, L.L., Dean. J.A. and Settle, E.A. (1982) *Instrumental Methods of Analysis.* Van Nostrand Reinhold, New York.
Williams, H.D. and Poole, R.K. (1987) *Current Microbiol.*, **15**, 319–24.
Williams, I.W. and Silver, S. (1984) *Enzy. Microbiol. Technol.*, **6**, 530–7.
Williams, R.J.P. (1981) *Proc. R. Soc. London B*, **213**, 361–97.
Williams, R.J.P. (1984) *Endeavour. New Series*, **8**, 65–70.
Williams. R.J.P. (1985) *Eur. J. Biochem.*, **150**, 231–48.
Wilson, D.O. and Reisenauer, H.M. (1970) *J. Bacteriol.*, **102**, 729–32.

Wilson, C.A.B., Hardman, N., Fothergill-Gilmore, L.A., Gamblin, S.J. *et al.* (1987) *Biochem. J.*, **241**, 609–14.
Winkelmann, G. (1979) *Arch. Microbiol.*, **121**, 43–51.
Winter, J., Lerp, C., Zabel, H.-P., Wildenauer, F.X. *et al.* (1984) *Syst. Applied Microbiol.*, **5**, 457–66.
Wolff, J., Newton, D.L. and Klee, C.B. (1986) *Biochemistry*, **25**, 7950–5.
Wong, M.H. (1980) *Microbios.*, **28**, 19–25.
Wong, C., Silver, M. and Kushner, D.J. (1982) *Can. J. Microbiol.*, **28**, 536–44.
Wood. J.M. (1974) *Science*, **183**, 1049–52.
Wood, J.M. (1984) in *Metal Ions in Biological Systems* **18**: *Circulation of Metals in the Environment* (Ed. H. Sigel) Marcel Dekker Inc., New York, pp. 333–51.
Wood, J.M., Moura, I., Moura, J.J.G., Santos, M.H. *et al.* (1982) *Science*, **216**, 303–5.
Wood, P.M. (1978) *Eur. J. Biochem.*, **87**, 9–13.
Wood, P.M. (1984) *Biochim. Biophys. Acta*, **768**, 293–317.
Woodland, M.P. and Cammack, R. (1985) in *Microbial Gas Metabolism: Mechanistic Metabolic and Biotechnological Aspects* (Eds. R.K. Poole and C.S. Dow), Academic Press, Orlando, 209–13.
Wright, C.T. and Klaenhammer, T.R. (1981) *Appl. Environ. Microbiol.*, **41**, 807–15.
Wright, C.T. and Klaenhammer, T.R. (1983) *Appl. Environ. Microbiol.*, **46**, 785–92.
Wu, H.M., Dattagupta, N., Hogan, M. and Crothers, D.M. (1979) *Biochemistry*, **18**, 3960.
Xavier, A.V., Moura, J.J.G. and Moura, I. (1980) *Structure and Bonding*, **43**, 187–93.
Yamada, J. and Watanabe, T. (1979a) *Agric. Biol. Chem.*, **43**, 1293–300.
Yamada, J. and Watanabe, T. (1979b) *Agric. Biol. Chem.*, **43**, 1681–6.
Yamanaka, K. (1961) *Agric. Biol. Chem.*, **25**, 272–81.
Yamasaki, K., Moriyama, Y., Futai, M. and Tsuchiya, T. (1980) *FEBS Letters*, **120**, 125–7.
Yariv, J. (1983) *Biochem, J.*, **211**, 527.
Yates, M.V., Brierley, J.A., Brierley, C.L. and Follin, S. (1983) *Appl. Environ. Microbiol.*, **46**, 779–84.
Yeh, T.Y., Godshalk, J.R., Olson, G.J. and Kelly, R.M. (1987) *Biotech. Bioeng.*, **30**, 138–46.
Yoch, D.C. and Carithers, R.P. (1979) *Microbiol. Rev.*, **43**, 384–421.
Yonetani, T. (1974) in *Microbial Iron Metabolism. A Comprehensive Treatise* (Ed. J.B. Neilands), Academic Press, New York, London, pp. 401–15.
Yoshimura, F. and Brodie, A.F. (1981) *J. Biol. Chem.*, **256**, 12239–42.
Young, J.M., Schray, K.J. and Mildvan, A.S. (1975) *J. Biol. Chem.*, **250**, 9021.
Zapkin, M.A. and Ehrlich, H.L. (1983) *Z. Allg. Mikrobiol.*, **23**, 447–55.
Zielinski, R.E. and Price, C.A. (1978) *Plant Physiol.*, **61**, 624–5.
Zillig, W., Stetter, K.O., Wunderl, S., Schulz, W. *et al.* (1980) *Arch. Microbiol.*, **125**, 259–69.
Zillig, W., Yeats, S. Holz, I., Bock, A. *et al.* (1985) *Nature*, **313**, 789–90.
Zimniak, P. and Barnes, E.M. (1980) *J. Biol. Chem.*, **255**, 10140–3.
Zonneveld, B.J.M. (1975a) *Arch. Microbiol.*, **105**, 101–4.
Zonneveld, B.J.M. (1975b) *Arch. Microbiol.*, **105**, 105–8.
Zumberge, J.E., Sigleo, A.C. and Nagy, B. (1978) *Mineral Sci. Eng.*, **10**, 223–9.

Index